AF389097

Advances in Molecular Breeding of Vegetable Crops

Advances in Molecular Breeding of Vegetable Crops

Editor

Yuyang Zhang

MDPI • Basel • Beijing • Wuhan • Barcelona • Belgrade • Manchester • Tokyo • Cluj • Tianjin

Editor
Yuyang Zhang
Huazhong Agricultural
University
China

Editorial Office
MDPI
St. Alban-Anlage 66
4052 Basel, Switzerland

This is a reprint of articles from the Special Issue published online in the open access journal *Horticulturae* (ISSN 2311-7524) (available at: https://www.mdpi.com/journal/horticulturae/special_issues/Molecular_Breeding_Vegetable_Crops).

For citation purposes, cite each article independently as indicated on the article page online and as indicated below:

LastName, A.A.; LastName, B.B.; LastName, C.C. Article Title. *Journal Name* **Year**, *Volume Number*, Page Range.

ISBN 978-3-0365-5789-2 (Hbk)
ISBN 978-3-0365-5790-8 (PDF)

Cover image courtesy of Yuyang Zhang

© 2022 by the authors. Articles in this book are Open Access and distributed under the Creative Commons Attribution (CC BY) license, which allows users to download, copy and build upon published articles, as long as the author and publisher are properly credited, which ensures maximum dissemination and a wider impact of our publications.
The book as a whole is distributed by MDPI under the terms and conditions of the Creative Commons license CC BY-NC-ND.

Contents

About the Editor

Yuyang Zhang

Dr. Yuyang Zhang is a Principal Investigator in Hubei Hongshan Laboratory, Wuhan, China. He has also been a professor in the College of Horticulture and Forestry Science at the Huazhong Agricultural University, Wuhan, China, since 2013. He obtained his bachelor's degree and PhD degree at the Huazhong Agricultural University in 2001 and 2006, respectively. He then carried out postdoctoral research at the Agricultural Research Organization, Israel, and worked as a Visiting Scientist at Wageningen University. His research areas cover the fields of the fruit development and quality, molecular biology, and molecular breeding of tomato.

 *horticulturae*

Editorial

Advances in Molecular Breeding of Vegetable Crops

Yaru Wang [1] and Yuyang Zhang [1,2,*]

1   Key Laboratory of Horticultural Plant Biology, Ministry of Education, Huazhong Agricultural University, Wuhan 430070, China
2   Hubei Hongshan Laboratory, Wuhan 430070, China
*   Correspondence: yyzhang@mail.hzau.edu.cn; Tel.: +86-27-8728-2010

Abstract: Molecular vegetable breeding has been progressed intensively in recent years. Huge advances have been made in germplasm evaluation, gene isolation, plant transformation, gene editing and molecular-marker-assisted breeding. The goal of this Special Issue is to highlight, through selected works, frontier research from basic to applied molecular vegetable breeding. The selected papers published in Special Issue of *Horticulturae* exhibit a diversity in molecular vegetable breeding. The papers listed in this editorial are especially noteworthy.

1. Advances in Gene Editing in Context of Vegetable Molecular Breeding

In recent years, it has become certain that genome editing is an efficient and powerful tool for precise genome manipulations in plants. For applications in molecular vegetable breeding, this new technique overcomes the shortcomings of conventional breeding, such as long-term artificial selection and limited genetic germplasm resources [1,2]. Wan et al. reviewed the development and application of CRISPR-Cas9 gene editing in vegetable crops. Currently, this system has been used to improve shelf life, fruit quality and stress resistance in major vegetable crops, such as tomato and cabbage. In the case of broccoli, genome editing has succeeded in limited *B. oleracea* crops [3]. Although the application of genome editing is extensive, how to obtain germplasm resources through gene editing of CREs (Cis-regulatory elements) and create a universal regeneration system for vegetable crops needs to be further studied and improved [1,4].

2. Germplasm Diversity Evaluation for Vegetable Improvement

In the modern breeding process, the evaluation of genetic diversity in agronomic and quality traits is still a fundamental method and approach for germplasm utilization and excavation. Uddin et al. performed phenotypic characterization and genetic diversity evaluation of 130 local eggplant germplasms [5]. Based on an analysis of trait variance, correlation matrix and MGIDI index, numerous traits were evaluated to determine the inherent variation and select applicable parents for eggplant improvement.

Simple sequence repeats (SSRs) are widely used genetic markers for genetic variation research in various crops due to co-dominance traits and high polymorphism. Zhong et al. employed this sequencing technology in *Capsicum frutescens* to provide resources of SSR molecular markers and analysis genetic diversity for pepper breeding [6]. Genome-wide identification of SSR markers revealed that trinucleotides were the dominant repeat motif. A total of 147 collected pepper cultivars were determined, clustered into seven main groups due to genetic diversity and phylogenetic relationships analysis. In *Cucurbita moschata*, 103,056 SSR loci were found by in silico PCR in which di-nucleotide motifs were the most common type [7]. Synteny analysis of cross-species SSR markers indicated that the main syntenic relationships between *Cucurbita* species were highly conserved during evolution.

3. Understanding the Genetic Basis of Biotic Resistance in Vegetable Crops

Fungal diseases remain challenges restricting the sustainable development of vegetable production. Although pesticides can prevent and control fungal diseases, excessive

Citation: Wang, Y.; Zhang, Y. Advances in Molecular Breeding of Vegetable Crops. *Horticulturae* **2022**, *8*, 821. https://doi.org/10.3390/horticulturae8090821

Received: 17 August 2022
Accepted: 25 August 2022
Published: 7 September 2022

Publisher's Note: MDPI stays neutral with regard to jurisdictional claims in published maps and institutional affiliations.

Copyright: © 2022 by the authors. Licensee MDPI, Basel, Switzerland. This article is an open access article distributed under the terms and conditions of the Creative Commons Attribution (CC BY) license (https://creativecommons.org/licenses/by/4.0/).

use of pesticides has brought great damages to the environment and human beings. Improving disease resistance has become an important breeding objective. The development and establishment of molecular marker technology makes it fast and effective to select germplasm resources directly.

The development of molecular markers associated with resistance to gray mold disease in onion (*Allium cepa* L.) through RAPD-PCR was assessed by Kim et al. [8]. RAPD analysis was performed to identify the genetic relationship between the resistant and susceptible lines and develop the SCAR marker. In addition, RNA-seq of the gray mold-resistant and -susceptible onion lines were analyzed to develop a selectable marker for the resistant line.

Phytophthora blight is a common disease that causes decreased yield and quality in pepper (*Capsicum annuum* L.). Li et al. generated a high-resolution genetic map of pepper associated with resistance to *Phytophthora capsici* by SLAF-seq and QTL analysis [9]. *CQPc5.1* was identified as a major quantitative trait locus (QTL) for the *P. capsici* resistance, including 23 candidate genes located within the interval.

4. Mining Genes Responsible for Abiotic Stresses for Vegetable Improvement

In vegetable crops, abiotic stresses cause serious damages, which limit growth and affect physiological metabolic processes. Therefore, screening genes responsible for abiotic stresses is essential to breeding resistant varieties. Pepper is sensitive to high temperatures, which leads to severe symptoms, such as pollination failure, growth defects and other aspects. Wang et al. identified differential expression genes in pepper leaves through a transcriptomics analysis of heat-tolerant and heat-sensitive varieties [10]. Heat shock (HS) proteins and HS transcription factors were identified as responsive to heat stress or recovery.

A small heat shock protein *CaHSP18.1a* was isolated and charactered from pepper [11]. Liu et al. demonstrated that *CaHSP18.1a* was sensitive to heat stress and showed high expression levels in thermo-tolerant line. The silencing of *CaHSP18.1a* caused elevated MDA contents and decreased resistance to heat, drought, and salt stresses, indicating that *CaHSP18.1a* positively regulates abiotic tolerance.

In eggplant, the genome-wide identification of *Hsf* and *Hsp* genes under heat stress was assessed by Gong et al. [12]. RNA-seq analysis showed that *Hsf* and *Hsp* genes exhibit different expression levels in the thermotolerant line 05-4 and the thermosensitive line 05-1, providing a basis for studying the relationship between thermotolerance and heat-response genes.

Nowadays, the irrational use of nitrogen fertilizer has resulted in undesirable growth and reduced yield in pepper. The molecular basis underlying the genetic variation in N-use efficiency (NUE) remains largely unknown. Based on comparative transcriptome analysis, Wang et al. selected two genotypes with contrasting low-N tolerance to explore the variation in NUEin pepper [13]. Numerous DEGs involved in N metabolism or other physiological processes were identified, providing candidate genes for improving N utilization in pepper.

5. Organic Compounds in Vegetables and Its Interaction with Environment

Flavonoids and volatile organic compounds act as important roles in the growth and developmental processes of vegetable crops, including the attraction of insect pollination, the inhibition of plants diseases and improvement in weed control.

In broccoli, anthocyanins contribute to the purple color and act as health-promoting antioxidants. Liu et al. identified major loci and candidate genes responsible for anthocyanin biosynthesis in broccoli [14]. Two QTLs on chromosomes 7 were identified to be tightly correlated with anthocyanin biosynthesis based on QTL-seq bulk segregant analysis. Further high-resolution mapping identified 14 candidate genes, providing a potential molecular marker into the breeding of novel varieties with abundant anthocyanins.

Volatile organic compounds released from plants are related to the allelopathy phenomenon, a chemical relationship of plant interaction. Xie et al. reviewed the recent advances in the allelopathy of volatile organic compounds (VOCs) of plants [15]. VOCs had multiple allelopathic effects on plants, such as enzyme activity, dormancy, diseases resistance, ROS scavenging, plant-to-plant communication, and other aspects [15]. The research suggested that the allelopathy of VOCs can be utilized in the development of economical and effective measures for sustainable agriculture [15].

With germplasm evaluation, gene isolation, and marker development, both scientists and breeders are working closely to generate more efficient breeding technology, e.g., gene editing, and to produce more elite cultivars. With emerging substantial genomic data and tools, further collaboration is worthwhile for next-generation breeding technology, e.g., genome-based breeding by design, to generate green, environmentally adaptive vegetable cultivars with high yield and quality.

Funding: In doing this work, the authors were sponsored by grants from the the National Key Research and Development Plan (2021YFD1200201; 2018YFD1000800); National Natural Science Foundation of China (31972426; 31991182); the Wuhan Biological Breeding Major Project (2022021302024852); the International Cooperation Promotion Plan of Shihezi University (GJHZ202104); the Key Project of Hubei Hongshan Laboratory (2021hszd007); and the Fundamental Research Funds for the Central Universities (2662022YLPY001).

Acknowledgments: We gratefully acknowledge all the authors that participated in this Special Issue.

Conflicts of Interest: The authors declare no conflict of interest.

References

1. Wan, L.; Wang, Z.; Tang, M.; Hong, D.; Sun, Y.; Ren, J.; Zhang, N.; Zeng, H. CRISPR-Cas9 Gene Editing for Fruit and Vegetable Crops: Strategies and Prospects. *Horticulturae* **2021**, *7*, 193. [CrossRef]
2. Chen, Y.; Mao, W.; Liu, T.; Feng, Q.; Li, L.; Li, B. Genome Editing as A Versatile Tool to Improve Horticultural Crop Qualities. *Hortic. Plant J.* **2020**, *6*, 372. [CrossRef]
3. Han, F.; Liu, Y.; Fang, Z.; Yang, L.; Zhuang, M.; Zhang, Y.; Lv, H.; Wang, Y.; Ji, J.; Li, Z. Advances in Genetics and Molecular Breeding of Broccoli. *Horticulturae* **2021**, *7*, 280. [CrossRef]
4. Li, X.; Li, H.; Zhao, Y.; Zong, P.; Zhan, Z.; Piao, Z. Establishment of a Simple and Efficient Agrobacterium-mediated Genetic Transformation System to Chinese Cabbage (*Brassica rapa* L. ssp. *pekinensis*). *Hortic. Plant J.* **2021**, *7*, 117. [CrossRef]
5. Uddin, M.; Billah, M.; Afroz, R.; Rahman, S.; Jahan, N.; Hossain, M.; Bagum, S.; Uddin, M.; Khaldun, A.; Azam, M.; et al. Evaluation of 130 Eggplant (*Solanum melongena* L.) Genotypes for Future Breeding Program Based on Qualitative and Quantitative Traits, and Various Genetic Parameters. *Horticulturae* **2021**, *7*, 376. [CrossRef]
6. Zhong, Y.; Cheng, Y.; Ruan, M.; Ye, Q.; Wang, R.; Yao, Z.; Zhou, G.; Liu, J.; Yu, J.; Wan, H. High-Throughput SSR Marker Development and the Analysis of Genetic Diversity in *Capsicum frutescens*. *Horticulturae* **2021**, *7*, 187. [CrossRef]
7. Zhu, L.; Zhu, H.; Li, Y.; Wang, Y.; Wu, X.; Li, J.; Zhang, Z.; Wang, Y.; Hu, J.; Yang, S.; et al. Genome Wide Characterization, Comparative and Genetic Diversity Analysis of Simple Sequence Repeats in Cucurbita Species. *Horticulturae* **2021**, *7*, 143. [CrossRef]
8. Kim, S.; Park, J.; Park, T.; Lee, H.; Choi, J.; Park, Y. Development of Molecular Markers Associated with Resistance to Gray Mold Disease in Onion (*Allium cepa* L.) through RAPD-PCR and Transcriptome Analysis. *Horticulturae* **2021**, *7*, 436. [CrossRef]
9. Li, Y.; Zhang, S.; Yang, X.; Wang, C.; Huang, Q.; Huang, R. Generation of a High-Density Genetic Map of Pepper (*Capsicum annuum* L.) by SLAF-seq and QTL Analysis of *Phytophthora capsici* Resistance. *Horticulturae* **2021**, *7*, 92. [CrossRef]
10. Wang, F.; Yin, Y.; Yu, C.; Li, N.; Shen, S.; Liu, Y.; Gao, S.; Jiao, C.; Yao, M. Transcriptomics Analysis of Heat Stress-Induced Genes in Pepper (*Capsicum annuum* L.) Seedlings. *Horticulturae* **2021**, *7*, 339. [CrossRef]
11. Liu, Y.; Liu, S.; Xiao, J.; Cheng, G.; Gong, Z. CaHSP18.1a, a Small Heat Shock Protein from Pepper (*Capsicum annuum* L.), Positively Responds to Heat, Drought, and Salt Tolerance. *Horticulturae* **2021**, *7*, 117. [CrossRef]
12. Gong, C.; Pang, Q.; Li, Z.; Li, Z.; Chen, R.; Sun, G.; Sun, B. Genome-Wide Identification and Characterization of *Hsf* and *Hsp* Gene Families and Gene Expression Analysis under Heat Stress in Eggplant (*Solanum melongema* L.). *Horticulturae* **2021**, *7*, 149. [CrossRef]
13. Wang, C.; Li, Y.; Bai, W.; Yang, X.; Wu, H.; Lei, K.; Huang, R.; Zhang, S.; Huang, Q.; Lin, Q. Comparative Transcriptome Analysis Reveals Different Low-Nitrogen-Responsive Genes in Pepper Cultivars. *Horticulturae* **2021**, *7*, 110. [CrossRef]
14. Liu, C.; Yao, X.; Li, G.; Huang, L.; Wu, X.; Xie, Z. Identification of Major Loci and Candidate Genes for Anthocyanin Biosynthesis in Broccoli Using QTL-Seq. *Horticulturae* **2021**, *7*, 246. [CrossRef]
15. Xie, Y.; Tian, L.; Han, X.; Yang, Y. Research Advances in Allelopathy of Volatile Organic Compounds (VOCs) of Plants. *Horticulturae* **2021**, *7*, 278. [CrossRef]

Review

Advances in Genetics and Molecular Breeding of Broccoli

Fengqing Han, Yumei Liu, Zhiyuan Fang, Limei Yang, Mu Zhuang, Yangyong Zhang, Honghao Lv , Yong Wang, Jialei Ji and Zhansheng Li *

Key Laboratory of Biology and Genetic Improvement of Horticultural Crops, Institute of Vegetables and Flowers, Chinese Academy of Agricultural Sciences, Ministry of Agriculture, #12 Zhong Guan Cun Nandajie Street, Beijing 100081, China; hanfengqing@caas.cn (F.H.); liuyumei@caas.cn (Y.L.); fangzhiyuan@caas.cn (Z.F.); yanglimei@caas.cn (L.Y.); zhuangmu@caas.cn (M.Z.); zhangyangyong@caas.cn (Y.Z.); lvhonghao@caas.cn (H.L.); wangyong03@caas.cn (Y.W.); jijialei@caas.cn (J.J.)
* Correspondence: lizhansheng@caas.cn; Tel.: +86-010-62135629

Abstract: Broccoli (*Brassica oleracea* L. var. *italica*) is one of the most important vegetable crops cultivated worldwide. The market demand for broccoli is still increasing due to its richness in vitamins, anthocyanins, mineral substances, fiber, secondary metabolites and other nutrients. The famous secondary metabolites, glucosinolates, sulforaphane and selenium have protective effects against cancer. Significant progress has been made in fine-mapping and cloning genes that are responsible for important traits; this progress provides a foundation for marker-assisted selection (MAS) in broccoli breeding. Genetic engineering by the well-developed *Agrobacterium tumefaciens*-mediated transformation in broccoli has contributed to the improvement of quality; postharvest life; glucosinolate and sulforaphane content; and resistance to insects, pathogens and abiotic stresses. Here, we review recent progress in the genetics and molecular breeding of broccoli. Future perspectives for improving broccoli are also briefly discussed.

Keywords: broccoli; progress; genetic researches; molecular breeding

Citation: Han, F.; Liu, Y.; Fang, Z.; Yang, L.; Zhuang, M.; Zhang, Y.; Lv, H.; Wang, Y.; Ji, J.; Li, Z. Advances in Genetics and Molecular Breeding of Broccoli. *Horticulturae* **2021**, *7*, 280. https://doi.org/10.3390/horticulturae7090280

Academic Editor: Yuyang Zhang

Received: 6 August 2021
Accepted: 1 September 2021
Published: 3 September 2021

Publisher's Note: MDPI stays neutral with regard to jurisdictional claims in published maps and institutional affiliations.

Copyright: © 2021 by the authors. Licensee MDPI, Basel, Switzerland. This article is an open access article distributed under the terms and conditions of the Creative Commons Attribution (CC BY) license (https://creativecommons.org/licenses/by/4.0/).

1. Introduction

Broccoli (*Brassica oleracea* L. var. *italica*) is a member of the Brassicaceae family and is widely cultivated as an important vegetable crop worldwide [1,2]. It produces edible hypertrophic reproductive organs (floral head and stalk), with rich health benefits and nutritious properties, such as vitamin A, vitamin K, calcium, magnesium and anticancer bioactive compounds, including glucosinolates, sulforaphane, selenium and flavonoids [3–5].

The *italica* group arises from the cultivation and domestication of *Brassica oleracea* (CC genome; 2n = 18) in the Mediterranean region. Accurate knowledge about the cultivation of *B. oleracea* mustard plants can be traced to the Hellenic culture, starting in approximately the 6th century BC [6]. By distinguishing the *B. oleracea* cultivars, 'Broccoli' is probably a colloquial Latin word for any projecting shoots of the cabbage family [6]. Broccoli-like varieties were developed from selections of desirable *B. oleracea* types during the past 2000 years and formed various broccoli landraces mainly in Italy [6–9]. The broccoli variety 'Vrocculi o Sparaceddi' is considered the first domesticated form of wild brassica from which broccoli originated [10]. During the past 300 years, the heading broccoli has greatly improved, largely attributed to selection by Danish and English horticulturists [6]. For a long time, the consumption of broccoli as a vegetable was confined to the Italian peninsula and it was grown mainly as sprouting broccoli cultivars [10]. With the breeding and improvement of calabrese broccoli varieties, a particular type producing large and compact heads more similar to cauliflower, broccoli spread and gained popularity worldwide [10,11]. Various broccoli landraces were introduced to the United Kingdom in the 1700s and to the United States in the 1800s and became popular after World War II [11]. Broccoli was initially introduced into several southern provinces of China in the 1980s and has been a popular vegetable widely grown in China. In recent years, China, with a cultivation area of over 80,000 ha, has become the largest producer of broccoli in the world [12]. Driven by scientific evidence that broccoli is beneficial to human health, the market demand is still increasing in the main broccoli-producing countries, including China, the US and India [12–14].

With the development of molecular biology technology and functional genomics, a large number of studies on broccoli have been performed. Marker-assisted selection (MAS) and genetic transformation were combined with conventional breeding to improve broccoli for high yield, quality, resistance to biotic and abiotic stresses, etc. We review the recent progress on the genetics and molecular breeding of broccoli, focusing on desirable agronomic traits, male sterility, abiotic stress resistance, disease resistance, secondary metabolites and genetic transformation. Postharvest yellowing (or prolonging shelf life) of broccoli is also a research hotspot that has been reviewed recently and thus is not included in this review [2]. Broccoli improvement by genetic engineering was reviewed in 2016 [1], so relevant advances in recent years from 2016 to 2021 are included in this review.

2. Genetics and Molecular Breeding of Broccoli

2.1. Abiotic Stress Resistance

2.1.1. Heat Stress

Broccoli production faces challenges of demand to extend plant areas and maintain production security under extreme weather brought by climate change [13,15]. Broccoli is suitable for growth in cool weather with optimal temperatures ranging from 15 to 23 °C during the early stages of floral development [16]. High temperatures above 25 severely reduce broccoli quality because (1) most broccoli germplasms require vernalization at temperatures below 23 °C and superoptimal temperatures would even result in no head formation; (2) some broccoli germplasms do not require vernalization, but floral development under high temperatures (e.g., above 30 °C) results in undesirable traits, such as bracting, uneven head surface and sizes of buds, discoloration or even brown bead, making the broccoli products unmarketable; and (3) high temperatures during the head maturity stage decrease broccoli yield [15,17,18]. In recent years, substantial progress has been made in creating heat-tolerant breeding lines and genetically controlling heat tolerance in broccoli. In the USA, researchers have made efforts to achieve sustainable broccoli production under heat tolerance in the main production area on the east coast, supported by projects (National Institute of Food and Agriculture (NIFA) Project No. 2010-51181-21062 and the USDA Vegetable Brassica Research Project (CRIS No. 6080-21000-019-00D)) [16,19,20]. In Asia, researchers are trying to introduce broccoli to subtropical and tropical regions, such as in Taiwan, China and Indonesia [21,22].

The ability to produce high-quality heads by several broccoli germplasms under heat stress is considered a quantitative trait controlling multiple positive loci [13,15]. Lin et al. identified 31 QTLs for head size and weight phenotypes of broccoli grown in high-temperature seasons (average 36.4 °C day/25.9 °C) [23]. Branham et al. constructed a high-density genetic map by genotyping-by-sequencing of a DH broccoli segregating population for heat tolerance and identified five QTLs and one positive epistatic interaction between *QHT_C03* and *QHT_C05*, explaining 62.1% of phenotypic variation [15]. Using a new DH population of broccoli, Branham et al. performed whole-genome resequencing of bulked segregants and identified two novel heat tolerance QTLs, of which *QHT_C09.2* may explain the negative correlation between maturity and heat tolerance [13].

Using reversed genetic approaches, a heat-stress-related broccoli catalase gene was cloned, and ectopic expression of this gene in *Arabidopsis* can enhance heat tolerance, but whether it plays a role in maintaining a high-quality head under high temperatures is still unknown [24,25]. In addition, benefiting from improved sequencing techniques and the release of reference genomes, some researchers performed omics-related studies and identified differentially expressed microRNAs/genes and potential pathways involved in heat tolerance [26,27].

2.1.2. Other Abiotic Stresses

Several studies have focused on broccoli resistance to other stresses, such as proteomic analysis for waterlogging stresses [28], microRNA analysis for salt stress [29] and transcriptome and metabolomics for wounding stress [30]; differentially expressed proteins/microRNAs/genes were identified as possibly related to resistance to these stresses [28–30].

In addition, cuticular waxes on the plant surface contribute to resistance to many environmental stresses, such as drought, UV light, high radiation and both bacterial and fungal pathogens [31]. Some loci and linked markers for this trait have been obtained. Using a natural glossy (cuticular wax defective) mutant, Branham and Farnham identified three candidates, *Bo3g001070*, *Bo3g122030* and *Bo3g008780*, for this trait on C03 [32]. In the broccoli × Chinese kale-derived BolTBDH population, leaf color was segregated, which resulted from the differences in cuticular waxes between broccoli and Chinese kale; a locus for this trait, *LC_C09@15.1*, was identified on C09, explaining 45.64% of the phenotypic variation [33].

2.2. Desirable Agronomic Traits

2.2.1. Heading

Broccoli produces edible reproductive organs characterized by proliferation and developmental arrest of floral buds [17]. Floral head quality is the most important agronomic trait selected by breeders. With forward and reversed genetic approaches, some genes/loci related to head formation have been identified, but the genetic basis remains elusive [34].

Some works tried to identify homologs of the Arabidopsis floral meristem identity genes *LEAFY* (*LFY*), *APETALA1* (*AP1*) and *CAULIFLOWER* (*CAL*) and implied that *BoCAL* and *BoAP1* are involved in curding in cauliflower, a subspecies similar to broccoli but different in the developmental stage of the reproductive meristem at harvest [35–37]. Subsequent studies suggest that heading is quite complex in both cauliflower and broccoli, which seems not to be controlled solely by these floral genes [34].

In the 1990s, researchers started to construct genetic maps by crossing broccoli cultivars/inbred lines with various materials, including broccoli cultivar/landrace, cabbage, cauliflower, kale and Chinese kale, to detect loci of important traits, such as disease resistance, head morphology, nutritional quality and flowering/maturation time [38–42]. Several quantitative trait loci influencing head traits, including head weight, head height/width and floret height/width, have been identified, but the early constructed genetic maps are hard to unify [23,42,43] due to the differences in plant germplasm, marker types and linkage group nomenclature and the lack of *B. oleracea* reference genomes before 2014. Using a double-haploid BolTBDH mapping population derived from Early Big (broccoli DH line) and TO1000DH3 (nonhead Chinese kale), Stansell et al. identified heading-quality QTLs, including *BU_C04@51.5*, *BR_C09@49.5*, *HC_C09@48.8*, *HU_C09@48.8*, *HE_C09@47.7* and *OQ_C09@49.5* (Table S1), and found genomic regions of approximately 49 Mb on C09 harboring *FLOWERING LOCUS C* (*FLC*) homologs *Bo9g173400* and *Bo9g173370*, as hotspots contributing largely to over 40% phenotypic variance of the heading phenotype [33]. In another study, three head quality QTLs, *qCQ-2*, *qCQ-3* and *qCQ-6*, associated with subtropical adaptation were identified [21]; and specific haplotype combinations of candidates *BoFLC3* in the interval containing *qCQ-3* and *PERIANTHIA* (*PAN*, a bZIP-transcription factor required for *AGAMOUS* activation) in the interval containing *qCQ-6*, were supposed to adapt broccoli to high ambient temperature and short daylength. Along with these key head-related traits, QTL mapping for bud morphology was also reported by Stansell et al. and Lin et al. [21,33]. These studies provide genetic information and breeding materials for improving broccoli varieties.

2.2.2. Flowering Time

Flowering is an important agronomic trait of broccoli, as it influences maturity, head quality, hybrid seed production and geographical region adaptation. Flowering time is considered to be controlled by multiple QTLs. To detect QTLs/genes associated with this trait, segregation populations were generated by crossing broccoli with different germplasms, such as broccoli × cabbage, broccoli × Chinese kale and broccoli × broccoli. Different QTLs were detected in these studies, even using similar populations, such as broccoli × cabbage and broccoli × Chinese kale [21,33,44–47]. Most studies have implied that flowering is largely controlled by one or a few major QTLs [21,33,44–47]. As early as the late 1990s, using populations of broccoli (nonvernalization type) × cabbage (vernalization type), broccoli (late flowering type) × Chinese kale (early flowering type), several QTLs for flowering time were mapped [44–47]. Two subsequent studies used a similar population derived from broccoli × cabbage but obtained different results [48,49], possibly due to the differences of the specific germplasms used and the planting environmental conditions. Okazaki et al. detected six QTLs controlling flowering time (from February to July, 2001, Niigata, Niigata Prefecture, Japan), among which the major QTL in the interval BRMS215–F2-R4b, accounting for 36.8% of the phenotypic variance and *BoFLC2* in the interval is thought to be the candidate control of flowering time [48]. Similarly, using a broccoli × cabbage population, Shu et al. combined QTL-seq and a traditional linkage map to detect flowering time loci (from the spring of 2013 to the winter of 2014, Beijing China). A major QTL *Ef2.1* is located on C02 2.65–2.68 Mb, responsible for early flowering and explaining 51.5% of the phenotypic variation, and a homolog of *GROWTH-REGULATING FACTOR 6* (*BolGRF6*) is a possible candidate [49]. Using DH populations of broccoli × Chinese kale, Stansell et al. (2019) identified two QTLs *DM_C03@6.4* and *DM_C09@50.0* for days to maturity, two QTLs *DF_C03@6.4* and *DF_C09@50.0* for days to flowering, and the major QTLs *DM_C09@50.0* and *DF_C09@50.0* on C09 at approximately 50 Mb, explaining approximately 50% of the phenotypic variation [33].

Broccoli is usually sensitive and not feasible in high-temperature areas/seasons, which are thought to impede vernalization, resulting in defects in floral meristem development. Using tropical accessions in Taiwan, China, Lin et al. 2018 identified nonvernalization-responsive QTLs that contribute to subtropical adaptation (high ambient temperature and short day length) [21]. The candidate gene *BoFLC3* identified in the major QTL *qDCI-3* may function as an alternative pathway for the control of flowering in temperate and tropical environments [21].

2.2.3. Plant Architecture

Plant architecture is a complex trait attributed to stem and leaf morphologies, including plant height, leaf size, leaf shape, leaf angle, petiole length and lateral shoot growth. It affects the planting density, yield and quality of broccoli [50]. Several QTLs for plant architecture-relevant stem and leaf traits have been reported [23,27,32,51,52]. Before the release of reference genomes, researchers mapped QTLs associated with leaf lamina width on linkage groups C01 and C07 [51], stem width on LG5 [23], leaf apex on linkage groups C06 and C07 [43], leaf shape on linkage group C3 [43], leaf length on linkage group C7 [43], wing petiole length on linkage group C7 [43] and lobe number, wing number, leaf shape and lamina petiole length on linkage group C3 [43]. In recent years, in addition to focusing on heading traits, Stansell et al. mapped several QTLs for leaf morphology and lateral shoots. Four QTLs for leaf apex, two QTLs for leaf margin and leaf-associated hotspot genomic regions, *Lea3* on C03 0.7–1.7 Mb and *Lea7* on C07 37.0–39.5 Mb were identified. A *GRF1-INTERACTING FACTOR 1* (*GIF1*) homolog (*Bo7g093130*) within major QTL *LA_C07@36.6* may be responsible for the narrow leaf phenotype, and a *LATE MERISTEM IDENTITY1* ortholog (*BoLMI1, Bo3g002560*) near the major *LM_C03@0.7*, explaining over 40% phenotype variation, may be responsible for leaf margin phenotype [33]. Three lateral shoot growth-associated QTLs, *LT_C03@5.9*, *LT_C04@15.0* and *LT_C09@9.0*, are located on C03, C04 and C09, although no likely candidates were predicted [33]. Huang et al. constructed a genetic linkage map using a broccoli DH population and identified QTLs for plant height (PH), maximum outer petiole length (PL) and leaf width (LW), including major QTLs *phc1* for PH on chromosome 1, *plc6−2* for PL on chromosome 6 and *lwc3−1* for LW on chromosome 3 [52].

2.2.4. Stem Development

Broccoli hollow stem is an undesirable phenotypic disorder showing symptoms of cracks in the internal stem tissue [53,54]. It reduces the quality of broccoli products because hollow stems can result in (1) yield reduction, as harvested broccoli comprise partially edible stalks; (2) secondary pathogen infection and rotting of stems and florets [38,39]. The incidence of hollow stems increases when plants grow rapidly, triggered by, for example, high levels of nitrogenous and warm weather but also varies in different broccoli accessions, indicating that this trait is largely genetically determined and can be controlled by breeding resistant varieties [53,54]. However, relevant studies on this trait are very limited. Yu et al. constructed a genetic map using specific locus-amplified fragment (SLAF) sequencing in a double-haploid segregation population of broccoli and defined nine QTLs on C02, C03, C05, C06 and C09 for hollow stems, among which *QHS. C09-2* could explain 14.1% of the phenotypic variation [55].

2.2.5. Head Color

Broccoli is rich in anthocyanin, an important nutritional value with antioxidant activity, can improve health, increase life expectancy and prevent diseases [56]. Anthocyanin accumulation in broccoli inflorescences, especially in septals, makes the appearance range from green/blue to purple. Some cultivars, such as 'Purple Sprouting Early', are selected for rich anthocyanin contents, producing obvious purple heads [56]. Purple traits in *B. oleracea* are attributed to the independent activation of *Brassica oleracea MYB DOMAIN PROTEIN 2* (*BoMYB2*) in subspecies of cabbage, cauliflower, kohlrabi and possibly broccoli [57].

On the other hand, broccoli cultivars producing heads with green-purple color are considered not beautiful and would be less attractive to consumers than the completely green type, especially in the market of China [58]. This green-purple type is sensitive to temperature, and cool weather would induce and deepen the purple degree. Yu et al. mapped this purple sepal trait using a DH population and SLAF sequencing; three QTLs were detected, with a major locus, *qPH. C01–2*, located on linkage group (LG)1, and two loci, *qPH. C01–4* and *qPH. C01–5*, located near *qPH. C01–2* [59].

2.3. Male Sterility and Fertility Restoration

Broccoli displays obvious heterosis and most commercial broccoli varieties are F1 hybrids. The production of broccoli F1 hybrids depends on self-incompatibility before the early 21st century and now nearly completely depends on male sterility-based breeding systems [11,12,59]. Male sterility comprises cytoplasmic male sterility (CMS) and genic male sterility (GMS) [60]. Among them, Ogura CMS, with the advantages of complete male gamete abortion, maternal inheritance and easy transfer, is now the most widely studied and applied male sterility source in broccoli seed production [60,61]. Ogura CMS is a natural mutation found in radish populations [62], which is caused by a mitochondrial gene named *orf138*, and can be fully restored by the nuclear gene *RFO (PPR-B)* [63,64].

Researchers have made efforts to introduce the CMS source to *B. oleracea* by distant hybridization and/or protoplast fusion, but the initially created CMSR1 and CMSR2 contain too much radish cytoplasm, displaying undesirable characteristics, including yellowing at low temperature, deformed flower shape and poor seed setting, which cannot be used in seed production [65–67]. Until the late 1990s, the American Asgrow company applied the method of asymmetric protoplast fusion to reduce the proportion of radish mitochondria, creating CMSR3 with normal fertility and pistil structure; this CMS has been transferred to many elite parent lines, playing a dominant role in the seed production of *B. oleracea* crops [67,68]. During the creation and transfer processes of Ogura CMS, specific *orf138* PCR markers were developed for MAS [69]. Additional mitochondrial markers were developed to distinguish the CMS types; detected by these six *orf138*-related and two simple sequence repeat markers in 2016, Shu et al. divided 39 CMS broccoli accessions into five groups, and observed that CMSR3 constituted 79.49% of the CMS accessions from China [67].

In addition to the Ogura CMS, GMS resources and GMS-based seed production systems were reported as promising alternatives [60,70–73]. A special dominant genic male sterility (DGMS) resource, 79–399–3, which arose in cabbage populations in China, has been successfully and widely applied in cabbage hybrid seed production [61,72]. The DGMS-based breeding system has been established in *B. oleracea* crops, including cabbage, broccoli and kohlrabi [61,72]. Compared with the Ogura CMS, the DGMS-based breeding system displayed advantages of much higher seed quality and yield [61]. However, its utilization is limited in broccoli, largely because homozygous DGMS plants must be preserved and reproduced by tissue culture, which is not effective for large-scale hybrid seed production [61]. Despite these disadvantages, this DGMS-based system has been preserved as an alternative for broccoli hybrid seed production. In recent decades, dozens of broccoli DGMS lines have been created, and several markers have been developed for MAS for the rapid creation of DGMS lines [61,74]. Shu et al. developed generic SSR markers linked to the male-sterile gene, with the marker scaffold10312a showed the highest accuracy of $\geq$96.43% [74]. By distinguishing the amplified products polyacrylamide gel, these markers were successfully used for identification of male and sterile plants in broccoli breeding lines DGMs8554, DGMs93219 and DGMs94174; enabled DGMS plants selection in the seedling stage. Han et al. developed a high-throughput kompetitive allele specific PCR (KASP) marker K6 with high accuracy and no genetic background bias applicable to all *B. oleracea* crops, including broccoli [61]. This marker was based on allele specific fluorescence on an Applied Biosystems Viia 7 real-time PCR system for high-throughput detection. In the DGMS-based breeding system, this marker was used for identifying homozygous DGMS plants from selfing progenies of heterozygous plants as an alternative to test crossing, which requires at least two years and additional labor in tissue culture [61]. These DGMS-specific markers enable effective selection in breeding programs.

On the other hand, there is increasing demand for the reutilization of CMS resources in *B. oleracea* crops. The Ogura CMS restorer gene *RFO* (*PPR-B*) was introduced from radish to rapeseed and recently to *B. oleracea* crops [75,76]. Liu et al. applied strategies of interspecific hybridization and backcrossing and introduced the *RFO* gene from rapeseed to broccoli. The foreground Rfo-specific markers BnRFO-AS2F/BnRFO-AS2F and BnRFO-AS2F/BnRFO-NEW-R, were used for detecting *Rfo*-positive interspecific hybrids; and 28 background SSR markers were used for detecting true intergeneric hybrids and assessing the genetic backgrounds of *Rfo*-positive interspecific hybrids. By evaluating polymorphism loci of the 28 background markers, the BC2 *Rfo*-positive individuals were found closer to the broccoli's genetic background [76].

2.4. Disease Resistance

2.4.1. Downy Mildew

Downy mildew, caused by the obligate fungus *Hyaloperonospora parasitica* (Pers. Fr.), is a destructive disease that affects brassica crops, including broccoli [77,78]. Broccoli plants are often stunned or killed when infected with downy mildew at the young seedling stage or infection can result in quality reduction and yield loss at the adult stage [79,80]. The disease is prevalent in cool weather, with initial symptoms of light green-yellow lesions on the upper leaf surface and later on the undersurface; the spot enlarges and turns yellow; white fungi are visible on the undersurface of leaves under high humidity conditions [79,80]. High resistance to downy mildew both at the young and adult stages is present in some broccoli germplasms and is controlled by a single dominant locus [79,81–84]. Resistance loci were mapped and linkage markers were developed for MAS, but the gene has not been cloned [79,82,83]. Giovannelli et al. identified 8 RAPD (random amplification of polymorphic DNA) markers linked to downy mildew resistance in broccoli (cotyledon and true leaf stage), among which two, $UBC3596_{620}$ and $OPM16_{750}$, were converted to SCAR (sequence characterized amplified regions) markers linked to the locus with 6.7 and

3.3 cM [82]. Farinhó et al. mapped the locus *Pp523* for downy mildew resistance to adult plants of broccoli and developed flanking RAPD markers OPK17_980 and AFLP marker AT. CTA_133/134, with genetic distances of 3.1 cM and 3.6 cM, respectively [83]; in a later study, new AFLP markers were developed and some of them were more user-friendly SCAR and CAPS (cleaved amplified polymorphic sequence) markers; sequencing indicated that *Pp523* is syntenic to the top arm end of *Arabidopsis thaliana* chromosome 1 [79]. We aligned the marker sequences to the broccoli HDEM reference genome [85] and found that the target *Pp523* region is 49.29–50.68 Mb on C8.

2.4.2. Clubroot

Clubroot, caused by the soil-borne pathogen *Plasmodiophora brassicae*, is one of the most devastating diseases of Brassica crops, including broccoli [86–88]. Plants infected by the pathogen form galls on roots, which prevent plant uptake of nutrients and water and become stunted and wilt under warm weather [89]. *B. olearcea* lacks germplasm highly resistant to clubroot, although it has been identified and studied for mining resistance loci/genes in its close relatives, such as turnip, radish and rapeseed [90–93]. The resistance gene *CRa* has been introduced from *B. rapa* to *B. olearcea* by distant hybridization and MAS [94]; in this process *CRa*-specific markers SC2930-Q-FW/SC2930-RV were applied for detection of *CRa* gene in the F1 and each backcross plants, enabled successful introgression of the *CRa* gene into the cabbage inbred lines. In recent years, commercial broccoli varieties with the *CRa* resistance gene, bred by the Syngenta Corporation, are available on the market of China, but the MAS process is not available.

While highly clubroot-resistant germplasms are lacking, some moderate clubroot resistance has been identified in *B. oleracea* [95–97]. There are two studies on genetic mapping for resistance loci related to broccoli, although both of them used broccoli as susceptible parents. These studies are useful for the rapid introduction of clubroot resistance from other subspecies/related species to broccoli with MAS [95,96]. Rocherieux et al. generated F2:3 segregation populations by crossing clubroot-resistant kale and clubroot-susceptible broccoli and constructed a restriction fragment length polymorphism (RFLP) based genetic map. The populations were infected by five isolates and two to five QTLs were identified depending on the isolates; one of these QTLs, *Pb-Bo1*, showed broad-spectrum resistance detected in all isolates [95]. Using populations of crossing resistant double-haploid line (Anju) with a susceptible double-haploid line (GC), Nagaoka et al. identified five CR-QTLs, *pb-Bo(Anju)1*, *PbBo(Anju)2*, *PbBo(Anju)3* and *PbBo(Anju)4* derived from Anju and *pb-Bo(GC)1* from the susceptible parent GC; this study also provided specific primer sequences linked to CR loci and a comparison with known *B. rapa* CR genes [96].

2.4.3. Black Rot

Black rot, caused by the bacterium *Xanthomonas campestris* pv. *campestris* (Pam.) Dowson (*Xcc*), is also one of the most destructive diseases of brassica crops in the world [98,99]. The pathogen often invades plants through hydathodes and spreads through vascular tissue, forming V-shaped lesions at the leaf margins, causing systemic infection and great loss of quality and yield [98–100]. While some resistant plant resources have been reported in *B. oleracea*, few loci/genes have been identified [101–103]. Camargo et al. identified genomic regions associated with young and adult plant resistance to black rot in linkage groups 1, 2 and 9 using a population of black rot-resistant cabbage line BI-16 and susceptible inbred broccoli line OSU Cr-7 [38]. Doullah et al. identified two genomic regions on LG 2 and LG 9 significantly associated with resistance to black rot, with a disease rating of populations from susceptible broccoli green commet P09 and resistant Reiho P01 [104]. In a later study using the same plant materials, Tonu et al. improved the previous genetic map and identified three QTLs, *XccBo(Reiho)1*, *XccBo(Reiho)2 and XccBo(Reiho)1*, for resistance to black rot, and the major QTL, *XccBo(Reiho)2*, was from parent Reiho [105]; comparison using common markers of the previous study by Camargo et al. revealed that *XccBo(Reiho)1* and *XccBo(GC)1* may be identical to the previously reported QTLs [104,105].

Iglesias-Bernabé et al. performed QTL analysis of black rot resistance (*Xcc* race 1) in the BolTBDH mapping population and identified four QTLs, including *Xcc1.1* showing overlap with the previously reported cabbage resistance locus *BRQTL-C1_1*, *BRQTL-C1_2* [106], *Xcc6.1* showing overlap with *BRQTL-C6*, *Xcc8.1* showing overlap with *XccBo(Reiho)2* [105] and a novel locus, *Xcc9.1* [107]; in addition, this study indicated that resistance might be related to the synthesis of secondary metabolites [107].

2.5. Secondary Metabolites

Broccoli contains a number of beneficial secondary metabolites, including glucosinolates/sulforaphane, carotenoids, phenolic acids and flavonoids. Several loci/genes regulating the accumulation of these compounds in broccoli have been identified. Genetic models of secondary metabolite biosynthesis in *Arabidopsis* provide a convenient tool for homologous studies in broccoli [108,109]. Via a homologous cloning strategy, some broccoli genes are isolated directly, including *cytochrome P450 79F1 (CYP79F1), cytochrome P450 83A1 (CYP83A1), UDP-glucosyltransferase 74B1 (UGT74B1), sulfotransferase 18 (ST5b)* and flavin-containing monooxygenase *GS-OX1 (FMOGS-OX1), cytochrome P45083B1 (BoCYP83B1), BoMYB51, GSL-PRO, GSL-ELONG, GSL-ALK, GSL-OH, Myb28* and *BoMYB51* for glucosinolate biosynthesis [109–113], and *BoPAL, BoDFR, BoTT8* and *BoTTG1* for anthocyanin biosynthesis [114]. Genetic loci determining the variation in these secondary metabolites were also detected by genetic mapping. Sotelo et al. performed genetic analysis to identify the genome regions regulating glucosinolate biosynthesis in the DH mapping population BolTBDH and detected eighty-two significant QTLs for individual and total glucosinolate synthesis in leaves, seeds and flower buds, and *QTL9.2* (proposed candidate as *GSL-ALK*) plays a central role in determining glucosinolate variation, showing epistatic interactions with other loci [115]. Brown et al. constructed a genetic linkage map with a broccoli mapping population, identified 14 QTLs associated with the accumulation of aliphatic, indolic or aromatic glucosinolates in florets, and a locus *GSL12* on C09 explains approximately 40% of the phenotypic variability of progoitrin [116]. Li et al. performed genetic mapping for sulforaphane metabolism with a DH population; 18 QTLs for sulforaphane metabolism in broccoli florets were identified, and six QTLs among them were detected in more than one environment [117]. Using the same population previously reported [116], Brown et al. constructed a genetic linkage map with an SNP array and identified three QTLs for carotenoid variation in broccoli florets [118]. Gardner et al. performed QTL analysis saturated with SNP markers in an Illumina 60 K array for total phenolic concentration and its individual components in the population previously reported by Brown et al. [118] and obtained twenty-three loci identified in at least two analyses [119]. In the BolTBDH mapping population, 33 QTLs were identified controlling phenolic concentrations in leaves, flower buds and seeds [120]. In addition, transcriptome analyses were performed to identify differentially expressed genes related to glucosinolate metabolism in broccoli seeds, sprouts and byproducts [121–123].

2.6. Development of Omics Research

Advances in techniques and reduced costs of high-throughput next- and third-generation sequencing have brought high-throughput tools for genomic-related studies and the improvement of broccoli. In 2014, *B. oleracea* draft genome-based short reads of the next generation were released [124,125]; in 2018, the first broccoli (HDEM) reference genome, a high-quality daft genome based on third-generation nanopore long reads and optical maps, was accessible [85]. These studies provided information on genome duplication and gene divergence and the direct prediction of genes related to phytochemicals and morphological variations and, as mentioned above, provided a reference for high density marker development [97,116,118]. Bulked-segregant analysis combined with whole genome resequencing (BSA-seq) for rapid gene/QTL mapping and candidate searching [13,32] and omics-related studies exploring differentially expressed genes/miRNAs related to important traits [26–30,121–123]. In addition, high-throughput strategies promote KASP marker-based fingerprinting for the essential broccoli germplasm [126], genetic diversity and population structure analysis for broccoli cultivars [11,127], and the genomic and morphological domestication syndrome of broccoli calabrese landraces, hybrids and sprouting broccoli [11].

2.7. Genome Editing

Genome editing is a powerful tool for efficient and targeted genome manipulations in living organisms. Depending on the genome editing tools, four engineered nucleases were developed: Meganucleases [128], zinc finger nucleases (ZFNs) [129], transcription activator-like effector-based nucleases (TALENs) [130] and short palindromic repeat (CRISPR)-associated protein (Cas9) systems [131,132]. CRISPR/Cas9 has proven to be a cost-effective and versatile tool for precise and efficient genome editing and in recent years, it has been extensively studied and applied to manipulate desired genes in plants [133]. While it has been realized in some *B. oleracea* crops [134,135], genome editing by CRISPR/Cas9 has not succeeded in complete broccoli background plants. Only one study applied this tool to broccoli-related plant material DH1012, a doubled haploid genotype from the crossing of *B. oleracea alboglabra* (A12DHd) with *B. oleracea italica* (Green Duke GDDH33), targeting *BolC.GA4.a* (*Bol038154*), resulting in dwarf stature [136,137].

2.8. Genetic Transformation

Agrobacterium-mediated transformation in broccoli was first reported by Metz et al. [138]. In the last decade, this genetic engineering tool has been applied for improving broccoli regarding (1) insect resistance by the genes *cry1A(c)*, *cry1C* and *cryIA(b)*; (2) fungal resistance by the Trichoderma harzianum endochitinase gene, *PR-1* and *PR-2*; (3) abiotic stress resistance by *AtHSP101*; (4) herbicide resistance by *Bar* gene; (5) prolonged shelf-life/delayed postharvest yellowing by *ipt* (isopentenyl transferase) gene, *ACC synthase 1*, *BoCLH1* and *ACC oxidase* gene; and (6) flowering control by *CYP86MF*, *SLG*, *FCA* and *CONSTANS*, which has been reviewed by Kumar and Srivastava in 2016 [1]. Thus, we review the advances of broccoli transgenic improvement in recent years (Table 1).

Table 1. Broccoli improvement by genetic transformation in recent years.

Gene Transferred	Origin	Recipient Plant	Performance	References
BoAPX	broccoli	broccoli	enhanced resistance to downy mildew enhanced tolerance to heat stress	[139]
BoWRKY6	broccoli	broccoli	enhanced resistance to downy mildew	[140]
BoiCesA (RNAi)	broccoli	broccoli	enhanced salt tolerance; dwarf and smaller leaves	[141]
BoC3H	broccoli	broccoli	enhanced salt stress tolerance	[142]
BoC3H4	broccoli	broccoli	enhanced salt stress tolerance; more susceptible to *S. sclerotiorum*	[143]
BoERF1	broccoli	broccoli	enhanced salt stress tolerance; enhanced resistance to Sclerotinia stem rot	[144]
cryIAa	Bacillus thuringiensis	broccoli	resistance to diamondback moth	[145]
BoMYB29	wild *B. oleracea*	DH line AG1012, (partial broccoli background)	increased glucosinolate content	[146]
BoTSB1, BoTSB2	broccoli	Arabidopsis	increased glucosinolate content	[147]
BroMYB28 (transient overexpression)	broccoli	broccoli	increased glucoraphanin content	[148]

Table 1. *Cont.*

Gene Transferred	Origin	Recipient Plant	Performance	References
MAM1	broccoli	broccoli	increased sulforaphane content	[149]
FMOGS–OX2	broccoli	broccoli	increased sulforaphane content	[149]
Myrosinase	broccoli	broccoli	increased sulforaphane content	[149]
BoiDAD1F (RNAi)	broccoli	broccoli	recoverable male sterility	[150]
bol-miR171b	broccoli	broccoli	nearly completely male sterile and increased the chlorophyll content	[151]

2.8.1. Transgenic Breeding for Fungal Resistance

In two independent studies, Jiang et al. generated transgenic broccoli plants overexpressing the cytosolic ascorbate peroxidase gene *BoAPX* and the WRKY transcription factor gene *BoWRKY6*; both of them obtained enhanced resistance to downy mildew [139,140]. *BoAPX*-overexpressing broccoli, with a lower level of electrical conductivity and a higher level of APX enzyme activity, exhibited significantly higher resistance to *Hyaloperonospora parasitica* infection, as well as to heat stress, than wild-type plants [139]. *BoWRKY6*-overexpressing broccoli exhibited significantly increased resistance to downy mildew but varied from low to very high [140]; two of them, lines BWK14 and BWK31, exhibited very high resistance to downy mildew [140].

2.8.2. Transgenic Breeding for Abiotic Stress Resistance

Li et al. generated RNAi transgenic broccoli lines targeting the cellulose synthase gene *BoiCesA*; the *BoiCesA* knockdown plants showed a loss of cellulose content and significantly enhanced salt tolerance, and the expression of related genes (*BoiProH, BoiPIP2;2, BoiPIP2;3*) was significantly changed but also displayed phenotypic defects characterized by dwarfs and smaller leaves [141].

In three independent studies, Jiang et al. reported that the overexpression of the C3H-type zinc finger genes *BoC3H* and *BoC3H4* and the ethylene response transcription factor gene *BoERF1* enhanced salt stress tolerance [142–144]. The BoC3H-overexpression lines exhibited higher germination rates, dry weight and chlorophyll content under salt stress and less cell death in the leaves due to the decreased hydrogen peroxide level, relative electrical conductivity and malondialdehyde contents but increased free proline content and catalase, peroxidase and superoxide dismutase enzyme activities [142]. The *BoC3H4*-overexpression lines exhibited increased salinity stress tolerance, with an increase in proline and H_2O_2 and a decrease in chlorophyll loss, MDA and REC compared with WT plants; however, the lines were more susceptible to *S. sclerotiorum*, possibly due to the inhibited expression of the *BoPDF1.2* gene [143]. The *BoERF1*-overexpression lines exhibited a higher seed germination rate and less chlorophyll loss under salt stress, with less cell death in the leaves similar to the BoC3H-overexpression lines; in addition, the transgenic lines showed enhanced resistance to *Sclerotinia* stem rot [144].

2.8.3. Transgenic Breeding for Insect Resistance

Transgenic broccoli for insect resistance was extensively studied in the late 1990s and the beginning of the 21st century [1], but in recent years there have been few related studies. Kumar et al. generated transgenic broccoli overexpressing *cryIAa*, which showed effective resistance to infestation by diamondback moth (*Plutella xylostella*) larvae [145].

2.8.4. Transgenic Breeding for Enriched Glucosinolate/Sulforaphane Content

In recent years, improving the anticancer metabolite glucosinolate/sulforaphane content in broccoli by the genetic engineering of biosynthesis-/regulation-related genes has increased [146–149,152]. Zuluaga et al. reported that the overexpression of *BoMYB29* in DH line AG1012 resulted in the upregulation of the aliphatic glucosinolate pathway and higher production of methylsulphinylalkyl glucosinolates, including glucoraphanin [146]. Li et al. isolated two tryptophan synthase beta subunit (TSB) genes from broccoli and generated overexpression lines of *BoTSB1* or *BoTSB2* in Arabidopsis, which showed accumulation of tryptophan, indole-3-acetic acid (IAA) and indole glucosinolates; this study provides a target for improving glucosinolates, but no broccoli transgenic plants were generated [147]. Studies on *BroMYB28* revealed its possible role in the biosynthesis of glucoraphanin [152], but its function was not proven in broccoli until 2019 [148]. *Agrobacterium*-mediated transient overexpression of *BroMYB28* in broccoli results in the accumulation of glucoraphanin [148]. Cao et al. generated transgenic broccoli by overexpressing *MAM1*, *FMO$_{GS-OX2}$* and *Myrosinase* independently or in triple [149]. Compared with wild-type plants, independent transgenes of *MAM1 FMO$_{GS-OX2}$* and *Myrosinase* enhanced sulforaphane content by 1.7–3.4-, 1.6–2.7- and 3.7-fold, while transgenic plants with the triple gene enhanced sulforaphane content by 1.86–5.5-fold [149].

2.8.5. Transgenic Breeding for Manipulating Male Fertility

Creation of a new male-sterile type by genetic engineering strategies can rapidly provide alternative resources for hybrid seed production. Male-sterile transgenic broccoli was reported by Chen et al. via RNAi of the jasmonic acid pathway gene *BoiDAD1F* [150]. These transgenic plants showed male sterility under normal conditions but recovered to fertility when treated with exogenous JA and were thus suitable for utilization in a two-line seed production system [150]. Li et al. reported that the overexpression of a microRNA *bol-miR171b* in broccoli resulted in nearly complete male sterility and increased the chlorophyll content [151].

3. Conclusions and Future Perspectives

In recent years, progress has been made in the molecular breeding of broccoli for agronomic traits, secondary metabolites, male sterility, abiotic stress resistance, disease resistance and insect resistance. MAS facilitates the breeding of heat-stress-resistant varieties and clubroot-resistant varieties. However, the molecular breeding of broccoli is still restrained by a lack of basic research and an unknown genetic basis of most desirable traits. Future research on the molecular breeding of broccoli may pay attention to the following aspects.

3.1. Mining Functional Loci/Genes

Some linked markers and mapped genes/QTLs for desirable traits have been reported, and with the development of sequencing technology in recent years, candidates have been predicted for mapped genes/QTLs. Omics technologies, such as transcriptomics, proteomics and metabolomics, have been employed to understand the mechanism of desirable traits. Despite efforts, quite a few genes in broccoli have been cloned and functionally verified. Further research should focus on mining and functionally verifying more genes/QTLs for guiding and promoting the breeding work of broccoli: (1) As the most desirable traits in broccoli are controlled by complex QTLs, secondary mapping populations, including near-isogenic lines, introgression lines and chromosome segment substitution lines, should be developed for fine mapping and isolation of these genes; (2) by reverse genetics approaches, released databases and advanced sequencing technology can be used to identify more functional genes; and (3) the obtained target genes, neither from fine mapping nor homology cloning, should be verified by transient expression or genetic transformation.

3.2. Improving Broccoli by Landraces or Other B. oleracea Subspecies

Modern broccoli has very narrow genetic diversity, which may cause undesirable quality, yield and resistance. Broccoli landraces (especially from Italy) and other subspecies provide diverse genetic resources with promising traits, such as differential heading type, differential flowering/maturation time, high glucosinolate content and strong disease resistance. Desirable genes can be introduced by MAS from landraces/different subspecies to breeding materials to improve the quality and extend the genetic diversity of broccoli. A particular case is the head compactness of calabrese broccoli, the most popular broccoli type; in recent decades, head compactness of this broccoli has been significantly enhanced for easy transport and storage, which may be improved via genomic fragment introgression from cauliflower (no published literature).

3.3. Introducing Disease Resistance Genes from Related Species

Broccoli lacks resistance to some devastating diseases, such as clubroot and black rot. To guard broccoli genotypes against these diseases, distant hybridization and MAS can be used to introduce pyramid resistance genes/loci from related species. The clubroot pathogen *P. brassicae* evolved many physiological races showing different infection responses on host plants. Only one resistant locus, *CRa* for race 4, has been introduced from *B. rapa*, which is not enough for sustainable production of broccoli under the threat of other *P. brassicae* races. More resistance genes/loci should be introduced from related species, such as turnip, radish and rapeseed, and pyramided in broccoli. For black rot disease, strong resistance sources have been reported in the A and B genomes of Brassica species, and moderate clubroot resistance has been reported in the C genome of cabbage. These resistance genes/loci can be introduced from *Brassica carinata* and cabbage to broccoli.

3.4. Improving the CRISPR/Cas9 Genome Editing System

The CRISPR/Cas9 system has been proven to be a highly efficient genome editing method in plants. In recent years, this genome editing system has been successfully applied in many crops, including rice, maize, soybean and tomato, for gene function studies and crop improvement, such as high yield, disease resistance, herbicide resistance, ideal plant architecture and other desirable traits. However, the CRISPR/Cas9 system has not been established in broccoli; thus, future studies should pay more attention to improving and employing the CRISPR/Cas9 system for broccoli improvement.

Supplementary Materials: The following are available online at https://www.mdpi.com/article/10.3390/horticulturae7090280/s1, Table S1: Genetic mapping of genes/QTLs in broccoli.

Author Contributions: Conceptualization, F.H.; writing—original draft preparation, F.H.; writing—review and editing, Y.L., Z.F., L.Y., M.Z., Y.Z., H.L., Y.W. and J.J.; supervision, F.H.; project administration, Z.L.; funding acquisition, Z.F. and Z.L. All authors have read and agreed to the published version of the manuscript.

Funding: This research was funded by grants from the Central Public-interest Scientific Institution Basal Research Fund (Grant No. IVF-BRF2020003, IVF-BRF2021003), the China Agriculture Research System (Grant No. CARS-23-08A), the Agricultural Science and Technology Innovation Program (Grant No. ASTIP), and the State Key Laboratory of Vegetable Germplasm Innovation (Grant No. SKL-VGI).

Institutional Review Board Statement: Not applicable.

Informed Consent Statement: Not applicable.

Conflicts of Interest: The authors declare no conflict of interest.

References

1. Kumar, P.; Srivastava, D.K. Biotechnological advancement in genetic improvement of broccoli (*Brassica oleracea* L. var. *italica*), an important vegetable crop. *Biotechnol. Lett.* **2016**, *38*, 1049–1063. [CrossRef] [PubMed]
2. Luo, F.; Fang, H.-X.; Wei, B.-D.; Cheng, S.-C.; Zhou, Q.; Zhou, X.; Zhang, X.; Zhao, Y.-B.; Ji, S.-J. Advance in yellowing mechanism and the regulation technology of post-harvested broccoli. *Food Qual. Saf.* **2020**, *4*, 107–113. [CrossRef]
3. Vallejo, F.; Garcia-Viguera, C.; Tomas-Barberan, F. Changes in Broccoli (*Brassica oleracea* L. var. *italica*) Health-Promoting Compounds with Inflorescence Development. *J. Agric. Food Chem.* **2003**, *51*, 3776–3782. [CrossRef] [PubMed]
4. Vallejo, F.; Tomas-Barberan, F.; Ferreres, F. Characterisation of flavonols in broccoli (*Brassica oleracea* L. var. *italica*) by liquid chromatography–UV diode-array detection–electrospray ionisation mass spectrometry. *J. Chromatogr. A* **2004**, *1054*, 181–193. [CrossRef] [PubMed]

5. Shapiro, T.A.; Fahey, J.W.; Wade, K.L.; Stephenson, K.K.; Talalay, P. Chemoprotective Glucosinolates and Isothiocyanates of Broccoli Sprouts. *Cancer Epidemiol. Prev. Biomark.* **2001**, *10*, 501–508.

6. Buck, P.A. Origin and taxonomy of broccoli. *Econ. Bot.* **1956**, *10*, 250–253. [CrossRef]

7. Massie, I.H.; Astley, D.; King, G.J. Patterns of genetic diversity and relationships between regional groups and populations of Italian landrace cauliflower and broccoli (*Brassica oleracea* L. var. *botrytis* L. and var. *italica* Plenck). *Acta Hort.* **1996**, *407*, 45–54. [CrossRef]

8. Ciancaleoni, S.; Chiarenza, G.L.; Raggi, L.; Branca, F.; Negri, V. Diversity characterisation of broccoli (*Brassica oleracea* L. var. *italica* Plenck) landraces for their on-farm (in situ) safeguard and use in breeding programs. *Genet. Resour. Crop. Evol.* **2013**, *61*, 451–464. [CrossRef]

9. Maggioni, L.; Von Bothmer, R.; Poulsen, G.; Lipman, E. Domestication, diversity and use of *Brassica oleracea* L., based on ancient Greek and Latin texts. *Genet. Resour. Crop. Evol.* **2017**, *65*, 137–159. [CrossRef]

10. Branca, F.; Chiarenza, G.L.; Cavallaro, C.; Gu, H.; Zhao, Z.; Tribulato, A. Diversity of Sicilian broccoli (*Brassica oleracea* var. *italica*) and cauliflower (*Brassica oleracea* var. *botrytis*) landraces and their distinctive bio-morphological, antioxidant, and genetic traits. *Genet. Resour. Crop. Evol.* **2017**, *65*, 485–502. [CrossRef]

11. Stansell, Z.; Björkman, T. From landrace to modern hybrid broccoli: The genomic and morphological domestication syndrome within a diverse *B. oleracea* collection. *Hortic. Res.* **2020**, *7*, 1–17. [CrossRef] [PubMed]

12. Li, Z.; Mei, Y.; Liu, Y.; Fang, Z.; Yang, L.; Zhuang, M.; Zhang, Y.; Lv, H. The evolution of genetic diversity of broccoli cultivars in China since 1980. *Sci. Hortic.* **2019**, *250*, 69–80. [CrossRef]

13. Branham, S.E.; Farnham, M.W. Identification of heat tolerance loci in broccoli through bulked segregant analysis using whole genome resequencing. *Euphytica* **2019**, *215*, 34. [CrossRef]

14. Kumar, R.; Kandpal, K. Influence of foliar fertilization of boron on broccoli (*Brassica oleracea* var. *italica*) in boron deficient soil of Doon Valley, India. *Progress. Hortic.* **2017**, *49*, 65. [CrossRef]

15. Branham, S.E.; Stansell, Z.J.; Couillard, D.M.; Farnham, M.W. Quantitative trait loci mapping of heat tolerance in broccoli (*Brassica oleracea* var. *italica*) using genotyping-by-sequencing. *Theor. Appl. Genet.* **2017**, *130*, 529–538. [CrossRef]

16. Farnham, M.W.; Bjorkman, T. Breeding Vegetables Adapted to High Temperatures: A Case Study with Broccoli. *HortScience* **2011**, *46*, 1093–1097. [CrossRef]

17. Björkman, T.; Pearson, K.J. High temperature arrest of inflorescence development in broccoli (*Brassica oleracea* var. *italica* L.). *J. Exp. Bot.* **1998**, *49*, 101–106. [CrossRef]

18. Heather, D.; Sieczka, J.; Dickson, M.; Wolfe, D. Heat Tolerance and Holding Ability in Broccoli. *J. Am. Soc. Hortic. Sci.* **1992**, *117*, 887–892. [CrossRef]

19. Atallah, S.S.; Gómez, M.I.; Björkman, T. Localization effects for a fresh vegetable product supply chain: Broccoli in the eastern United States. *Food Policy* **2014**, *49*, 151–159. [CrossRef]

20. Ward, B.; Smith, P.; James, S.; Stansell, Z.; Farnham, M. Increasing Plant Density in Eastern United States Broccoli Production Systems to Maximize Marketable Head Yields. *Horttechnology* **2015**, *25*, 330–334. [CrossRef]

21. Lin, Y.-R.; Lee, J.-Y.; Tseng, M.-C.; Lee, C.-Y.; Shen, C.-H.; Wang, C.-S.; Liou, C.-C.; Shuang, L.-S.; Paterson, A.H.; Hwu, K.-K. Subtropical adaptation of a temperate plant (*Brassica oleracea* var. *italica*) utilizes non-vernalization-responsive QTLs. *Sci. Rep.* **2018**, *8*, 13609. [CrossRef]

22. Astarini, I.A.; Defiani, M.R.; Suriani, N.L.; Griffiths, P.; Stefanova, K.; Siddique, K. Adaptation of broccoli (*Brassica oleracea* var. *italica* L.) to high and low altitudes in Bali, Indonesia. *Biodiversitas J. Biol. Divers.* **2020**, *21*, 5263–5269. [CrossRef]

23. Lin, K.H.; Chang, L.C.; Lai, C.D.; Lo, H.F. AFLP mapping of quantitative trait loci influencing seven head-related traits in broccoli (*Brassica oleracea* var. *italica*). *J. Hortic. Sci. Biotechnol.* **2013**, *88*, 257–268. [CrossRef]

24. Lin, K.-H.; Huang, H.-C.; Lin, C.-Y. Cloning, expression and physiological analysis of broccoli catalase gene and Chinese cabbage ascorbate peroxidase gene under heat stress. *Plant Cell Rep.* **2010**, *29*, 575–593. [CrossRef]

25. Chiang, C.-M.; Chen, S.-P.; Chen, L.-F.O.; Chiang, M.-C.; Chien, H.-L.; Lin, K.-H. Expression of the broccoli catalase gene (*BoCAT*) enhances heat tolerance in transgenic *Arabidopsis*. *J. Plant Biochem. Biotechnol.* **2013**, *23*, 266–277. [CrossRef]

26. Chen, C.-C.; Fu, S.-F.; Norikazu, M.; Yang, Y.-W.; Liu, Y.-J.; Ikeo, K.; Gojobori, T.; Huang, H.-J. Comparative miRNAs analysis of Two contrasting broccoli inbred lines with divergent head-forming capacity under temperature stress. *BMC Genom.* **2015**, *16*, 1026. [CrossRef] [PubMed]

27. Lin, C.-W.; Fu, S.-F.; Liu, Y.-J.; Chen, C.-C.; Chang, C.-H.; Yang, Y.-W.; Huang, H.-J. Analysis of ambient temperature-responsive transcriptome in shoot apical meristem of heat-tolerant and heat-sensitive broccoli inbred lines during floral head formation. *BMC Plant Biol.* **2019**, *19*, 3. [CrossRef]

28. Lin, H.-H.; Lin, K.-H.; Chen, S.-C.; Shen, Y.-H.; Lo, H.-F. Proteomic analysis of broccoli (*Brassica oleracea*) under high temperature and waterlogging stresses. *Bot. Stud.* **2015**, *56*, 1–11. [CrossRef] [PubMed]

29. Tian, Y.; Tian, Y.; Luo, X.; Zhou, T.; Huang, Z.; Liu, Y.; Qiu, Y.; Hou, B.; Sun, D.; Deng, H.; et al. Identification and characterization of microRNAs related to salt stress in broccoli, using high-throughput sequencing and bioinformatics analysis. *BMC Plant Biol.* **2014**, *14*, 1–13. [CrossRef]

30. Torres-Contreras, A.M.; Senés-Guerrero, C.; Pacheco, A.; González-Agüero, M.; Ramos-Parra, P.A.; Cisneros-Zevallos, L.; Jacobo-Velázquez, D.A. Genes differentially expressed in broccoli as an early and late response to wounding stress. *Postharvest Biol. Technol.* **2018**, *145*, 172–182. [CrossRef]

31. Lee, J.; Yang, K.; Lee, M.; Kim, S.; Kim, J.; Lim, S.; Kang, G.-H.; Min, S.R.; Kim, S.-J.; Park, S.U.; et al. Differentiated cuticular wax content and expression patterns of cuticular wax biosynthetic genes in bloomed and bloomless broccoli (*Brassica oleracea* var. *italica*). *Process. Biochem.* **2015**, *50*, 456–462. [CrossRef]

32. Branham, S.E.; Farnham, M.W. Genotyping-by-sequencing of waxy and glossy near-isogenic broccoli lines. *Euphytica* **2017**, *213*, 84. [CrossRef]

33. Stansell, Z.; Farnham, M.; Björkman, T. Complex Horticultural Quality Traits in Broccoli Are Illuminated by Evaluation of the Immortal BolTBDH Mapping Population. *Front. Plant Sci.* **2019**, *10*, 1104. [CrossRef] [PubMed]

34. Duclos, D.V.; Bjoerkman, T. Meristem identity gene expression during curd proliferation and flower initiation in *Brassica oleracea*. *J. Exp. Bot.* **2008**, *59*, 421–433. [CrossRef]

35. Fujime, Y.; Okuda, N. The physiology of flowering in brassicas, especially about cauliflower and broccoli. *Acta Hortic.* **1996**, 247–254. [CrossRef]

36. Carr, S.M.; Irish, V.F. Floral homeotic gene expression defines developmental arrest stages in *Brassica oleracea* L. vars. *botrytis* and *italica*. *Planta* **1997**, *201*, 179–188. [CrossRef]

37. Smith, L.B.; King, G.J. The distribution of BoCAL-a alleles in *Brassica oleracea* is consistent with a genetic model for curd development and domestication of the cauliflower. *Mol. Breed.* **2000**, *6*, 603–613. [CrossRef]

38. Camargo, L.; Williams, P.; Osborn, T. Mapping of quantitative trait loci controlling resistance of *Brassica oleracea* to *Xanthomonas campestris* pv. *campestris* in the field and greenhouse. *Phytopathology* **1995**, *85*, 1296–1300. [CrossRef]

39. Bohuon, E.J.R.; Keith, D.J.; Parkin, I.A.P.; Sharpe, A.G.; Lydiate, D.J. Alignment of the conserved C genomes of *Brassica oleracea* and Brassica napus. *Theor. Appl. Genet.* **1996**, *93*, 833–839. [CrossRef]

40. Kianian, S.F.; Quiros, C.F. Generation of a *Brassica oleracea* composite RFLP map: Linkage arrangements among various populations and evolutionary implications. *Theor. Appl. Genet.* **1992**, *84*, 544–554. [CrossRef] [PubMed]

41. Slocum, M.K.; Figdore, S.S.; Kennard, W.C.; Suzuki, J.Y.; Osborn, T.C. Linkage arrangement of restriction fragment length polymorphism loci in *Brassica oleracea*. *Theor. Appl. Genet.* **1990**, *80*, 57–64. [CrossRef]

42. Brown, A.F.; Jeffery, E.; Juvik, J.A. A Polymerase Chain Reaction-based Linkage Map of Broccoli and Identification of Quantitative Trait Loci Associated with Harvest Date and Head Weight. *J. Am. Soc. Hortic. Sci.* **2007**, *132*, 507–513. [CrossRef]

43. Walley, P.G.; Carder, J.; Skipper, E.; Mathas, E.; Lynn, J.; Pink, D.; Buchanan-Wollaston, V. A new broccoli x broccoli immortal mapping population and framework genetic map: Tools for breeders and complex trait analysis. *Theor. Appl. Genet.* **2012**, *124*, 467–484. [CrossRef]

44. Kennard, W.C.; Slocum, M.K.; Figdore, S.S.; Osborn, T.C. Genetic analysis of morphological variation in *Brassica oleracea* using molecular markers. *Theor. Appl. Genet.* **1994**, *87*, 721–732. [CrossRef] [PubMed]

45. Camargo, L.; Osborn, T. Mapping loci controlling flowering time in *Brassica oleracea*. *Theor. Appl. Genet.* **1996**, *92*, 610–616. [CrossRef] [PubMed]

46. Bohuon, E.J.R.; Ramsay, L.D.; Craft, J.A.; Arthur, A.E.; Marshall, D.; Lydiate, D.J.; Kearsey, M.J. The Association of Flowering Time Quantitative Trait Loci with Duplicated Regions and Candidate Loci in *Brassica oleracea*. *Genetics* **1998**, *150*, 393–401. [CrossRef] [PubMed]

47. Rae, A.M.; Howell, E.C.; Kearsey, M.J. More QTL for flowering time revealed by substitution lines in *Brassica oleracea*. *Heredity* **1999**, *83*, 586–596. [CrossRef]

48. Okazaki, K.; Sakamoto, K.; Kikuchi, R.; Saito, A.; Togashi, E.; Kuginuki, Y.; Matsumoto, S.; Hirai, M. Mapping and characterization of FLC homologs and QTL analysis of flowering time in *Brassica oleracea*. *Theor. Appl. Genet.* **2006**, *114*, 595–608. [CrossRef] [PubMed]

49. Shu, J.S.; Liu, Y.M.; Zhang, L.L.; Li, Z.S.; Fang, Z.Y.; Yang, L.M.; Zhuang, M.; Zhang, Y.Y.; Lv, H.H. QTL-seq for rapid identification of candidate genes for flowering time in broccoli x cabbage. *Theor. Appl. Genet.* **2018**, *131*, 917–928. [CrossRef] [PubMed]

50. Hale, A.L.; Farnham, M.W.; Nzaramba, M.N.; Kimbeng, C.A. Heterosis for horticultural traits in Broccoli. *Theor. Appl. Genet.* **2007**, *115*, 351–360. [CrossRef] [PubMed]

51. Lan, T.-H.; Paterson, A.H. Comparative Mapping of Quantitative Trait Loci Sculpting the Curd of *Brassica oleracea*. *Genetics* **2000**, *155*, 1927–1954. [CrossRef]

52. Huang, J.; Sun, J.; Liu, E.; Yuan, S.; Liu, Y.; Han, F.; Li, Z.; Fang, Z.; Yang, L.; Zhuang, M.; et al. Mapping of QTLs detected in a broccoli double diploid population for planting density traits. *Sci. Hortic.* **2021**, *277*, 109835. [CrossRef]

53. Boersma, M.; Gracie, A.; Brown, P. Evidence of mechanical tissue strain in the development of hollow stem in broccoli. *Sci. Hortic.* **2013**, *164*, 353–358. [CrossRef]

54. Boersma, M.; Gracie, A.J.; Brown, P.H. Relationship between growth rate and the development of hollow stem in broccoli. *Crop. Pasture Sci.* **2009**, *60*, 995–1001. [CrossRef]

55. Yu, H.; Wang, J.; Zhao, Z.; Sheng, X.; Shen, Y.; Branca, F.; Gu, H. Construction of a High-Density Genetic Map and Identification of Loci Related to Hollow Stem Trait in Broccoli (*Brassic oleracea* L. *italica*). *Front. Plant Sci.* **2019**, *10*, 10. [CrossRef] [PubMed]

56. Chaudhary, A.; Choudhary, S.; Sharma, U.; Vig, A.P.; Singh, B.; Arora, S. Purple head broccoli (*Brassica oleracea* L. var. *italica* Plenck), a functional food crop for antioxidant and anticancer potential. *J. Food Sci. Technol.* **2018**, *55*, 1806–1815. [CrossRef] [PubMed]

57. Yan, C.; An, G.; Zhu, T.; Zhang, W.; Zhang, L.; Peng, L.; Chen, J.; Kuang, H. Independent activation of the BoMYB2 gene leading to purple traits in *Brassica oleracea*. *Theor. Appl. Genet.* **2018**, *132*, 895–906. [CrossRef] [PubMed]

58. Yu, H.; Wang, J.; Sheng, X.; Zhao, Z.; Shen, Y.; Branca, F.; Gu, H. Construction of a high-density genetic map and identification of loci controlling purple sepal trait of flower head in *Brassica oleracea* L. italica. *BMC Plant Biol.* **2019**, *19*, 228. [CrossRef] [PubMed]
59. Reshma, V.; Adarsh, M.; Dhanush, K. Development of hybrids in cole crops: A review. *Plant Arch.* **2018**, *18*, 1–11.
60. Han, F.; Yuan, K.; Kong, C.; Zhang, X.; Yang, L.; Zhuang, M.; Zhang, Y.; Li, Z.; Wang, Y.; Fang, Z.; et al. Fine mapping and candidate gene identification of the genic male-sterile gene ms3 in cabbage 51S. *Theor. Appl. Genet.* **2018**, *131*, 2651–2661. [CrossRef]
61. Han, F.; Zhang, X.; Yuan, K.; Fang, Z.; Yang, L.; Zhuang, M.; Zhang, Y.; Wang, Y.; Liu, Y.; Li, Z.; et al. A user-friendly KASP molecular marker developed for the DGMS-based breeding system in *Brassica oleracea* species. *Mol. Breed.* **2019**, *39*, 90. [CrossRef]
62. Ogura, H. Studies on the new male sterility in Japanese radish, with special references on the utilization of this sterility towards the practical raising of hybrid seeds. *Mem. Fac. Agric. Kagoshima Univ.* **1968**, *6*, 40–75.
63. Uyttewaal, M.; Arnal, N.; Quadrado, M.; Martin-Canadell, A.; Vrielynck, N.; Hiard, S.; Gherbi, H.; Bendahmane, A.; Budar, F.; Mireau, H. Characterization of *Raphanus sativus* Pentatricopeptide Repeat Proteins Encoded by the Fertility Restorer Locus for Ogura Cytoplasmic Male Sterility. *Plant Cell* **2009**, *20*, 3331–3345. [CrossRef]
64. Yamagishi, H.; Jikuya, M.; Okushiro, K.; Hashimoto, A.; Fukunaga, A.; Takenaka, M.; Terachi, T. A single nucleotide substitution in the coding region of Ogura male sterile gene, orf138, determines effectiveness of a fertility restorer gene, Rfo, in radish. *Mol. Genet. Genom.* **2021**, *296*, 705–717. [CrossRef]
65. Bannerot, H.; Boulidard, L.; Cauderon, Y.; Tempe, J. Transfer of cytoplasmic male sterility from *Raphanus sativus* to *Brassica oleracea*. *Proc. Eucarpia Meet. Crucif.* **1974**, *25*, 52–54.
66. Sigareva, M.A.; Earle, E.D. Direct transfer of a cold-tolerant Ogura male-sterile cytoplasm into cabbage (*Brassica oleracea* ssp. *capitata*) via protoplast fusion. *Theor. Appl. Genet.* **1997**, *94*, 213–220. [CrossRef]
67. Shu, J.; Liu, Y.; Li, Z.; Zhang, L.; Fang, Z.; Yang, L.; Zhuang, M.; Zhang, Y.; Lv, H. Detection of the Diversity of Cytoplasmic Male Sterility Sources in Broccoli (*Brassica oleracea* var. *Italica*) Using Mitochondrial Markers. *Front. Plant Sci.* **2016**, *7*, 927. [CrossRef] [PubMed]
68. Wang, Q.; Zhang, Y.; Fang, Z.; Liu, Y.; Yang, L.; Zhuang, M. Chloroplast and mitochondrial SSR help to distinguish allo-cytoplasmic male sterile types in cabbage (*Brassica oleracea* L. var. *capitata*). *Mol. Breed.* **2011**, *30*, 709–716. [CrossRef]
69. Yao, X.; Li, Y.; Xie, Z.; Liu, L. Identification of specific SRAP marker associated with cytoplasmic male sterility gene of broccoli. *Mol. Plant Breed.* **2009**, *7*, 941–947.
70. Sampson, D.R. Linkage of genetic male sterility with a seedling marker and its use in producing f1 hybrid seed of *brassica oleracea* (cabbage, broccoli, kale, etc.). *Can. J. Plant Sci.* **1966**, *46*, 703. [CrossRef]
71. Ruffio-Chable, V.; Bellis, H.; Herve, Y. A dominant gene for male sterility in cauliflower (*Brassica oleracea* var. *botrytis*): Phenotype expression, inheritance, and use in F1 hybrid production. *Euphytica* **1993**, *67*, 9–17. [CrossRef]
72. Fang, Z.; Sun, P.; Liu, Y.; Yang, L.; Wang, X.; Hou, A.; Bian, C. A male sterile line with dominant gene (Ms) in cabbage (*Brassica oleracea* var. *capitata*) and its utilization for hybrid seed production. *Euphytica* **1997**, *97*, 265–268. [CrossRef]
73. Ji, J.-L.; Yang, L.-M.; Fang, Z.-Y.; Zhuang, M.; Zhang, Y.-Y.; Lv, H.-H.; Liu, Y.-M.; Li, Z.-S. Recessive male sterility in cabbage (*Brassica oleracea* var. *capitata*) caused by loss of function of BoCYP704B1 due to the insertion of a LTR-retrotransposon. *Theor. Appl. Genet.* **2017**, *130*, 1441–1451. [CrossRef] [PubMed]
74. Shu, J.; Liu, Y.; Li, Z.; Zhang, L.; Fang, Z.; Yang, L.; Zhuang, M.; Lv, H. A generic SSR marker closely linked to a dominant genic male sterility gene (DGMs79-399-3) in broccoli (*Brassica oleracea* var. *italica*). *Mol. Breed.* **2016**, *36*, 86. [CrossRef]
75. Yu, H.-L.; Fang, Z.-Y.; Liu, Y.-M.; Yang, L.-M.; Zhuang, M.; Lv, H.; Li, Z.-S.; Han, F.-Q.; Liu, X.-P.; Zhang, Y.-Y. Development of a novel allele-specific Rfo marker and creation of Ogura CMS fertility-restored interspecific hybrids in *Brassica oleracea*. *Theor. Appl. Genet.* **2016**, *129*, 1625–1637. [CrossRef] [PubMed]
76. Liu, C.-Q.; Li, G.-Q.; Yao, X.-Q.; Huang, L.; Wu, X.-Y.; Xie, Z.-J. Characterization of Ogura CMS fertility-restored interspecific hybrids and backcross progenies from crosses between broccoli and rapeseed. *Euphytica* **2020**, *216*, 1–12. [CrossRef]
77. Monot, C.; Pajot, E.; Le Corre, D.; Silué, D. Induction of systemic resistance in broccoli (*Brassica oleracea* var. *botrytis*) against downy mildew (*Peronospora parasitica*) by avirulent isolates. *Biol. Control.* **2002**, *24*, 75–81. [CrossRef]
78. Monot, C.; Penguilly, D.; Silué, D. First confirmed report of downy mildew caused by *Hyaloperonospora parasitica* on broccoli, cauliflower and Romanesco-type cauliflower heads in France. *Plant Pathol.* **2010**, *59*, 1165. [CrossRef]
79. Farinhó, M.; Coelho, P.; Carlier, J.; Svetleva, D.; Monteiro, A.; Leitão, J.M. Mapping of a locus for adult plant resistance to downy mildew in broccoli (*Brassica oleracea* convar. *italica*). *Theor. Appl. Genet.* **2004**, *109*, 1392–1398. [CrossRef]
80. Wang, M.; Farnham, M.W.; Thomas, C.E. Phenotypic Variation for Downy Mildew Resistance among Inbred Broccoli. *HortScience* **2000**, *35*, 925–929. [CrossRef]
81. Wang, M.; Farnham, M.W.; Thomas, C.E. Inheritance of True Leaf Stage Downy Mildew Resistance in Broccoli. *J. Am. Soc. Hortic. Sci.* **2001**, *126*, 727–729. [CrossRef]
82. Giovannelli, J.L.; Farnham, M.W.; Wang, M.; Strand, A.E. Development of Sequence Characterized Amplified Region Markers Linked to Downy Mildew Resistance in Broccoli. *J. Am. Soc. Hortic. Sci.* **2002**, *127*, 597–601. [CrossRef]
83. Farnham, M.; Wang, M.; Thomas, C. A single dominant gene for downy mildew resistance in broccoli. *Euphytica* **2002**, *128*, 405–407. [CrossRef]
84. Coelho, P.S.; Monteiro, A.A. Inheritance of downy mildew resistance in mature broccoli plants. *Euphytica* **2003**, *131*, 65–69. [CrossRef]

85. Belser, C.; Istace, B.; Denis, E.; Dubarry, M.; Baurens, F.-C.; Falentin, C.; Genete, M.; Berrabah, W.; Chèvre, A.-M.; Delourme, R.; et al. Chromosome-scale assemblies of plant genomes using nanopore long reads and optical maps. *Nat. Plants* **2018**, *4*, 879–887. [CrossRef] [PubMed]

86. Lovelock, D.A.; Donald, C.E.; Conlan, X.; Cahill, D.M. Salicylic acid suppression of clubroot in broccoli (Brassicae oleracea var. *italica*) caused by the obligate biotroph *Plasmodiophora brassicae*. *Australas. Plant Pathol.* **2012**, *42*, 141–153. [CrossRef]

87. Crute, I.; Gray, A.; Crisp, P.; Buczacki, S. Variation in *Plasmodiophora brassicae* and resistance to clubroot disease in brassicas and allied crops-a critical review. *Plant Breed. Abstr.* **1980**, *50*, 91–104.

88. Hirai, M. Genetic Analysis of Clubroot Resistance in Brassica Crops. *Breed. Sci.* **2006**, *56*, 223–229. [CrossRef]

89. Voorrips, R.E. *Plasmodiophora brassicae*: Aspects of pathogenesis and resistance in *Brassica oleracea*. *Euphytica* **1995**, *83*, 139–146. [CrossRef]

90. Diederichsen, E.; Frauen, M.; Linders, E.G.A.; Hatakeyama, K.; Hirai, M. Status and Perspectives of Clubroot Resistance Breeding in Crucifer Crops. *J. Plant Growth Regul.* **2009**, *28*, 265–281. [CrossRef]

91. Hirani, A.; Gao, F.; Liu, J.; Fu, G.; Wu, C.; McVetty, P.B.E.; Duncan, R.W.; Li, G. Combinations of Independent Dominant Loci Conferring Clubroot Resistance in All Four Turnip Accessions (*Brassica rapa*) from the European Clubroot Differential Set. *Front. Plant Sci.* **2018**, *9*, 1628. [CrossRef]

92. Gan, C.; Deng, X.; Cui, L.; Yu, X.; Yuan, W.; Dai, Z.; Yao, M.; Pang, W.; Ma, Y.; Yu, X.; et al. Construction of a high-density genetic linkage map and identification of quantitative trait loci associated with clubroot resistance in radish (*Raphanus sativus* L.). *Mol. Breed.* **2019**, *39*, 1–12. [CrossRef]

93. Li, L.; Luo, Y.; Chen, B.; Xu, K.; Zhang, F.; Li, H.; Huang, Q.; Xiao, X.; Zhang, T.; Hu, J.; et al. A Genome-Wide Association Study Reveals New Loci for Resistance to Clubroot Disease in Brassica napus. *Front. Plant Sci.* **2016**, *7*, 1483. [CrossRef]

94. Ren, W.; Li, Z.; Han, F.; Zhang, B.; Li, X.; Fang, Z.; Yang, L.; Zhuang, M.; Lv, H.; Liu, Y.; et al. Utilization of Ogura CMS germplasm with the clubroot resistance gene by fertility restoration and cytoplasm replacement in *Brassica oleracea* L. *Hortic. Res.* **2020**, *7*, 1–10. [CrossRef] [PubMed]

95. Rocherieux, J.; Glory, P.; Giboulot, A.; Boury, S.; Barbeyron, G.; Thomas, G.; Manzanares-Dauleux, M.J. Isolate-specific and broad-spectrum QTLs are involved in the control of clubroot in *Brassica oleracea*. *Theor. Appl. Genet.* **2004**, *108*, 1555–1563. [CrossRef]

96. Nagaoka, T.; Doullah, M.A.U.; Matsumoto, S.; Kawasaki, S.; Ishikawa, T.; Hori, H.; Okazaki, K. Identification of QTLs that control clubroot resistance in *Brassica oleracea* and comparative analysis of clubroot resistance genes between B. rapa and B. oleracea. *Theor. Appl. Genet.* **2010**, *120*, 1335–1346. [CrossRef]

97. Peng, L.; Zhou, L.; Li, Q.; Wei, D.; Ren, X.; Song, H.; Mei, J.; Si, J.; Qian, W. Identification of Quantitative Trait Loci for Clubroot Resistance in *Brassica oleracea* With the Use of Brassica SNP Microarray. *Front. Plant Sci.* **2018**, *9*, 822. [CrossRef]

98. Mirik, M.; Selcuk, F.; Aysan, Y.; Sahin, F. First Outbreak of Bacterial Black Rot on Cabbage, Broccoli, and Brussels Sprouts Caused by *Xanthomonas campestris* pv. *campestris* in the Mediterranean Region of Turkey. *Plant Dis.* **2008**, *92*, 176. [CrossRef] [PubMed]

99. Nagai, H.; Miyake, N.; Kato, S.; Maekawa, D.; Inoue, Y.; Takikawa, Y. Improved control of black rot of broccoli caused by *Xanthomonas campestris* pv. *campestris* using a bacteriophage and a nonpathogenic *Xanthomonas* sp. strain. *J. Gen. Plant Pathol.* **2017**, *83*, 373–381. [CrossRef]

100. Sakai, H.; Ikeda, K.; Urushibara, T.; Shiraishi, T. Black rot of broccoli (*Brassica oleracea* var. *italica*) caused by *Xanthomonas campestris* pv. *campestris*. *Jpn. J. Phytopathol.* **2006**, *72*, 116–119. [CrossRef]

101. Saha, P.; Kalia, P.; Sharma, M.; Singh, D. New source of black rot disease resistance in *Brassica oleracea* and genetic analysis of resistance. *Euphytica* **2016**, *207*, 35–48. [CrossRef]

102. Afrin, K.S.; Rahim, A.; Park, J.-I.; Natarajan, S.; Kim, H.-T.; Nou, I.-S. Identification of NBS-encoding genes linked to black rot resistance in cabbage (*Brassica oleracea* var. *capitata*). *Mol. Biol. Rep.* **2018**, *45*, 773–785. [CrossRef]

103. Afrin, K.S.; Rahim, A.; Park, J.-I.; Natarajan, S.; Rubel, M.H.; Kim, H.-T.; Nou, A.I.-S. Screening of Cabbage (*Brassica oleracea* L.) Germplasm for Resistance to Black Rot. *Plant Breed. Biotechnol.* **2018**, *6*, 30–43. [CrossRef]

104. Doullah, M.; Mohsin, G.; Ishikawa, K.; Hori, H.; Okazaki, K. Construction of a Linkage Map and QTL analysis for Black Rot Resistance in *Brassica oleracea* L. *Int. J. Nat. Sci.* **2011**, *1*, 1–6. [CrossRef]

105. Tonu, N.N.; Doullah, A.-U.; Shimizu, M.; Karim, M.; Kawanabe, T.; Fujimoto, R.; Okazaki, K. Comparison of Positions of QTLs Conferring Resistance to *Xanthomonas campestris* pv. *campestris* in *Brassica oleracea*. *Am. J. Plant Sci.* **2013**, *4*, 11–20. [CrossRef]

106. Lee, Y.H.; Hong, J.K. Differential defence responses of susceptible and resistant kimchi cabbage cultivars to anthracnose, black spot and black rot diseases. *Plant Pathol.* **2014**, *64*, 406–415. [CrossRef]

107. Iglesias-Bernabé, L.; Madloo, P.; Rodríguez, V.M.; Francisco, M.; Soengas, P. Dissecting quantitative resistance to *Xanthomonas campestris* pv. *campestris* in leaves of *Brassica oleracea* by QTL analysis. *Sci. Rep.* **2019**, *9*, 2015. [CrossRef] [PubMed]

108. Frerigmann, H.; Gigolashvili, T. MYB34, MYB51, and MYB122 Distinctly Regulate Indolic Glucosinolate Biosynthesis in *Arabidopsis thaliana*. *Mol. Plant* **2014**, *7*, 814–828. [CrossRef] [PubMed]

109. Guo, L.; Yang, R.; Gu, Z. Cloning of genes related to aliphatic glucosinolate metabolism and the mechanism of sulforaphane accumulation in broccoli sprouts under jasmonic acid treatment. *J. Sci. Food Agric.* **2016**, *96*, 4329–4336. [CrossRef]

110. Li, G.; Riaz, A.; Goyal, S.; Abel, S.; Quiros, C. Inheritance of Three Major Genes Involved in the Synthesis of Aliphatic Glucosinolates in *Brassica oleracea*. *J. Am. Soc. Hortic. Sci.* **2001**, *126*, 427–431. [CrossRef]

111. Traka, M.H.; Saha, S.; Huseby, S.; Kopriva, S.; Walley, P.G.; Barker, G.C.; Moore, J.; Mero, G.; van den Bosch, F.; Constant, H.; et al. Genetic regulation of glucoraphanin accumulation in Beneforte® broccoli. *New Phytol.* **2013**, *198*, 1085–1095. [CrossRef] [PubMed]

112. Xu, R.; Kong, W.W.; Peng, Y.F.; Zhang, K.X.; Li, R.; Li, J. Identification and expression pattern analysis of the glucosinolate biosynthetic gene BoCYP83B1 from broccoli. *Biol. Plant.* **2018**, *62*, 521–533. [CrossRef]

113. Yu, Q.; Hao, G.; Zhou, J.; Wang, J.; Evivie, E.R.; Li, J. Identification and expression pattern analysis of BoMYB51 involved in indolic glucosinolate biosynthesis from broccoli (*Brassica oleracea* var. *italica*). *Biochem. Biophys. Res. Commun.* **2018**, *501*, 598–604. [CrossRef] [PubMed]

114. Rahim, A.; Afrin, K.S.; Jung, H.-J.; Kim, H.-T.; Park, J.-I.; Hur, Y.; Nou, I.S. Molecular analysis of anthocyanin biosynthesis-related genes reveal BoTT8 associated with purple hypocotyl of broccoli (*Brassica oleracea* var. *italica* L.). *Genome* **2019**, *62*, 253–266. [CrossRef] [PubMed]

115. Sotelo, T.; Soengas, P.; Velasco, P.; Rodríguez, V.M.; Cartea, M.E. Identification of Metabolic QTLs and Candidate Genes for Glucosinolate Synthesis in *Brassica oleracea* Leaves, Seeds and Flower Buds. *PLoS ONE* **2014**, *9*, e91428. [CrossRef]

116. Brown, A.F.; Yousef, G.G.; Reid, R.W.; Chebrolu, K.K.; Thomas, A.; Krueger, C.; Jeffery, E.; Jackson, E.; Juvik, J.A. Genetic analysis of glucosinolate variability in broccoli florets using genome-anchored single nucleotide polymorphisms. *Theor. Appl. Genet.* **2015**, *128*, 1431–1447. [CrossRef]

117. Li, Z.; Liu, Y.; Yuan, S.; Han, F.; Fang, Z.; Yang, L.; Zhuang, M.; Zhang, Y.; Lv, H.; Wang, Y.; et al. Fine mapping of the major QTLs for biochemical variation of sulforaphane in broccoli florets using a DH population. *Sci. Rep.* **2021**, *11*, 9004. [CrossRef]

118. Brown, A.F.; Yousef, G.G.; Chebrolu, K.K.; Byrd, R.W.; Everhart, K.W.; Thomas, A.; Reid, R.; Parkin, I.A.P.; Sharpe, A.G.; Oliver, R.; et al. High-density single nucleotide polymorphism (SNP) array mapping in *Brassica oleracea*: Identification of QTL associated with carotenoid variation in broccoli florets. *Theor. Appl. Genet.* **2014**, *127*, 2051–2064. [CrossRef]

119. Gardner, A.M.; Brown, A.F.; Juvik, J.A. QTL analysis for the identification of candidate genes controlling phenolic compound accumulation in broccoli (*Brassica oleracea* L. var. *italica*). *Mol. Breed.* **2016**, *36*, 81. [CrossRef]

120. Francisco, M.; Ali, M.A.A.; Ferreres, F.; Moreno-Fernández, D.; Ángel; Velasco, P.; Soengas, P. Organ-Specific Quantitative Genetics and Candidate Genes of Phenylpropanoid Metabolism in *Brassica oleracea*. *Front. Plant Sci.* **2016**, *6*, 1240. [CrossRef]

121. Gao, J.; Yu, X.; Ma, F.; Li, J. RNA-Seq Analysis of Transcriptome and Glucosinolate Metabolism in Seeds and Sprouts of Broccoli (*Brassica oleracea* var. italic). *PLoS ONE* **2014**, *9*, e88804. [CrossRef]

122. Lee, Y.-S.; Ku, K.-M.; Becker, T.M.; Juvik, J.A. Chemopreventive glucosinolate accumulation in various broccoli and collard tissues: Microfluidic-based targeted transcriptomics for by-product valorization. *PLoS ONE* **2017**, *12*, e0185112. [CrossRef]

123. Li, Z.; Liu, Y.; Li, L.; Fang, Z.; Yang, L.; Zhuang, M.; Zhang, Y.; Lv, H. Transcriptome reveals the gene expression patterns of sulforaphane metabolism in broccoli florets. *PLoS ONE* **2019**, *14*, e0213902. [CrossRef]

124. Parkin, I.A.P.; Koh, C.; Tang, H.; Robinson, S.J.; Kagale, S.; Clarke, W.E.; Town, C.D.; Nixon, J.; Krishnakumar, V.; Bidwell, S.L.; et al. Transcriptome and methylome profiling reveals relics of genome dominance in the mesopolyploid *Brassica oleracea*. *Genome Biol.* **2014**, *15*, R77. [CrossRef]

125. Liu, S.; Liu, Y.; Yang, X.; Tong, C.; Edwards, D.; Parkin, I.A.P.; Zhao, M.; Perumal, S.; Yu, J.; Huang, S.; et al. The *Brassica oleracea* genome reveals the asymmetrical evolution of polyploid genomes. *Nat. Commun.* **2014**, *5*, 3930. [CrossRef] [PubMed]

126. Shen, Y.; Wang, J.; Shaw, R.K.; Yu, H.; Sheng, X.; Zhao, Z.; Li, S.; Gu, H. Development of GBTS and KASP Panels for Genetic Diversity, Population Structure, and Fingerprinting of a Large Collection of Broccoli (*Brassica oleracea* L. var. *italica*) in China. *Front. Plant Sci.* **2021**, *12*, 655254. [CrossRef] [PubMed]

127. Huang, J.; Liu, Y.; Han, F.; Fang, Z.; Yang, L.; Zhuang, M.; Zhang, Y.; Lv, H.; Wang, Y.; Jialei, J.; et al. Genetic Diversity and Population Structure Analysis of 161 Broccoli Cultivars Based on SNP Markers. *Hortic. Plant J.* **2021**. [CrossRef]

128. Silva, G.; Poirot, L.; Galetto, R.; Smith, J.; Montoya, G.; Duchateau, P.; Paques, F. Meganucleases and Other Tools for Targeted Genome Engineering: Perspectives and Challenges for Gene Therapy. *Curr. Gene Ther.* **2011**, *11*, 11–27. [CrossRef]

129. Kim, Y.G.; Cha, J.; Chandrasegaran, S. Hybrid restriction enzymes: Zinc finger fusions to Fok I cleavage domain. *Proc. Natl. Acad. Sci. USA* **1996**, *93*, 1156. [CrossRef]

130. Christian, M.; Cermak, T.; Doyle, E.L.; Schmidt, C.; Zhang, F.; Hummel, A.; Bogdanove, A.J.; Voytas, D.F. Targeting DNA Double-Strand Breaks with TAL Effector Nucleases. *Genetics* **2010**, *186*, 757–761. [CrossRef]

131. Mali, P.; Esvelt, K.M.; Church, G. Cas9 as a versatile tool for engineering biology. *Nat. Methods* **2013**, *10*, 957–963. [CrossRef] [PubMed]

132. Ma, X.; Zhang, X.; Liu, H.; Li, Z. Highly efficient DNA-free plant genome editing using virally delivered CRISPR–Cas9. *Nat. Plants* **2020**, *6*, 773–779. [CrossRef] [PubMed]

133. Liu, X.; Wu, S.; Xu, J.; Sui, C.; Wei, J. Application of CRISPR/Cas9 in plant biology. *Acta Pharm. Sin. B* **2017**, *7*, 292–302. [CrossRef] [PubMed]

134. Ma, C.; Zhu, C.; Zheng, M.; Liu, M.; Zhang, D.; Liu, B.; Li, Q.; Si, J.; Ren, X.; Song, H. CRISPR/Cas9-mediated multiple gene editing in *Brassica oleracea* var. *capitata* using the endogenous tRNA-processing system. *Hortic. Res.* **2019**, *6*, 20. [CrossRef]

135. Sun, B.; Jiang, M.; Zheng, H.; Jian, Y.; Huang, W.-L.; Yuan, Q.; Zheng, A.-H.; Chen, Q.; Zhang, Y.-T.; Lin, Y.-X.; et al. Color-related chlorophyll and carotenoid concentrations of Chinese kale can be altered through CRISPR/Cas9 targeted editing of the carotenoid isomerase gene BoaCRTISO. *Hortic. Res.* **2020**, *7*, 161. [CrossRef] [PubMed]

136. Lawrenson, T.; Shorinola, O.; Stacey, N.; Li, C.; Østergaard, L.; Patron, N.; Uauy, C.; Harwood, W. Induction of targeted, heritable mutations in barley and *Brassica oleracea* using RNA-guided Cas9 nuclease. *Genome Biol.* **2015**, *16*, 258. [CrossRef]

137. Sparrow, P.A.C.H.N.; Irwin, J.A. *Brassica oleracea* and *B. napus*. In *Methods in Molecular Biology*; Springer: Berlin/Heidelberg, Germany, 2014; Volume 1223, pp. 287–297.

138. Metz, T.D.; Dixit, R.; Earle, E.D. *Agrobacterium tumefaciens*-mediated transformation of broccoli (*Brassica oleracea* var. *italica*) and cabbage (*B. oleracea* var. *capitata*). *Plant Cell Rep.* **1995**, *15*, 287–292. [CrossRef]

139. Jiang, M.; Jiang, J.; He, C.; Guan, M. Broccoli plants over-expressing a cytosolic ascorbate peroxidase gene increase resistance to downy mildew and heat stress. *J. Plant Pathol.* **2016**, *98*, 413–420.

140. Jiang, M.; Liu, Q.-E.; Liu, Z.-N.; Li, J.-Z.; He, C.-M. Over-expression of a WRKY transcription factor gene BoWRKY6 enhances resistance to downy mildew in transgenic broccoli plants. *Australas. Plant Pathol.* **2016**, *45*, 327–334. [CrossRef]

141. Li, S.; Zhang, L.; Wang, Y.; Xu, F.; Liu, M.; Lin, P.; Ren, S.; Ma, R.; Guo, Y.-D. Knockdown of a cellulose synthase gene BoiCesA affects the leaf anatomy, cellulose content and salt tolerance in broccoli. *Sci. Rep.* **2017**, *7*, 41397. [CrossRef]

142. Jiang, M.; Jiang, J.-J.; Miao, L.-X.; He, C.-M. Over-expression of a C3H-type zinc finger gene contributes to salt stress tolerance in transgenic broccoli plants. *Plant Cell Tissue Organ Cult. (PCTOC)* **2017**, *130*, 239–254. [CrossRef]

143. Jiang, M.; Miao, L.; Zhang, H.; Zhu, X. Over-Expression of a Transcription Factor Gene BoC3H4 Enhances Salt Stress Tolerance but Reduces Sclerotinia Stem Rot Disease Resistance in Broccoli. *J. Plant Growth Regul.* **2019**, *39*, 1162–1176. [CrossRef]

144. Jiang, M.; Ye, Z.-H.; Zhang, H.-J.; Miao, L.-X. Broccoli Plants Over-expressing an ERF Transcription Factor Gene BoERF1 Facilitates Both Salt Stress and Sclerotinia Stem Rot Resistance. *J. Plant Growth Regul.* **2018**, *38*, 1–13. [CrossRef]

145. Kumar, P.; Gambhir, G.; Gaur, A.; Sharma, K.C.; Thakur, A.K.; Srivastava, D.K. Development of transgenic broccoli with cryIAa gene for resistance against diamondback moth (*Plutella xylostella*). *3 Biotech* **2018**, *8*, 299. [CrossRef]

146. Zuluaga, D.L.; Graham, N.S.; Klinder, A.; Kloeke, A.E.E.V.O.; Marcotrigiano, A.R.; Wagstaff, C.; Verkerk, R.; Sonnante, G.; Aarts, M.G.M. Overexpression of the MYB29 transcription factor affects aliphatic glucosinolate synthesis in *Brassica oleracea*. *Plant Mol. Biol.* **2019**, *101*, 65–79. [CrossRef] [PubMed]

147. Li, R.; Jiang, J.; Jia, S.; Zhu, X.; Su, H.; Li, J. Overexpressing broccoli tryptophan biosynthetic genes BoTSB1 and BoTSB2 promotes biosynthesis of IAA and indole glucosinolates. *Physiol. Plant.* **2019**, *168*, 174–187. [CrossRef]

148. Kim, Y.-C.; Cha, A.; Hussain, M.; Lee, K.; Lee, S. Impact of *Agrobacterium*-infiltration and transient overexpression of BroMYB28 on glucoraphanin biosynthesis in broccoli leaves. *Plant Biotechnol. Rep.* **2020**, *14*, 373–380. [CrossRef]

149. Cao, H.; Liu, R.; Zhang, J.; Liu, Z.; Fan, S.; Yang, G.; Jin, Z.; Pei, Y. Improving sulforaphane content in transgenic broccoli plants by overexpressing MAM1, FMOGS–OX2, and Myrosinase. *Plant Cell Tissue Organ Cult. (PCTOC)* **2021**, *146*, 461–471. [CrossRef]

150. Chen, G.J.; Cao, B.H.; Xu, F.; Lei, J.J. Development of adjustable male sterile plant in broccoli by antisense DAD1 fragment transformation. *Afr. J. Biotechnol.* **2010**, *9*, 4534–4541.

151. Li, H.; Zhang, Q.L.; Li, L.H.; Yuan, J.Y.; Wang, Y.; Wu, M.; Han, Z.P.; Liu, M.; Chen, C.B.; Song, W.Q.; et al. Ectopic overexpression of bol-mir171b increases chlorophyll content and results in sterility in broccoli (*Brassica oleracea* L. var. *italica*). *J. Agric. Food Chem.* **2018**, *66*, 9588–9597. [CrossRef] [PubMed]

152. Baskar, V.; Park, S.W. Molecular characterization of BrMYB28 and BrMYB29 paralogous transcription factors involved in the regulation of aliphatic glucosinolate profiles in *Brassica rapa* ssp. pekinensis. *Comptes Rendus Biol.* **2015**, *338*, 434–442. [CrossRef] [PubMed]

Review

CRISPR-Cas9 Gene Editing for Fruit and Vegetable Crops: Strategies and Prospects

Lili Wan [1], Zhuanrong Wang [1], Mi Tang [1,*], Dengfeng Hong [2], Yuhong Sun [1], Jian Ren [1], Na Zhang [1] and Hongxia Zeng [1]

1   Institute of Crop, Wuhan Academy of Agricultural Sciences, Wuhan 430065, China; wanlili13226@163.com (L.W.); zhuanrongwang@163.com (Z.W.); sunyh68@163.com (Y.S.); 18571636516@163.com (J.R.); zhangna@wuhanagri.com (N.Z.); 18971053198@163.com (H.Z.)
2   National Key Laboratory of Crop Genetic Improvement, Huazhong Agricultural University, Wuhan 430070, China; dfhong@mail.hzau.edu.cn
*   Correspondence: xtgyjs@wuhanagri.com; Tel.: +86-027-62302993

Abstract: Fruit and vegetable crops are rich in dietary fibre, vitamins and minerals, which are vital to human health. However, many biotic stressors (such as pests and diseases) and abiotic stressors threaten crop growth, quality, and yield. Traditional breeding strategies for improving crop traits include a series of backcrosses and selection to introduce beneficial traits into fine germplasm, this process is slow and resource-intensive. The new breeding technique known as clustered regularly interspaced short palindromic repeats (CRISPR)-CRISPR-associated protein-9 (Cas9) has the potential to improve many traits rapidly and accurately, such as yield, quality, disease resistance, abiotic stress tolerance, and nutritional aspects in crops. Because of its simple operation and high mutation efficiency, this system has been applied to obtain new germplasm resources via gene-directed mutation. With the availability of whole-genome sequencing data, and information about gene function for important traits, CRISPR-Cas9 editing to precisely mutate key genes can rapidly generate new germplasm resources for the improvement of important agronomic traits. In this review, we explore this technology and its application in fruit and vegetable crops. We address the challenges, existing variants and the associated regulatory framework, and consider future applications.

Keywords: CRISPR/Cas; gene knockout; genome editing; germplasm resource; precision editing; regulatory framework; trait improvement

Citation: Wan, L.; Wang, Z.; Tang, M.; Hong, D.; Sun, Y.; Ren, J.; Zhang, N.; Zeng, H. CRISPR-Cas9 Gene Editing for Fruit and Vegetable Crops: Strategies and Prospects. *Horticulturae* **2021**, *7*, 193. https://doi.org/10.3390/horticulturae7070193

Academic Editor: Yuyang Zhang

Received: 1 June 2021
Accepted: 10 July 2021
Published: 14 July 2021

Publisher's Note: MDPI stays neutral with regard to jurisdictional claims in published maps and institutional affiliations.

Copyright: © 2021 by the authors. Licensee MDPI, Basel, Switzerland. This article is an open access article distributed under the terms and conditions of the Creative Commons Attribution (CC BY) license (https://creativecommons.org/licenses/by/4.0/).

1. Introduction

Fruit and vegetable crops are rich in cellulose, vitamins, trace elements, minerals, and other important nutrients, which are essential in the human diet [1]. However, climate and environmental changes potentially threaten the production and supply of fruits and vegetables [2]. Humans have long domesticated and cultivated wild species. Cross-breeding technology enables breeders to improve varieties by crossing selected dominant varieties [3]. However, with long-term artificial selection, the shortcomings of conventional breeding become increasingly prominent, mainly in the excessive dependence on naturally occurring allelic variation. There are limited genetic germplasm resources for improving target traits, and conventional breeding can expose many adverse traits, thereby reducing breeding efficiency [4]. Although traditional breeding can produce new vegetable cultivars with high yield, good quality and disease resistance, with the increasing global population and continuous food-supply demands, it is important to rapidly select new varieties to meet market demands [5]. The development and application of emerging methods in crop biotechnology can promote high-efficiency and precise varietal breeding [6].

Genetic engineering has been used to improve the responses to biotic and abiotic stress, and to improve the quality of fruits and vegetables. In 1994, a storage resistant transgenic tomato was approved by the Food and Drug Administration (FDA) [7]. For papaya, 80%

of the market was supplied with high-yield transgenic papaya with high resistance to the cyclic spot virus [8]. However, in order to ensure safety in planting processes and product consumption, genetically modified (GM) plant development and application are strictly legislated and regulated, greatly delaying the development to market of transgenic cultivars [9]. In 2013, CRISPR-Cas-mediated gene editing was developed as a tool to study plant gene function. Over the next two years, many new gene-edited crop germplasm resources emerged. In 2016, the US FDA approved the CRISPR gene editing of a waxy corn null segregant line and an anti-browning mushroom (*Agaricus bisporus*) for the market, without applying the strict regulatory process required for GM crops [10,11]. This indicates that CRISPR gene editing has already succeeded in promoting the development of crop cultivars.

In this review, we summarise the mechanisms underlying CRISPR technology, recent applications in fruit and vegetable crops, and improvements in CRISPR-Cas systems. We further outline CRISPR-associated regulatory frameworks that enable commercialisation of gene edited crops in different countries. Finally, we discuss the future challenges and opportunities for introducing desirable alleles and improving many traits.

1.1. The Discovery and Development of CRISPR Technology

CRISPR-associated (Cas) genes were first discovered in the *Escherichia coli* genome in 1987 and were officially named by the Dutch scientist who identified them [12]. In 2005, it was discovered that many CRISPR spacers consist of short sequences that are highly homologous with sequences originating from extrachromosomal DNA. The *Cas*-encoded protein can combine with the CRISPR transcription products and with the homologous foreign DNA sequences to form a protein–RNA complex, which can cut the foreign DNA fragments. The primary function of the CRISPR complex in bacteria and archaea is to integrate specific fragments of exogenous DNA (from invading phages or other sources) into their own genomes to become interval sequences. During subsequent invasion by foreign DNA, the specific recognition system is then activated, providing an acquired immune defence function [13–15].

CRISPR-Cas technology has been successfully applied to the editing of human, animal, and plant genomes, and has been developed for use in drug screening, animal domestication, and food science research [16–18]. There are three main types of CRISPR-Cas systems. Types I and III use a large multi-Cas protein complex for interference [19]. Type II requires only a simple effector-module architecture to accomplish interference via its two signature nuclease domains, RuvC and HNH [20]. Among various CRISPR nucleases, type II Cas9 from *Streptococcus pyogenes* (SpCas9) is the most widely used in CRISPR-Cas technology [21]. The sgRNA-Cas complex recognises the protospacer adjacent motif (PAM) and Cas9 cleaves the target DNA to generate a double-strand break (DSB), triggering cellular DNA repair mechanisms (Figure 1). In eukaryotes, DSBs have two main repair mechanisms. The first is nonhomologous end joining (NHEJ). In the absence of a homologous repair template, the NHEJ repair pathway is activated at the DSB site, thus disrupting gene function. The second is homology directed repair (HDR). If a donor DNA template homologous to the sequence surrounding the DSB site is available, the HDR pathway is initiated, precisely introducing specific mutations such as insertion or replacement of desired sequences into the break sites [22]. Using a donor DNA as a template, gene targeting (GT) can precisely modify a target locus to repair DNA DSBs.

Several strategies are used to improve the homologous recombination frequency between a genomic target and an exogenous homologous template donor. Most of the strategies focus on enhancing the number of donor repair templates using virus replicons [23], suppressing the NHEJ pathway [24], and timing DSB induction at target sites to coincide with donor repair template delivery in plant cells [25]. Finally, the recombination frequency can be enhanced by treatment with Rad51-stimulatory compound1 (RS-1) [26].

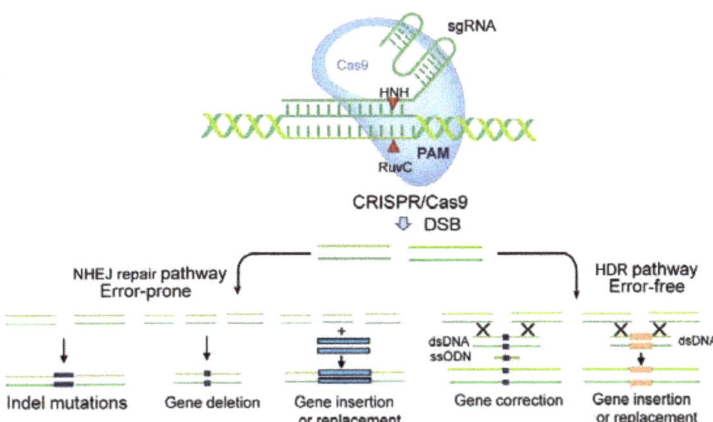

Figure 1. The potential applications of CRISPR-Cas systems in genome editing. CRISPR-Cas systems mediated genome modification depending on the two main double-strand break (DSB) repair pathways. Indel mutation and gene deletion are outcomes of the dominant nonhomologous end joining (NHEJ) repair pathway. Gene insertion, correction, and replacement, using a DNA donor template, are outcomes of the homology directed repair (HDR) pathway.

1.2. Development of the CRISPR-Cas System in Plant Studies

Since the CRISPR-Cas system was first adopted for plant genetic engineering in 2013, numerous efforts have been made to develop it into a more powerful tool, for instance, to enable precisely targeted DNA mutations or genetic modifications [27]. CRISPR-Cas can now target the open reading frame, untranslated region, and promoter region of a target coding gene, as well as noncoding RNAs [28–30]. Single-base mutations at genomic targets have also been achieved by nickase Cas9 (nCas9) or catalytically inactive Cas9 (dead Cas9; dCas9) variants fused with cytosine or adenine deaminases, without inducing DSBs [31]. Cas9 proteins have been developed extensively to broaden PAM preferences. Cas9 orthologs which possess not only the canonical NGG PAM, but also NG and other PAMs, will expand the repertoire of CRISPR-Cas9 genome editing in plants [32].

1.3. CRISPR-Cas9 in Fruit and Vegetable Crop Improvement

In 2014, CRSPR-Cas9 was used to create the first needle-leaf mutant in tomato, by knocking out *Argonaute 7* [33]. Many studies have since been published on its possible applications in protecting plants against biotic and abiotic stresses, and improving fruit quality, plant architecture, and shelf life [34]. Currently, the system is in the research stage for many fruits and vegetables crops, such as cabbage, mustard, tomato, and watermelon.

Most gene-editing studies have evaluated mutation efficiency in terms of the number of albino plants obtained after mutation of the endogenous phytoene desaturase (*PDS)* gene. The disruption of *PDS* impairs the production of chlorophyll and carotenoid, generating an easily identifiable albinism phenotype in plants. However, the products of gene editing obtained in this way have no economic value [35–37]. Because of its high economic value and the availability of Agrobacterium-mediated transformation, tomato has become a model crop for testing CRISPR-Cas9 applications (Figure 2).

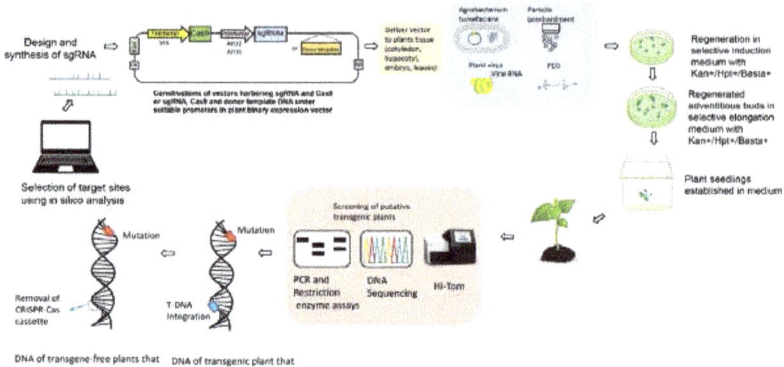

Figure 2. CRISPR-Cas9 mediated genome editing. (I) Selection of the desired genomic DNA target, and recognition of protospacer adjacent motif (PAM) sequences before 20 bp sequences. Design of the sgRNA using online bioinformatics tools. (II) Cloning of designed sgRNAs, and binary vector construction using promoters. (III) The delivery of CRISPR-Cas editing reagents into plant cells. The vector can be transferred into the plant via *Agrobacterium tumefaciens*, nanoparticles, biolistic bombardment, or polyethylene glycol (PEG). Alternatively, plant RNA viruses have been used to induce heritable genome editing. When the cassette harbouring the sgRNA, RNA mobile element, and tobacco rattle virus (TRV) is transformed into the Cas9 expressing plants, the systemic spread of sgRNA will introduce heritable genome editing. (IV) Plant transformation and development of transgenic plants. (V) Genotyping of transgenic plants. (VI) Transgene-free plants with the desired mutation are obtained.

1.3.1. Improvement of Biotic Stress Resistance

Two strategies have been used to improve plant resistance to viruses: (1) designing sgRNAs and targeting the virus genome; or (2), modifying the fruit crop genes in the antiviral pathway. The binding of virus genome linked protein (VPg) to the plant protein 'eukaryotic translation initiation factor 4E' (eIF4E) is key in Y virus infection of plants. Mutation of a key site of eIF4E can affect the virus–plant interaction, and mediate plant resistance to this virus [38]. In cucumbers, using CRISPR-Cas to target the N′ and C′ ends of eIF4E-produced nontransgenic homozygous plants in the T3 generation; these showed immunity to cucumber vein yellow virus and pumpkin mosaic virus, and resistance to papaya ring spot mosaic virus (PRSV-W) [39].

CRISPR-Cas9 can generate mutations in the coding and noncoding regions of geminivirus, effectively reducing its pathogenicity. In *Nicotiana benthamiana*, sgRNA-Cas9 constructs target beet severe curly top virus (a geminivirus), inhibiting its accumulation in leaves [40]. Geminivirus noncoding-region mutations are believed to reduce or even inhibit its replication ability. Compared with coding-region mutations, noncoding-region mutations generate fewer viral variants [41].

Fungi cause many diseases, potentially causing severe losses in crop yield and quality. For instance, downy and powdery mildews cause serious economic losses in tomato [42]. *Arabidopsis thaliana DMR6* (down mildew resistant) is a member of the 2-oxoglutarate oxygenase Fe(II)-dependent superfamily and is involved in salicylic acid homeostasis. Overexpression of *DMR6* in plants can reduce susceptibility to downy mildew [43]. The *DMR6* mutation obtained using CRISPR-Cas9 to knock out the homologous genes in tomato showed resistance to *Pseudomonas syringae*, *Phytophthora* and *Xanthomonas* spp. [44]. *Mlo1* (Mildew resistant locus 1) encodes a membrane-associated protein and is a powdery mildew disease-sensitivity gene. In tomato, *Mlo1* mutants obtained via gene editing exhibited resistance to the powdery mildew *Oidium neolycopersici*. Further, a mutant free of mlo1 T-DNA was obtained by selfing T0 generation plants [45].

The fungal pathogen *Fusarium oxysporum* can cause *Fusarium* wilt disease in fruit and vegetable crops [46]. In tomatoes, *Solyc08g075770*-knockout via CRISPR-Cas9 resulted in sensitivity to *Fusarium* wilt disease [47]. In watermelons, the knockout of *Clpsk1*, encoding the Phytosulfokine (PSK) precursor, confers enhanced resistance to *Fusarium oxysporum* f.sp.*niveum* (FON) [48]. *Botrytis cinerea*, an airborne plant pathogen that infects fruit and vegetable crops, causes great economic losses. Its initial symptoms are not obvious, and the lack of effective pesticides makes its prevention and control difficult. Pathogens can be effectively controlled in crops by the use of genetic resources that convey heritable resistance. In tomatoes, mutations in *MAPK3* (mitogen-activated protein kinase 3) produced using CRISPR-Cas9 induce resistance to *Botrytis cinerea* [49].

The bacterial pathogen *Pseudomonas syringae* causes leaf spot diseases in crops, severely impacting the yield and sensory qualities of fruits and vegetables. In *Arabidopsis thaliana*, CRISPR-Cas9 was used to mutate the C-terminal jasmonate domain (JAZ2Δjas) of *JAZ2* (jasmonate ZIM domain protein 2), causing expression of JAZ2 repressors; these repressors confer resistance to *Pseudomonas syringae* [50].

1.3.2. Abiotic Stress Resistance Improvement

With climate change, crop production is exposed to increased potential risks of abiotic stress. Although traditional breeding can to some extent ensure stable crop production, the application of new technologies to rapidly obtain new crop germplasm resources capable of responding to abiotic stress is essential for accelerating the cultivation of new varieties [51]. The emergence of CRISPR-Cas9 gene editing has shortened the time required to create new varieties. Brassinazole-resistant 1 gene (*BZR1*) participates in various brassinosteroid (BR) mediated development processes. The CRISPR mediated mutation in *BZR1* impaired the induction of *RESPIRATORY BURST OXIDASE HOMOLOG1(RBOH1)* and the production of H_2O_2. Exogenous H_2O_2 recovered the heat tolerance in tomato bzr1 mutant [52]. Further, new cold- and drought-tolerant germplasms can be created using gene-editing, for instance, of *CBF1* (C-repeat binding factor 1), which regulates cold tolerance in plants, and *MAPK3*, which participates in the drought stress response to protect plant cell membranes from peroxidative damage in tomatoes [53,54].

1.3.3. Herbicide Resistance Improvement

Weeds are an important cause of stress that affect vegetable yield and quality, and selective herbicides are often used to control weed growth during cultivation. To obtain herbicide-resistant fruits and vegetables for field production, CRISPR-Cas9 gene editing was used for site-directed mutagenesis of the herbicide target gene acetolactate synthase (ALS) in watermelon, yielding a herbicide-resistant watermelon germplasm [55]. Cytidine base editing (CBE) was used for cytidine editing of key *ALS* sites in tomato and potato, resulting in amino acid mutations. Up to 71% of edited tomato plants exhibited resistance to the pesticide chlorsulfuron, and of the edited tomato and potato plants, 12% and 10%, respectively, were free of GM components [56]. *Phelipanche aegyptiaca*, an obligate weedy plant parasite, requires the host roots to release the plant hormone strigolactone (SL) to promote seed germination; CRISPR-Cas9 was used to mutate carotenoid dioxygenase 8 (*CCD8*), a key enzyme in the carotenoid synthesis pathway that produces SLs in tomato, and More Axillary Growth1 (*MAX1*), which is involved in the synthesis of SLs, thereby significantly reducing SL content, and creating *P. aegyptiaca*-resistant tomato plants [57,58].

1.3.4. Fruit and Vegetable Quality Improvement

The primary goal in fruit and vegetable breeding is to improve quality and prolong shelf life after harvest. Quality refers to both external and internal factors. External quality refers to fruit size, colour, and texture, which can be discerned by the naked eye. Internal quality must be measured using equipment, and includes the levels of nutrients such as sugars, vitamins, and bioactive compounds including lycopene, anthocyanins, and malate. For example, in tomato, the ovary locule number, which determines 50% of the genetic variation in fruit size, is determined by multiple QTLs [59]. Researchers at Cold Spring Harbor Laboratory designed eight sgRNAs and used CRISPR-Cas9 to edit the promoter region of the tomato CLAVATA-WUSCHEL (CLV-WUS) stem cell gene *CLV3* to obtain fruits that are larger and more numerous than wild-type fruits [60]; editing of fruit-size determining QTLs, such as the QTLs for locule number (lc) and fasciated number (fas), generated germplasm resources with an increased number of locules [61].

Fruit and vegetable colour and texture are important traits for consumers. For example, European and American consumers prefer red tomatoes, whereas Asian consumers prefer pink tomatoes [62,63]. CRISPR-Cas was used to modify phytoene synthase 1 (*PSY1*), MYB transcription factor 12 (*MYB12*), and anthocyanin 2 (*ANT2*) to obtain yellow, pink, and purple tomatoes, respectively [64–66]. The carotenoid isomerase gene of Chinese kale (*BoaCRTISO*) is responsible for catalysis, then conversion of lycopene precursors to lycopene. When *BoaCRTISO* was targeted and edited, the colour of mutants changed from green to yellow [67]. The primary goal of improving the intrinsic quality of fruits and vegetables is to improve their nutrient and bioactive compound content. Carbohydrates and vitamins are essential nutrients. Many genes are involved in the synthesis and metabolism of sucrose and carotenoids. One of the carotenoids, provitamin A, can be absorbed by the human body and converted into vitamin A. For example, CRISPR-Cas was used to knock out *MPK20* (mitogen-activated protein kinase 20), blocking the transcription and protein products of multiple genes in the sucrose metabolism pathway [68]. Biofortification, the biotechnological improvement of the absorption, transport, and metabolism of minerals by plants, increases the levels of micronutrients that are beneficial to human health; long-term consumption of these micronutrients can effectively prevent cardiovascular disease and cancer [69].

Anthocyanins [70], malate [71], γ-aminobutyric acid (GABA) [72], and lycopene [73] are bioactive compounds. Adjusting key metabolic-pathway-related genes via CRISPR-Cas9 can enrich these nutrients in fruits. For example, in tomatoes, butylamine content was increased 19-fold through editing multiple genes in the GABA synthesis pathway, and malate content was improved by regulating aluminium-activated malate transporter (*ALMT9*) [72].

CRISPR-Cas9 can also be used to reduce the content of substances in vegetables that are not conducive to human health, by targeting mutations that inactivate genes in biosynthetic pathways. In potato tubers, for example, excessive content of steroidal glycoalkaloids (SGAs), such as α-solanine and α-chaconine, affects their taste and makes them less safe for human consumption, hence low content is an indicator of high quality. CRISPR-Cas9 was used to delete *St16DOX* (steroid 16α-hydroxylase) in the potato SGA biosynthetic pathway, resulting in SGA-free potato lines [74].

Prolonged shelf-life is an important breeding goal in fruit and vegetable production. CRISPR was used to knock out ripening inhibitor (*RIN*) or DNA demethylase (DNA demethylase 2, *DML2*) to slow fruit ripening, thereby prolonging their shelf life. However, regulating these two genes in fruit alters peel colour and reduces flavour and nutritional value, severely reducing the fruit's palatability and sensory qualities [75,76]. In tomatoes, inhibiting the expression of the pectate lyase (*PL*) and alcobaca (*ALC*) genes effectively extended shelf life, without affecting the sensory qualities or nutritional value [77,78].

1.3.5. Application of CRISPR-Cas9 to Crop Domestication

The domestication of wild species into commercial cultivated species requires changes in numerous crop traits, including seed setting, size, consistency of maturation, flowering, photoperiod sensitivity, and the nutritional value of the fruit [79]. Plant domestication mostly affects the genes controlling plant morphology, plant growth habit, floral induction, fruit size and number, dispersal, and architecture, as well as the nutritional composition. To achieve the ideotype, alleles controlling favourable nutritional attributes and stress resilience from wild relatives are introduced into cultivated species via traditional domestication technology, but this process is very time-consuming in bringing about changes to many loci. With its ability to precisely manipulate the genome, CRISPR-Cas9 can substantially accelerate de novo domestication.

The tomato is a model crop for artificial domestication using CRISPR-Cas9. In tomato plants, the joint is a weak region of the stem which allows the fruit to drop from the plant, making the fruit prone to fall after ripening, thus improving seed dispersal. Many years of artificial domestication based on harvesting habits has generated cultivars with jointless fruit stems, in which the fruit do not fall after maturation [79,80]. Roldan et al. [81] used CRISPR-Cas to mutate *MBP21* (MADS-box protein 21), obtaining a new jointless germplasm resource.

Parthenocarpy (fertilisation-independent seedless fruit development) is an important agronomic fruit and vegetable trait and can help ensure stable yield in fluctuating environments. It satisfies consumer preferences for seedless over seeded fruits and provides savings in energy consumption when separating the seeds for industrial production. In tomatoes, *SlAGL6* (SlAGAMOUS-like 6) is essential for parthenocarpy during high temperature stress. *SlAGL6*-mutant plants grow normally and have the same fruit weight and morphology as wild-type plants. Therefore, this gene is an important resource for creating new parthenocarpic germplasms. Homozygous or biallelic mutant plants obtained by modifying *SlAGL6* produced parthenocarpic fruits and fruits with a maximum of 10 seeds, respectively [82]. CRISPR-Cas has also been used to knock out *SlARF7* (auxin response factor 7) and *SlIAA9* (indole-3-acetic acid inducible 9) to obtain seedless tomatoes. Seedless tomatoes are obtained from the T0 generation of the biallelic and homozygous *SlIAA9*-mutant Micro-Tom cultivar and the commercial Ailsa Craig cultivar [83,84].

Plant yield depends primarily on the number of flowers, which in turn is determined by inflorescence structure. *BOP* (blade-on-petiole) is homologous to genes associated with leaf complexity and silique dehiscence in tomato and *Arabidopsis*. Knocking out *BOP* via CRISPR-Cas9 altered inflorescence morphology. CRISPR-Bop1/2/3 triple mutants flower faster and have simpler inflorescence structure than wild-type plants [85]. The site-directed editing of six key genes that determine yield in wild tomatoes (*Solanum pimpinellifolium*) has resulted in morphological changes in aspects such as size, fruit number, and nutritional composition [86]. In domesticated wild tomatoes, genes associated with morphology, number of flowers, fruit yield, and vitamin C synthesis have been improved by editing their coding sequences, cis-regulatory sequences, and upstream open reading frames [81]. APETALA2a (*AP2a*), NON-RIPENING (*NOR*), and FRUITFULL (*FUL1/TDR4* and *FUL2/MBP7*) have been modified to accelerate tomato maturation, producing plants that mature earlier in natural environments [87].

Crop sensitivity to photoperiod restricts their planting areas and regulating the photoperiod-associated genes can accelerate domestication. The disruption of self-pruning 5G (*SP5G*) generated a rapid surge in flowering that leads to an early fruit harvest [88].

Dwarf-crop breeding is an important direction in domestication research, as dwarf plants are resistant to lodging under high wind conditions. Compared with normal plants, it is more convenient to pick fruits from dwarfed plants [89]. Dwarf plants transport nutrients more readily over the shorter distances from the roots to the leaves [89]. The application of CRISPR-Cas9 technology in the genomes of several commercially important fruit and vegetables has been achieved as outlined in Table 1.

Table 1. List of target genes and traits modified via CRISPR-Cas9 technology in fruit and vegetable crops.

Crop Species	Target	Mutation	Transformation Method	Trait Modification	References
Tomato	*SlAGO7*	Loss of function	*Agrobacterium*-mediated transformation	Wiry phenotype	[33]
Cabbage	*BoPDS*	Loss of function	*Agrobacterium*-mediated hypocotyl transformation	Albino phenotype	[35]
Watermelon	*ClPDS*	Loss of function	*Agrobacterium*-mediated cotyledon transformation	Albino phenotype	[36]
Chinese kale	*BaPDS1BaPDS2*	Loss of function	*Agrobacterium*-mediated cotyledon transformation	Albino phenotype	[37]
Cucumber	*eIF4E*	Loss of function	*Agrobacterium*-mediated cotyledon transformation	Resistance against cucumber vein yellowing virus, zucchini yellow mosaic virus, and papaya ring spot mosaic virus	[38]
Tomato	*ClDMR6*	Loss of function	*Agrobacterium*-mediated cotyledon transformation	Resistance against downy mildew	[44]
Tomato	*ClMlo1*	Loss of function	*Agrobacterium*-mediated cotyledon transformation	Resistance against downy mildew	[45]
Tomato	*Solyc08g075770*	Loss of function	*Agrobacterium*-mediated cotyledon transformation	Fusarium wilt susceptibility	[47]
Watermelon	*ClPSK1*	Loss of function	Agrobacterium-mediated cotyledon transformation	Resistance to *Fusarium oxysporum* f.sp. *niveum*	[48]
Tomato	*MAPK3*	Loss of function	*Agrobacterium*-mediated cotyledon transformation	Resistance to *Botrytis cinerea*	[49]
Tomato	*BZR1*	Loss of function	*Agrobacterium*-mediated cotyledon transformation	Decrease in heat stress tolerance	[52]
Tomato	*CBF1*	Loss of function	*Agrobacterium*-mediated cotyledon transformation	Decrease in chilling stress tolerance	[53]
Tomato	*SlMAPK3*	Loss of function	*Agrobacterium*-mediated cotyledon transformation	Decrease in drought stress tolerance	[54]
Watermelon	*ClALS*	Site-directed mutagenesis	*Agrobacterium*-mediated cotyledon transformation	Herbicide resistance	[55]
Tomato and Potato	*StALS2*	Site-directed mutagenesis	*Agrobacterium*-mediated transformation	Herbicide resistance	[56]
Tomato	Carotenoid cleavage dioxygenase8 (*CCD8*)	Loss of function	*Agrobacterium*-mediated transformation	Resistance against *Phelipanche aegytiaca*	[57]
Tomato	More Axillary Growth1 (*MAX1*)	Loss of function	*Agrobacterium*-mediated transformation	Resistance against *Phelipanche aegytiaca*	[58]
Tomato	*SP,SP5G,CLV3, WUS, GGP1*	Cis-regulatory variation and loss of function	*Agrobacterium*-mediated transformation	Introduction of desirable traits with morphology, flower number, fruit size and number, and ascorbic acid synthesis	[60]

Table 1. *Cont.*

Crop Species	Target	Mutation	Transformation Method	Trait Modification	References
Tomato	*SlWUS* CarG element, *SlCLV3* promoter	Cis-regulatory variation	*Agrobacterium*-mediated transformation	Fruit size, inflorescence branching, and plant architecture	[61]
Tomato	*PSY1*	Different mutations in alleles	*Agrobacterium*-mediated cotyledon transformation	Yellow-coloured tomato	[64]
Tomato	*MYB12*	Different mutations in alleles	*Agrobacterium*-mediated cotyledon transformation	Pink-coloured tomato	[65]
Tomato	*ANT2*	Gene insertion	*Agrobacterium*-mediated cotyledon transformation	Purple-coloured tomato	[66]
Chinese kale	*BoaCRTISO*	Gene insertion and replacement	*Agrobacterium*-mediated transformation	Colour of mutants changed from green to yellow	[67]
Tomato	*MPK20*	Loss of function	*Agrobacterium*-mediated transformation	Repression of genes controlling sugar and auxin metabolism	[68]
Tomato	*GAD2,GAD3,SlyGABA-TP1,SlyGABA-TP2,SlyGABA-TP3,SlyCAT9,SlySSADH*	Autoinhibitory domain deletion	*Agrobacterium*-mediated transformation	Increase in γ-aminobutyric acid (GABA) content	[72]
Tomato	*SGR1,lycopene ε-cyclase (LCY-E), beta-lycopene cyclase(Blc), lycopene β-cyclase1(LCY-B1) and LCY-B2*	Loss of function	*Agrobacterium*-mediated apical segments of hypocotyls transformation	Lycopene content	[73]
Potato	*16α-hydroxylation* (St16DOX)	Loss of function	*Agrobacterium*-mediated shoot transformation	Steroidal glycoalkaloid (SGA) biosynthesis	[74]
Tomato	Ripening inhibitor (*RIN*)	Single base insertion or deletion of more than three bases	*Agrobacterium*-mediated transformation	MADS-box transcription factor regulating fruit ripening	[75]
Tomato	DNA demethylases (*SlDML2*)	Loss of function	*Agrobacterium*-mediated cotyledon transformation	Activation and inhibition of fruit ripening	[76]
Tomato	Enzymes pectate lyase (*PL*), Polygalacturonase 2a (*PG2a*), and β-galactanase (*TBG4*)	Generation of a range of CRISPR alleles	*Agrobacterium*-mediated transformation	Pectin degradation control	[77]
Tomato	Alcobaca (*SLALC*)	Loss of function	*Agrobacterium*-mediated hypocotyls transformation	Long shelf-life	[78]

Table 1. *Cont.*

Crop Species	Target	Mutation	Transformation Method	Trait Modification	References
Tomato	*SlMBP21*	Loss of function	*Agrobacterium*-mediated transformation	Jointless fruit stem	[81]
Tomato	*SlAGAMOUS-LIKE 6 (Sl AGL6)*	Loss of function	*Agrobacterium*-mediated transformation	Parthenocarpic	[82]
Tomato	*ARF7*	Loss of function	*Agrobacterium*-mediated transformation	Parthenocarpic	[83]
Tomato	*SlIAA9*	Loss of function	*Agrobacterium*-mediated leaf disk transformation	Parthenocarpic	[84]
Tomato	*Blade-on-petiole (SlBOP)*	Loss of function	*Agrobacterium*-mediated cotyledon segments transformation	Early flowering with simplified inflorescence architecture	[85]
Tomato	*Self-pruning 5G(SlSP5G)*	cis-regulatory variation	*Agrobacterium*-mediated transformation	Day-length-sensitive flowering	[88]

1.4. Improvements to CRISPR-Cas9 Gene-Editing Systems

1.4.1. Production of Non-GM Plants Using CRISPR-Cas9 Gene Editing

Most fruit crops have heterozygous genotypes, hybrid incompatibility, and long growth periods. Some have complex triploid or polyploid genomes [90]. During the last 20 years, innovative breeding technology employing genetic engineering has provided a favourable way to accelerate crop improvement for such species [91]. For example, the introduction of foreign DNA fragments when creating transgenic lines may block the function of endogenous genes and affect the expression of adjacent genes. In contrast to transgenic approaches, CRISPR/Cas9 technology is able to generate nontransgenic plants. Because CRISPR/Cas9 expression cassettes and their target sites are located at different positions of the genome, segregation and removal of the CRISPR/Cas9 cassettes is possible via subsequent selfing or crossing; however, this is not feasible in most fruit crops, because of their complex, highly heterozygous, and polyploid genomes, and because they are usually vegetatively propagated.

Fruit trees have a long juvenile stage and take several years to reach the reproductive stage. In such cases, the CRISPR/Cas9 components can be transiently expressed in the nucleus and function for a short time to induce precise mutations. This means that transgene-free edited plants can be generated, since the CRISPR/Cas9 expression cassette is not integrated into the genome. Currently, the overall efficiency of the transient system for the production of T-DNA-free edited apple lines is very low (0.4%) [92]; thus, the next step is to improve editing efficiency and to make this system suitable for other crops.

Preassembled CRISPR-Cas-sgRNA ribonucleoproteins (RNPs) can be delivered into plant cells and used for genome editing without the integration of foreign DNA because of degradation by endogenous proteases [93]. The protoplast transformation technique has been used to transform grape, apple, and lettuce with purified Cas9 RNPs. Sequencing analysis of transformed cells revealed mutagenesis efficiencies of 0.1 to 6.9% in grapevine and apple; however, due to the poor regeneration ability of protoplasts, no plants were regenerated [93,94].

Two methods have been used to obtain transgene-free plants with mutations via CRISPR/Cas9 gene editing. The first is based on the site-specific recombinase flippase (Flp) [95], which recognises 34 bp-long flippase recognition target site (FRT) sequences. The Flp/FRT system has been extensively used to remove undesired transgenic components in transgenic apple [96,97]. The second removal method relies on the Cas9 enzyme cleavage mechanism. Two additional synthetic target sites, referred to as cleavage target sites, were added next to the left border (LB) and right border (RB) sites of the CRISPR/Cas9 vector. When plants are transformed using CRISPR/Cas9, the Cas9 cleavage activity not only edits the endogenous target site, but also removes T-DNA by inserting two additional cleavage target sites, thereby resulting in T-DNA-free plants [98].

1.4.2. Novel Variants of Cas Protein and Applications

In commonly used CRISPR-Cas9 systems, the *Streptococcus pyogenes* (SpCas9)-gRNA complex generally recognises the region 20 nt upstream of the PAM sequence (5′-NGG-3′). To broaden the Cas9 protein recognition sequence and reduce the off-target editing rate, several approaches have been used to broaden PAM compatibility and enhance specificity. These approaches are based on the structural characteristics of SpCas9 binding to gRNA and target DNA. For example, the Cas9 variants SpCas9-VQR (NGA-PAM), SpCas9-EQR (NGAG-PAM), and SpCas9-VRER (NGCG-PAM) functioned, but their cleavage activity levels were lower compared to that of the wild-type SpCas9 in *Arabidopsis* and rice [99–102]. SpCas9-NG has a broader recognition sequence with enhanced compatibility, recognising NG-PAM, and has successfully generated targeted mutations in rice and *Arabidopsis* [103–105].

The variants SpCas9-HF1, eSpCas9, and HypaCas9 have been developed to enhance Cas9 protein-cleavage specificity. They show reduced off-target editing activities, indicating high specificity in plant cells [106–108]. Cas9 protein-directed evolution has been developed for Cas9 engineering, conferring high-specificity engineered SpCas9 proteins such as xCas9 [109], evoCas9 [110], and Sniper-Cas9 [111]. xCas9 recognises the NG, GAA, and GAT PAM sequences. Although the gRNA containing these PAM sequences can be mutated in plants, the mutation efficiencies and the preferences for different corresponding PAM sequences differ between cells. For example, the cleavage activity of xCas9 is lower in rice callus than in mammalian cells [112], and xCas9 does not recognise the NG-PAM sequence in tomatoes [32]. Cas9-NG has a stronger cleavage activity than xCas9, especially at CGG, AGC, TGA, and CGT sequence recognition sites [98], making Cas9-NG more suitable for genome editing at the NG-PAM site in plants. The single-base editing system developed based on Cas9 variants (SpCas9-NG and SpCas9-VQR) has been applied to precise base-editing of plant genomes [103,113].

At present, the most commonly used Cas9 protein comes from *Streptococcus pyogenes*, in order to broaden the Cas9 protein recognition sequence, orthologous Cas9 proteins have been isolated from other bacteria; for instance, NmCas9 has been isolated from *Neisseria meningitidis* [114], SaCas9 from *Staphylococcus aureus* [115], StCas9 from *Streptococcus thermophilus* [116], FnCas9 from *Francisella novicida* [117], and CjCas9 from *Campylobacter jejuni* [118]. These proteins are smaller than SpCas9, which is an advantage in cassette delivery. In *Arabidopsis*, the SaCas9 and SpCas9 systems do not interfere with each other [119], so they can fully utilise Cas9 orthologues that recognise different PAM sequences. Such simultaneous targeting by Cas9 orthologues with different PAM sequences would enable multiplex genome engineering by simultaneously targeting more than one site.

With the continued discovery and investigation of the functions of CRISPR protein family members, new types of Cas proteins have been discovered including the type VI CRISPR-Cas system Cas13 (C2c2) protein, which recognises RNA sequences and exhibits RNA editing activity without altering the genome sequence [120,121]. This system has been successfully applied to knockout gene function in rice and tobacco and promote resistance to RNA viruses in *Arabidopsis* [121,122]. Likewise, the CRISPR-Cas13 system created RNA-guided immunity against RNA viruses in plants. Type V CRISPR-Cas systems, such as Cas12c, Cas12g, Cas12h, Cas12i, and Cas14, are distinguished according to the type of their target template (ssRNA, ssDNA, dsDNA, or ssDNA) and cleavage activity strength. Their functions range from dsDNA nicking and cleavage, and can have collateral cleavage activity on ssRNA, ssDNA, dsDNA, and ssDNA [123,124]. In short, the functional differentiation of these Cas protein variants can be used to target mutations in different nucleic acid types, induce a small number of 100 kb sequence deletions, and expand the repertoire of plant genome-editing tools.

1.5. Regulatory Framework of CRISPR-Cas-Edited Crops

During CRISPR-Cas gene editing of plants, Cas cuts the target sequence to produce double-strand breaks, resulting in the loss of gene function. The CRISPR-Cas technique has been utilised to create modifications in the genome that are identical to natural genetic variation [94]. Similarly, in HDR (homologous DNA repair) of CRISPR-Cas-mediated dsDNA fragmentation, exogenously provided homologous DNA sequences are deemed transgenic; however, when the repair template is derived from the genes from the same species and related interbreeding species, the resultant crops are not regarded as transgenic crops [125]. Nevertheless, the regulatory framework regarding NHEJ and HDR-mediated gene editing contains differing definitions.

The United States Department of Agriculture (USDA), FDA, and Environmental Protection Agency state that the removal of transgenic elements in plants by CRISPR-mediated editing is equivalent to crop improvement by conventional breeding programs, and crops generated in this way are thus not considered GMOs for regulatory purposes [126]. In 2016, the USDA approved the marketing of gene-edited waxy corn without exogenous transgenic elements [127], and an *Agaricus bisporus* mushroom with an anti-browning trait obtained from CRISPR-Cas9 editing was exempted from GMO regulatory procedures [128]. This definition of gene-edited crops by the United States regulatory agencies promotes the genetic improvement of crops and accelerates the introduction of gene-edited crops to the market. Genome editing has emerged as a powerful and elegant technology to develop novel varieties or organisms with desirable traits in tomato, citrus, soybean, sugarcane, camelina, and rice. In May 2020, the USDA–APHIS issued the latest edition of biotechnology regulations that provide three exemptions for genetic modifications in any plant species: (i) resultant changes in DNA after DSB in the absence of an external repair donor template; (ii) single base pair substitution in targeted loci; and (iii) introduction of a known gene that exists in the plant's gene pool.

As a major producer of GM crops, Canada considers gene-edited products such as plants, animal feed, or human food as different from nonedited products, so they must undergo a premarket assessment. In Europe, there are stringent regulations regarding CRISPR/Cas9, and the European Court of Justice has included gene-edited crops in the scope of GM crop regulation [129]. Australia has taken a milder approach, allowing gene-editing without the introduction of any foreign genetic material [130]. In Asia, the attitude towards gene-edited crops has eased in China and Japan, and cases of gene-edited crops being planted in the field have been reported [131]. In addition, some countries have regulatory frameworks that are applied case-by-case, considering the breeding methodology used, new traits or characteristics introduced, and evidence of the genetic changes in the final product.

2. Future Challenges in the Application of CRISPR-Cas Gene Editing

CRISPR-Cas9 genome editing has been introduced and used to obtain abundant germplasm resources with genetic variation, thanks to the results of whole genome sequencing and functional genomics studies in fruit and vegetable crops. Nonetheless, its future application faces two major challenges. The first is the accurate selection of key genes for targeted mutations and the corresponding types of mutations. Important agronomic traits are often complex quantitative traits and editing a single gene does not produce phenotypic changes. Therefore, efficient CRISPR-Cas-mediated target site-specific insertion and chromosome recombination methods can be used to accumulate mutant alleles [132].

The use of gene-editing technology to inhibit the expression of specific genes in plants reduces their adaptability. Therefore, precise genome editing requires efficient and specific regulation of gene functions. Mutations in the exons of genes can change the function of proteins, and mutations in the exon-intron splice sites can result in different alternative splicing variants. Cis-regulatory elements (CREs) are noncoding DNA sequences that often serve to regulate gene transcription. In CREs, single-nucleotide mutations, insertions, deletions, inversions, and epigenetic variations in gene regulatory regions are closely associated with crop domestication [133]. The CRISPR-Cas system has been used to create mutations in the regulatory regions of promoters, and to generate a number of alleles with variable phenotypes, which serve as excellent genetic resources in breeding programs. Currently, the gene editing of CREs in fruit and vegetable crops is still in its preliminary stages. Future studies are aimed at associating changes in expression produced by different mutant CREs with the corresponding phenotypes and obtaining abundant breeding resources through gene editing of CREs. Target induction of DSBs using CRISPR-Cas results not only in small mutations, such as base substitutions, insertions, and deletions, but also large rearrangements of the genome, including large deletions, chromosomal translocations, and inversion, which is an efficient way to completely delete undesired genes, such as those encoding allergens, in fruit and vegetable crops [134–136]. Although such chromosomal rearrangements occur at a lower frequency than with conventional targeted mutagenesis, a deficiency of Ku70, which is involved in the NHEJ pathway, increases the frequency of inversion and translocation (Figure 3) [135].

The second challenge in the application of gene-editing technology is transforming the CRISPR-Cas gene-editing system into plant cells and obtaining regenerated plants. Genetic mutation vector systems mediated by conventional Cas9 or Cas variants, and precise gene-editing systems mediated by CBE, ABE, and prime editing, have been successfully applied in model crops such as rice, corn, tobacco, and tomatoes, via *Agrobacterium*, gene gun, polyethylene glycol (PEG), and electroporation-based methods, providing a foundation for application in other crops [137]. However, developing a universal and efficient genetic transformation and regeneration system for fruit and vegetable crops is more difficult. In particular, the genetic transformation efficiency of CRISPR-Cas cassettes, and the ability of transformed tissues to regenerate plants, have become limiting factors. To improve explant regeneration ability after transformation, and meristem induction activity, the plant morphogenesis regulatory genes *Bbm* (baby boom) and *Wus2* (Wuschel2) can be overexpressed at the same time as CRISPR-Cas expression cassette transformation [138]. Using plant RNA and DNA viruses as vectors to transform plant cells via CRISPR-Cas can provide sufficient sgRNAs and shorten and simplify the process of genetic transformation and regeneration, making this method suitable for in situ transformation, and facilitating the production of gene-edited plants without transgenic elements [139].

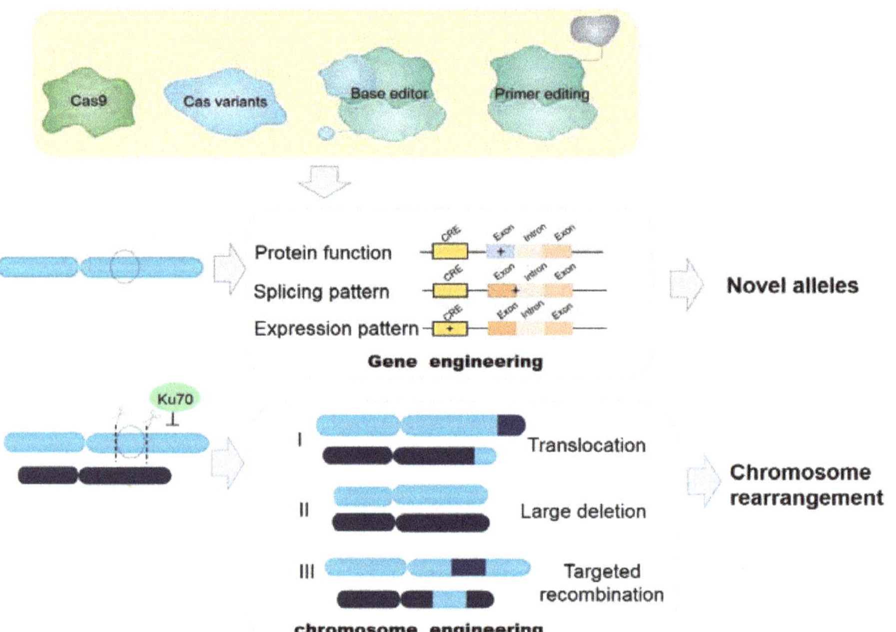

Figure 3. Genome-editing toolbox for trait improvement. CRISPR-Cas9, Cas9 variants, and base and primer editing enable precise gene modification and predictable rearrangement of genomes and chromosomes.

Author Contributions: This review was primarily conducted by L.W., Z.W., D.H., M.T., J.R., Y.S., N.Z. and H.Z., who were the contributors for the collection of the materials and revision of the manuscript. All authors have read and agreed to the published version of the manuscript.

Funding: This review was funded by the Natural Science Foundation of Hubei Province (2020CFB845), Innovation Research Foundation of Wuhan Academy of Agricultural Sciences (QNCX202110).

Institutional Review Board Statement: Not applicable.

Informed Consent Statement: Not applicable.

Acknowledgments: We appreciate internal review by Mao Donghai in Key Laboratory of Agro-Ecological Processes in Subtropical Region, Institute of Subtropical Agriculture, Chinese Academy of Sciences.

Conflicts of Interest: The authors declare no conflict of interest.

References

1. Giovannoni, J.; Nguyen, C.; Ampofo, B.; Zhong, S.; Fei, Z. The Epigenome and Transcriptional Dynamics of Fruit Ripening. *Ann. Rev. Plant Biol.* **2017**, *68*, 61–84. [CrossRef]
2. Karkute, S.G.; Singh, A.K.; Gupta, O.P.; Singh, P.M.; Singh, B. CRISPR/Cas9 Mediated Genome Engineering for Improvement of Horticultural Crops. *Front. Plant Sci.* **2017**, *8*, 1635. [CrossRef]
3. Meyer, R.S.; Purugganan, M.D. Evolution of crop species: Genetics of domestication and diversification. *Nat. Rev. Genet.* **2013**, *14*, 840–852. [CrossRef]
4. Tester, M.; Langridge, P. Breeding technologies to increase crop production in a changing world. *Science* **2010**, *327*, 818–822. [CrossRef] [PubMed]
5. Bigliardi, B.; Galati, F. Innovation trends in the food industry: The case of functional foods. *Trends Food Sci. Technol.* **2013**, *31*, 118–129. [CrossRef]
6. Parmar, N.; Singh, K.H.; Sharma, D.; Singh, L.; Kumar, P.; Nanjundan, J.; Khan, Y.J.; Chauhan, D.K.; Thakur, A.K. Genetic engineering strategies for biotic and abiotic stress tolerance and quality enhancement in horticultural crops: A comprehensive review. *3 Biotech.* **2017**, *7*, 239. [CrossRef] [PubMed]
7. Kramer, M.G.; Redenbaugh, K. Commercialization of a tomato with an antisense polygalacturonase gene: The FLAVR SAVR™ tomato story. *Euphytica* **1994**, *79*, 293–297. [CrossRef]
8. Millstone, E.; Stirling, A.; Glover, D. Regulating Genetic Engineering: The Limits and Politics of Knowledge. *Issues Sci. Technol.* **2015**, *31*, 23–26.
9. Bawa, A.S.; Anilakumar, K.R. Genetically modified foods: Safety, risks and public concerns-a review. *J. Food Sci. Technol.* **2013**, *50*, 1035–1046. [CrossRef]
10. Waltz, E. Gene-edited CRISPR mushroom escapes US regulation. *Nature* **2016**, *7599*, 293. [CrossRef]
11. Waltz, E. With a free pass, CRISPR-edited plants reach market in record time. *Nat. Biotechnol.* **2018**, *36*, 6–7. [CrossRef]
12. Jansen, R.; Embden, J.D.V.; Gaastra, W.; Schouls, L.M. Identification of genes that are associated with DNA repeats in prokaryotes. *Mol. Microbiol.* **2002**, *43*, 1565–1575. [CrossRef] [PubMed]
13. Bolotin, A. Clustered regularly interspaced short palindrome repeats (CRISPRs) have spacers of extrachromosomal origin. *Microbiology* **2005**, *151*, 2551–2561. [CrossRef]
14. Mojica, F.J.; Díez-Villaseñor, C.; García-Martínez, J.; Soria, E. Intervening sequences of regularly spaced prokaryotic repeats derive from foreign genetic elements. *J. Mol. Evol.* **2005**, *60*, 174–182. [CrossRef] [PubMed]
15. Pourcel, C. CRISPR elements in Yersinia pestis acquire new repeats by preferential uptake of bacteriophage DNA, and provide additional tools for evolutionary studies. *Microbiology* **2005**, *151*, 653–663. [CrossRef] [PubMed]
16. Cong, L.; Ran, F.A.; Cox, D.; Lin, S.; Barretto, R. Multiplex Genome Engineering Using CRISPR/Cas Systems. *Science* **2013**, *339*, 819–823. [CrossRef] [PubMed]
17. Kaboli, S.; Babazada, H. CRISPR Mediated Genome Engineering and its Application in Industry. *Curr. Issues Mol. Biol.* **2018**, *26*, 81–92. [CrossRef]
18. Shan, Q.; Wang, Y.; Li, J.; Zhang, Y.; Chen, K.; Liang, Z.; Zhang, K.; Liu, J.; Xi, J.J.; Qiu, J.L.; et al. Targeted genome modification of crop plants using a CRISPR-Cas system. *Nat. Biotechnol.* **2013**, *31*, 686–688. [CrossRef]
19. Rouillon, C.; Zhou, M.; Zhang, J.; Politis, A.; Beilsten-Edmands, V.; Cannone, G.; Graham, S.; Robinson, C.V.; Spagnolo, L.; White, M.F. Structure of the CRISPR interference complex CSM reveals key similarities with cascade. *Mol. Cell* **2013**, *52*, 124–134. [CrossRef]
20. Gasiunas, G.; Barrangou, R.; Horvath, P.; Siksnys, V. Cas9-crRNA ribonucleoprotein complex mediates specific DNA cleavage for adaptive immunity in bacteria. *Proc. Natl. Acad. Sci. USA* **2012**, *109*, 15539–15540. [CrossRef] [PubMed]
21. Doudna, J.A.; Charpentier, E. Genome editing. The new frontier of genome engineering with CRISPR-Cas9. *Science* **2014**, *346*, 1258096. [CrossRef]
22. Yin, K.; Gao, C.; Qiu, J.L. Progress and prospects in plant genome editing. *Nat. Plants* **2017**, *3*, 17107. [CrossRef] [PubMed]
23. Baltes, N.J.; Gil-Humanes, J.; Cermak, T.; Atkins, P.A.; Voytas, D.F. DNA replicons for plant genome engineering. *Plant cell* **2014**, *26*, 151–163. [CrossRef]
24. Endo, M.; Ishikawa, Y.; Osakabe, K.; Nakayama, S.; Kaya, H.; Araki, T.; Shibahara, K.-I.; Abe, K.; Ichikawa, H.; Valentine, L.; et al. Increased frequency of homologous recombination and T-DNA integration in Arabidopsis CAF-1 mutants. *EMBO J.* **2006**, *25*, 5579–5590. [CrossRef] [PubMed]
25. Gil-Humanes, J.; Wang, Y.; Liang, Z.; Shan, Q.; Ozuna, C.V.; Sánchez-León, S.; Baltes, N.J.; Starker, C.; Barro, F.; Gao, C.; et al. High-efficiency gene targeting in hexaploid wheat using DNA replicons and CRISPR/Cas9. *Plant J.* **2017**, *89*, 1251–1262. [CrossRef] [PubMed]
26. Ayathilaka, K.; Sheridan, S.D.; Bold, T.D.; Bochenska, K.; Logan, H.L.; Weichselbaum, R.R.; Bishop, D.K.; Connell, P.P. A chemical compound that stimulates the human homologous recombination protein RAD51. *Proc. Natl. Acad. Sci. USA* **2008**, *105*, 15848–15853. [CrossRef]
27. Donohoue, P.D.; Barrangou, R.; May, A.P. Advances in Industrial Biotechnology Using CRISPR-Cas Systems. *Trends Biotechnol.* **2018**, *36*, 134–146. [CrossRef]
28. Li, R.; Fu, D.; Zhu, B.; Luo, Y.; Zhu, H. CRISPR/Cas9-mediated mutagenesis of lncRNA1459 alters tomato fruit ripening. *Plant J.* **2019**, *97*, 795. [CrossRef]

29. Liang, W.; van Wersch, S.; Tong, M.; Li, X. TIR-NB-LRR immune receptor SOC3 pairs with truncated TIR-NB protein CHS1 or TN2 to monitor the homeostasis of E3 ligase SAUL1. *New Phytol.* **2019**, *221*, 2054–2066. [CrossRef]

30. Mao, Y.; Yang, X.; Zhou, Y.; Zhang, Z.; Botella, J.R.; Zhu, J.-K. Manipulating plant RNA-silencing pathways to improve the gene editing efficiency of CRISPR/Cas9 systems. *Genome Biol.* **2018**, *19*, 149. [CrossRef]

31. Hess, G.T.; Tycko, J.; Yao, D.; Bassik, M.C. Methods and Applications of CRISPR-Mediated Base Editing in Eukaryotic Genomes. *Mol. Cell* **2017**, *68*, 26–43. [CrossRef]

32. Niu, Q.; Wu, S.; Li, Y.; Yang, X.; Liu, P.; Xu, Y.; Lang, Z. Expanding the scope of CRISPR/Cas9-mediated genome editing in plants using an xCas9 and Cas9-NG hybrid. *J. Integr. Plant Biol.* **2020**, *62*, 398–402. [CrossRef]

33. Brooks, C.; Nekrasov, V.; Lippman, Z.B.; Van Eck, J. Efficient gene editing in tomato in the first generation using the clustered regularly interspaced short palindromic repeats/CRISPR-associated9 system. *Plant Physiol.* **2014**, *166*, 1292–1297. [CrossRef] [PubMed]

34. Kulus, D. Genetic resources and selected conservation methods of tomato. *J. Appl. Bot. Food Qual.* **2018**, *91*, 135–144.

35. Ma, C.; Liu, M.; Li, Q.; Si, J.; Ren, X.; Song, H. Efficient BoPDS Gene Editing in Cabbage by the CRISPR/Cas9 System. *Hortic. Plant J.* **2019**, *5*, 164–169. [CrossRef]

36. Sun, B.; Zheng, A.; Jiang, M.; Xue, S.; Yuan, Q.; Jiang, L.; Chen, Q.; Li, M.; Wang, Y.; Zhang, Y.; et al. CRISPR/Cas9-mediated mutagenesis of homologous genes in Chinese kale. *Sci. Rep.* **2018**, *8*, 16786. [CrossRef] [PubMed]

37. Tian, S.; Jiang, L.; Gao, Q.; Zhang, J.; Zong, M.; Zhang, H.; Ren, Y.; Guo, S.; Gong, G.; Liu, F.; et al. Efficient CRISPR/Cas9-based gene knockout in watermelon. *Plant Cell Rep.* **2017**, *36*, 399–406. [CrossRef]

38. Bastet, A.; Zafirov, D.; Giovinazzo, N.; Guyon-Debast, A.; Nogué, F.; Robaglia, C.; Gallois, J.L. Mimicking natural polymorphism in eIF4E by CRISPR-Cas9 base editing is associated with resistance to potyviruses. *Plant Biotechnol. J.* **2019**, *17*, 1736–1750. [CrossRef]

39. Chandrasekaran, J.; Brumin, M.; Wolf, D.; Leibman, D.; Klap, C.; Pearlsman, M.; Sherman, A.; Arazi, T.; Gal-On, A. Development of broad virus resistance in non-transgenic cucumber using CRISPR/Cas9 technology. *Mol. Plant Pathol.* **2016**, *17*, 1140–1153. [CrossRef] [PubMed]

40. Ji, X.; Zhang, H.; Zhang, Y.; Wang, Y.; Gao, C. Establishing a CRISPR–Cas-like immune system conferring DNA virus resistance in plants. *Nat. Plants* **2015**, *1*, 15144. [CrossRef] [PubMed]

41. Ali, Z.; Ali, S.; Tashkandi, M.; Zaidi, S.S.E.A.; Mahfouz, M.M. CRISPR/Cas9-Mediated Immunity to Geminiviruses: Differential Interference and Evasion. *Sci. Rep.* **2016**, *6*, 26912. [CrossRef]

42. Borrelli, V.M.G.; Brambilla, V.; Rogowsky, P.; Marocco, A.; Lanubile, A. The Enhancement of Plant Disease Resistance Using CRISPR/Cas9 Technology. *Frontiers in Plant Science.* **2018**, *9*, 1245. [CrossRef] [PubMed]

43. Zeilmaker, T.; Ludwig, N.R.; Elberse, J.; Seidl, M.F.; Berke, L.; Van Doorn, A.; Schuurink, R.C.; Snel, B.; Van den Ackerveken, G. DOWNY MILDEW RESISTANT 6 and DMR6-LIKE OXYGENASE 1 are partially redundant but distinct suppressors of immunity in Arabidopsis. *Plant J.* **2015**, *81*, 210–222. [CrossRef] [PubMed]

44. Paula de Toledo Thomazella, D.; Brail, Q.; Dahlbeck, D.; Staskawicz, B. CRISPR-Cas9 mediated mutagenesis of a DMR6 ortholog in tomato confers broad-spectrum disease resistance. *BioRxiv* **2016**, 064824. [CrossRef]

45. Nekrasov, V.; Wang, C.; Win, J.; Lanz, C.; Weigel, D.; Kamoun, S. Rapid generation of a transgene-free powdery mildew resistant tomato by genome deletion. *Sci. Rep.* **2017**, *7*, 1–6. [CrossRef]

46. Chaudhary, R.; Hs, A. Resistance-Gene-Mediated Defense Responses against Biotic Stresses in the Crop Model Plant Tomato. *J. Plant Pathol. Microbiol.* **2017**, *8*, 2.

47. Cahya, P.; Barbetti, M.J.; Barker, S.J. A Novel Tomato Fusarium Wilt Tolerance Gene. *Front. Microbiol.* **2018**, *9*, 1226.

48. Zhang, M.; Liu, Q.; Yang, X.; Xu, J.; Liu, G.; Yao, X.; Ren, R.; Xu, J.; Lou, L. CRISPR/Cas9-mediated mutagenesis of Clpsk1 in watermelon to confer resistance to Fusarium oxysporum f.sp. niveum. *Plant Cell Rep.* **2020**, *39*, 589–595. [CrossRef] [PubMed]

49. Shujuan, Z.; Liu, W.; Ruirui, Z.; Wenqing, Y.; Rui, L.; Yujing, L.; Jiping, S.; Lin, S. Knockout of SlMAPK3 Reduced Disease Resistance to Botrytis cinerea in Tomato Plants. *J. Agric. Food Chem.* **2018**, *66*, 8949–8956.

50. Andrés, O.; Selena, G.I.; Nathalie, L.; Roberto, S. Design of a bacterial speck resistant tomato by CRISPR/Cas9-mediated editing of SlJAZ2. *Plant Biotechnol. J.* **2018**, *17*, 665–673.

51. Haque, E.; Taniguchi, H.; Hassan, M.M.; Bhowmik, P.; Karim, M.R.; Śmiech, M.; Zhao, K.; Rahman, M.; Islam, T. Application of CRISPR/Cas9 Genome Editing Technology for the Improvement of Crops Cultivated in Tropical Climates: Recent Progress, Prospects, and Challenges. *Front. Plant Sci.* **2018**, *9*, 617. [CrossRef]

52. Yin, Y.; Qin, K.; Song, X.; Zhang, Q.; Zhou, Y.; Xia, X.; Yu, J. BZR1 Transcription Factor Regulates Heat Stress Tolerance Through FERONIA Receptor-Like Kinase-Mediated Reactive Oxygen Species Signaling in Tomato. *Plant Cell Physiol.* **2018**, *59*, 2239–2254. [CrossRef]

53. Li, R.; Zhang, L.; Wang, L.; Chen, L.; Zhao, R.; Sheng, J.; Shen, L. Reduction of Tomato-Plant Chilling Tolerance by CRISPR-Cas9-Mediated SlCBF1 Mutagenesis. *J. Agric. Food Chem.* **2018**, *66*, 9042–9051. [CrossRef] [PubMed]

54. Wang, L.; Chen, L.; Li, R.; Zhao, R.; Yang, M.; Sheng, J.; Shen, L. Reduced drought tolerance by CRISPR/Cas9-mediated SlMAPK3 mutagenesis in tomato plants. *J. Agric. Food Chem.* **2017**, *65*, 8674–8682. [CrossRef] [PubMed]

55. Tian, S.; Jiang, L.; Cui, X.; Zhang, J.; Guo, S.; Li, M.; Zhang, H.; Ren, Y.; Gong, G.; Zong, M. Engineering herbicide-resistant watermelon variety through CRISPR/Cas9-mediated base-editing. *Plant Cell Rep.* **2018**, *37*, 1353–1356. [CrossRef]

56. Butler, N.M.; Atkins, P.A.; Voytas, D.F.; Douches, D.S. Generation and Inheritance of Targeted Mutations in Potato (Solanum tuberosum L.) Using the CRISPR/Cas System. *PLoS ONE* **2015**, *10*, e0144591. [CrossRef]

57. Bari, V.K.; Nassar, J.A.; Aly, R. CRISPR/Cas9 mediated mutagenesis of MORE AXILLARY GROWTH 1 in tomato confers resistance to root parasitic weed Phelipanche aegyptiaca. *Sci. Rep.* **2021**, *11*, 3905. [CrossRef] [PubMed]

58. Bari, V.K.; Nassar, J.A.; Kheredin, S.M.; Gal-On, A.; Ron, M.; Britt, A.; Steele, D.; Yoder, J.; Aly, R. CRISPR/Cas9-mediated mutagenesis of CAROTENOID CLEAVAGE DIOXYGENASE 8 in tomato provides resistance against the parasitic weed Phelipanche aegyptiaca. *Sci. Rep.* **2019**, *9*, 11438. [CrossRef] [PubMed]

59. Li, H.; Qi, M.; Sun, M.; Liu, Y.; Liu, Y.; Xu, T.; Li, Y.; Li, T. Tomato Transcription Factor SlWUS Plays an Important Role in Tomato Flower and Locule Development. *Front. Plant Sci.* **2017**, *8*, 457. [CrossRef]

60. Li, T.; Yang, X.; Yu, Y.; Si, X.; Zhai, X.; Zhang, H.; Dong, W.; Gao, C.; Xu, C. Domestication of wild tomato is accelerated by genome editing. *Nat. Biotechnol.* **2018**, *36*, 1160–1163. [CrossRef]

61. Rodríguez-Leal, D.; Lemmon, Z.H.; Man, J.; Bartlett, M.E.; Lippman, Z.B. Engineering Quantitative Trait Variation for Crop Improvement by Genome Editing. *Cell* **2017**, *171*, 470–480. [CrossRef]

62. Lin, T.; Zhu, G.; Zhang, J.; Xu, X.; Yu, Q.; Zheng, Z.; Zhang, Z.; Lun, Y.; Li, S.; Wang, X.; et al. Genomic analyses provide insights into the history of tomato breeding. *Nat. Genet.* **2014**, *46*, 1220–1226. [CrossRef] [PubMed]

63. Ballester, A.R.; Molthoff, J.; de Vos, R.; Hekkert, B.T.L.; Orzaez, D.; Fernaʊndez-Moreno, J.P.; Tripodi, P.; Gran-dillo, S.; Martin, C.; Heldens, J.; et al. Biochemical and Molecular Analysis of Pink Tomatoes: Deregulated Expression of the Gene Encoding Transcription Factors SlMYB12 Leads to Pink Tomato Fruit Color. *Plant Physiol.* **2010**, *152*, 71–84. [CrossRef] [PubMed]

64. Čermák, T.; Baltes, N.J.; Čegan, R.; Zhang, Y.; Voytas, D.F. High-frequency, precise modification of the tomato genome. *Genome Biol.* **2015**, *16*, 232. [CrossRef] [PubMed]

65. Deng, L.; Wang, H.; Sun, C.; Li, Q.; Jiang, H.; Du, M.; Li, C.B.; Li, C. Efficient generation of pink-fruited tomatoes using CRISPR/Cas9 system. *J. Genet. Genom.* **2018**, *45*, 51–54. [CrossRef]

66. Filler Hayut, S.; Melamed Bessudo, C.; Levy, A.A. Targeted recombination between homologous chromosomes for precise breeding in tomato. *Nat. Commun.* **2017**, *8*, 15605. [CrossRef]

67. Sun, B.; Jiang, M.; Zheng, H.; Jian, Y.; Huang, W.L.; Yuan, Q.; Zheng, A.H.; Chen, Q.; Zhang, Y.T.; Lin, Y.X.; et al. Color-related chlorophyll and carotenoid concentrations of Chinese kale can be altered through CRISPR/Cas9 targeted editing of the carotenoid isomerase gene BoaCRTISO. *Hortic. Res.* **2020**, *7*, 161. [CrossRef]

68. Chen, L.; Yang, D.; Zhang, Y. Evidence for a specific and critical role of mitogen-activated protein kinase 20 in uni-to-binucleate transition of microgametogenesis in tomato. *Plant Biol.* **2018**, *219*, 176–194. [CrossRef]

69. Kris-Etherton, P.M.; Hecker, K.D.; Bonanome, A.; Coval, S.M.; Binkoski, A.E.; Hilpert, K.F.; Griel, A.E.; Etherton, T.D. Bioactive compounds in foods: Their role in the prevention of cardiovascular disease and cancer. *Am. J. Med.* **2002**, *113*, 71–88. [CrossRef]

70. Meng, X.; Yang, D.; Li, X.; Zhao, S.; Sui, N.; Meng, Q. Physiological changes in fruit ripening caused by overexpression of tomato SlAN2, an R2R3-MYB factor. *Plant Physiol. Biochem.* **2015**, *89*, 24–30. [CrossRef]

71. Ye, J.; Wang, X.; Hu, T.; Zhang, F.; Wang, B.; Li, C.; Yang, T.; Li, H.; Lu, Y. An InDel in the Promoter of Al-ACTIVATED MALATE TRANSPORTER9 Selected during Tomato Domestication Determines Fruit Malate Contents and Aluminum Tolerance. *Plant Cell* **2017**, *29*, 2249. [CrossRef] [PubMed]

72. Nonaka, S.; Arai, C.; Takayama, M.; Matsukura, C.; Ezura, H. Efficient increase of γ-aminobutyric acid (GABA) content in tomato fruits by targeted mutagenesis. *Sci. Rep.* **2017**, *7*, 7057. [CrossRef]

73. Li, X.; Wang, Y.; Chen, S.; Tian, H.; Fu, D.; Zhu, B.; Luo, Y.; Zhu, H. Lycopene Is Enriched in Tomato Fruit by CRISPR/Cas9-Mediated Multiplex Genome Editing. *Front. Plant Sci.* **2018**, *9*, 559. [CrossRef] [PubMed]

74. Nakayasu, M.; Akiyama, R.; Lee, H.J.; Osakabe, K.; Osakabe, Y.; Watanabe, B.; Sugimoto, Y.; Umemoto, N.; Saito, K.; Muranaka, T.; et al. Generation of α-solanine-free hairy roots of potato by CRISPR/Cas9 mediated genome editing of the St16DOX gene. *Plant Physiol. Biochem.* **2018**, *131*, 70–77. [CrossRef] [PubMed]

75. Ito, Y.; Nishizawa-Yokoi, A.; Endo, M.; Mikami, M.; Toki, S. CRISPR/Cas9-mediated mutagenesis of the RIN locus that regulates tomato fruit ripening. *Biochem. Biophys. Res. Commun.* **2015**, *467*, 76–82. [CrossRef]

76. Lang, Z.; Wang, Y.; Tang, K.; Tang, D.; Datsenka, T.; Cheng, J.; Zhang, Y.; Handa, A.K.; Zhu, J.K. Critical roles of DNA demethylation in the activation of ripening-induced genes and inhibition of ripening-repressed genes in tomato fruit. *Proc. Natl. Acad. Sci. USA* **2017**, *114*, E4511–E4519. [CrossRef]

77. Wang, D.; Samsulrizal, A.; Yan, B.C.; Allcock, C.; Craigon, D.J. Characterization of CRISPR Mutants Targeting Genes Modulating Pectin Degradation in Ripening Tomato. *Plant Physiol.* **2019**, *179*, 544–557.

78. Yu, Q.H.; Wang, B.; Li, N.; Tang, Y.; Yang, S.; Yang, T.; Xu, J.; Guo, C.; Yan, P.; Wang, Q.; et al. CRISPR/Cas9-induced Targeted Mutagenesis and Gene Replacement to Generate Long-shelf Life Tomato Lines. *Sci. Rep.* **2017**, *7*, 11874. [CrossRef]

79. Ledford, H. Fixing the tomato: CRISPR edits correct plant-breeding snafu. *Nature* **2017**, *545*, 394–395. [CrossRef]

80. Soyk, S.; Lemmon, Z.H.; Oved, M.; Fisher, J.; Liberatore, K.L.; Park, S.J.; Goren, A.; Jiang, K.; Ramos, A.; van der Knaap, E.; et al. Bypassing Negative Epistasis on Yield in Tomato Imposed by a Domestication Gene. *Cell* **2017**, *169*, 1142–1155. [CrossRef]

81. Roldan, M.V.G.; Périlleux, C.; Morin, H.; Huerga-Fernandez, S.; Latrasse, D.; Benhamed, M.; Bendahmane, A. Natural and induced loss of function mutations in SlMBP21 MADS-box gene led to jointless-2 phenotype in tomato. *Sci. Rep.* **2017**, *7*, 1–10. [CrossRef]

82. Klap, C.; Yeshayahou, E.; Bolger, A.M.; Arazi, T.; Gupta, S.K.; Shabtai, S.; Usadel, B.R.; Salts, Y.; Barg, R. Tomato facultative parthenocarpy results from SlAGAMOUS-LIKE 6 loss of function. *Plant Biotechnol. J.* **2016**, *15*, 634–647. [CrossRef] [PubMed]

83. Hu, J.; Alon, I.; Naomi, O.; Sun, T.P. DELLA-ARF/IAA Interaction Mediates Crosstalk between Gibberellin and Auxin Signaling in Controlling Fruit Initiation in Solanum lycopersicum. *Plant Cell* **2018**, *30*, 1710–1728. [CrossRef] [PubMed]

84. Ueta, R.; Abe, C.; Watanabe, T.; Sugano, S.S.; Ishihara, R.; Ezura, H.; Osakabe, Y.; Osakabe, K. Rapid breeding of parthenocarpic tomato plants using CRISPR/Cas9. *Sci. Rep.* **2017**, *7*, 507. [CrossRef] [PubMed]

85. Xu, C.; Park, S.J.; Van Eck, J.; Lippman, Z.B. Control of inflorescence architecture in tomato by BTB/POZ transcriptional regulators. *Genes Dev.* **2016**, *30*, 2048–2061. [CrossRef]

86. Zsögön, A.; Čermák, T.; Naves, E.R.; Notini, M.M.; Edel, K.H.; Weinl, S.; Freschi, L.; Voytas, D.F.; Kudla, J.; Peres, L.E.P. De novo domestication of wild tomato using genome editing. *Nat. Biotechnol.* **2018**, *36*, 1211–1216. [CrossRef] [PubMed]

87. Wang, R.; Tavano, E.C.d.R.; Lammers, M.; Martinelli, A.P.; Angenent, G.C.; de Maagd, R.A. Re-evaluation of transcription factor function in tomato fruit development and ripening with CRISPR/Cas9-mutagenesis. *Sci. Rep.* **2019**, *9*, 1696. [CrossRef] [PubMed]

88. Soyk, S.; Müller, N.A.; Park, S.J.; Schmalenbach, I.; Jiang, K.; Hayama, R.; Zhang, L.; Van Eck, J.; Jiménez-Gómez, J.M.; Lippman, Z.B. Variation in the flowering gene SELF PRUNING 5G promotes day-neutrality and early yield in tomato. *Nat. Genet.* **2017**, *49*, 162–168. [CrossRef]

89. Zsögön, A.; Cermak, T.; Voytas, D.; Peres, L.E. Genome editing as a tool to achieve the crop ideotype and de novo domestication of wild relatives: Case study in tomato. *Plant Sci.* **2017**, *256*, 120–130. [CrossRef] [PubMed]

90. Bisognin, D.A. Breeding vegetatively propagated horticultural crops. *Crop Breed. Appl. Biotechnol.* **2011**, *11*, 35–43. [CrossRef]

91. Corte, E.D.; MMahmoud, L.; SMoraes, T.; Mou, Z.; WGrosser, J.; Dutt, M. Development of Improved Fruit, Vegetable, and Ornamental Crops Using the CRISPR/Cas9 Genome Editing Technique. *Plants* **2019**, *8*, 601. [CrossRef]

92. Charrier, A.; Vergne, E.; Dousset, N.; Richer, A.; Petiteau, A.; Chevreau, E. Efficient Targeted Mutagenesis in Apple and First Time Edition of Pear Using the CRISPR-Cas9 System. *Front. Plant Sci.* **2019**, *10*, 40. [CrossRef]

93. Malnoy, M.; Viola, R.; Jung, M.H.; Koo, O.J.; Kim, S.; Kim, J.S.; Velasco, R.; Nagamangala Kanchiswamy, C. DNA-Free Genetically Edited Grapevine and Apple Protoplast Using CRISPR/Cas9 Ribonucleoproteins. *Front. Plant Sci.* **2016**, *7*, 1904. [CrossRef]

94. Woo, J.W.; Kim, J.; Kwon, S.I.; Corvalán, C.; Cho, S.W.; Kim, H.; Kim, S.G.; Kim, S.T.; Choe, S.; Kim, J.S. DNA-free genome editing in plants with preassembled CRISPR-Cas9 ribonucleoproteins. *Nat. Biotechnol.* **2015**, *33*, 1162–1164. [CrossRef]

95. Pompili, V.; Dalla Costa, L.; Piazza, S.; Pindo, M.; Malnoy, M. Reduced fire blight susceptibility in apple cultivars using a high-efficiency CRISPR/Cas9-FLP/FRT-based gene editing system. *Plant Biotechnol. J.* **2020**, *18*, 845–858. [CrossRef] [PubMed]

96. Dalla Costa, L.; Piazza, S.; Campa, M.; Flachowsky, H.; Hanke, M.-V.; Malnoy, M. Efficient heat-shock removal of the selectable marker gene in genetically modified grapevine. *Plant Cell Tissue Organ Cult.* **2016**, *124*, 471–481. [CrossRef]

97. Herzog, K.; Flachowsky, H.; Deising, H.B.; Hanke, M.V. Heat-shock-mediated elimination of the nptII marker gene in transgenic apple (Malus×domestica Borkh.). *Gene* **2012**, *498*, 41–49. [CrossRef]

98. Dalla Costa, L.; Piazza, S.; Pompili, V.; Salvagnin, U.; Cestaro, A.; Moffa, L.; Vittani, L.; Moser, C.; Malnoy, M. Strategies to produce T-DNA free CRISPRed fruit trees via Agrobacterium tumefaciens stable gene transfer. *Sci. Rep.* **2020**, *10*, 1–14. [CrossRef] [PubMed]

99. Akihiro, Y.; Takashi, I.; Mika, Y.; Yuri, K.; Shinichiro, S. Developing Heritable Mutations in Arabidopsis thaliana Using a Modified CRISPR/Cas9 Toolkit Comprising PAM-Altered Cas9 Variants and gRNAs. *Plant Cell Physiol.* **2019**, *60*, 2255–2262.

100. Hu, X.; Meng, X.; Liu, Q.; Li, J.; Wang, K. Increasing the efficiency of CRISPR-Cas9-VQR precise genome editing in rice. *Plant Biotechnol. J.* **2018**, *16*, 292–297. [CrossRef] [PubMed]

101. Hu, X.; Wang, C.; Fu, Y.; Liu, Q.; Jiao, X.; Wang, K. Expanding the Range of CRISPR/Cas9 Genome Editing in Rice. *Mol. Plant* **2016**, *9*, 943–945. [CrossRef]

102. Kleinstiver, B.P.; Prew, M.S.; Tsai, S.Q.; Topkar, V.V.; Nguyen, N.T.; Zheng, Z.; Gonzales, A.P.W.; Li, Z.; Peterson, R.T.; Yeh, J.-R.J.; et al. Engineered CRISPR-Cas9 nucleases with altered PAM specificities. *Nature* **2015**, *523*, 481–485. [CrossRef] [PubMed]

103. Hua, K.; Tao, X.; Han, P.; Wang, R.; Zhu, J.K. Genome Engineering in Rice Using Cas9 Variants that Recognize NG PAM Sequences. *Mol. Plant* **2019**, *12*, 1003–1014. [CrossRef]

104. Endo, M.; Mikami, M.; Endo, A.; Kaya, H.; Itoh, T.; Nishimasu, H.; Nureki, O.; Toki, S. Genome editing in plants by engineered CRISPR-Cas9 recognizing NG PAM. *Nat. Plants* **2019**, *5*, 14–17. [CrossRef]

105. Zhong, Z.; Sretenovic, S.; Ren, Q.; Yang, L.; Bao, Y.; Qi, C.; Yuan, M.; He, Y.; Liu, S.; Liu, X.; et al. Improving Plant Genome Editing with High-Fidelity xCas9 and Non-canonical PAM-Targeting Cas9-NG. *Mol. Plant* **2019**, *12*, 1027–1036. [CrossRef]

106. Kleinstiver, B.P.; Pattanayak, V.; Prew, M.S.; Tsai, S.Q.; Nguyen, N.T.; Zheng, Z.; Joung, J.K. High-fidelity CRISPR-Cas9 nucleases with no detectable genome-wide off-target effects. *Nature* **2016**, *529*, 490–495. [CrossRef] [PubMed]

107. Slaymaker, I.M.; Gao, L.; Zetsche, B.; Scott, D.A.; Yan, W.X.; Zhang, F. Rationally engineered Cas9 nucleases with improved specificity. *Science* **2016**, *351*, 84–88. [CrossRef] [PubMed]

108. Zhang, D.; Zhang, H.; Li, T.; Chen, K.; Qiu, J.L.; Gao, C. Perfectly matched 20-nucleotide guide RNA sequences enable robust genome editing using high-fidelity SpCas9 nucleases. *Genome Biol.* **2017**, *18*, 191. [CrossRef] [PubMed]

109. Hu, J.H.; Miller, S.M.; Geurts, M.H.; Tang, W.; Chen, L.; Sun, N.; Zeina, C.M.; Gao, X.; Rees, H.A.; Lin, Z.; et al. Evolved Cas9 variants with broad PAM compatibility and high DNA specificity. *Nature* **2018**, *556*, 57–63. [CrossRef] [PubMed]

110. Casini, A.; Olivieri, M.; Petris, G.; Bianchi, A.; Montagna, C.; Reginato, G.; Maule, G.; Lorenzin, F.; Prandi, D.; Romanel, A.; et al. Evocas9, a highly specific SpCas9 variant from a yeast in vivo screening. *Nat. Biotechnol.* **2018**, *36*, 265–271. [CrossRef]

111. Lee, J.K.; Jeong, E.; Lee, J.; Jung, M.; Shin, E.; Kim, Y.-H.; Lee, K.; Jung, I.; Kim, D.; Kim, S.; et al. Directed evolution of CRISPR-Cas9 to increase its specificity. *Nat. Commun.* **2018**, *9*, 3048. [CrossRef]
112. Wang, J.; Meng, X.; Hu, X.; Sun, T.; Li, J.; Wang, K.; Yu, H. xCas9 expands the scope of genome editing with reduced efficiency in rice. *Plant Biotechnol. J.* **2019**, *17*, 709–711. [CrossRef] [PubMed]
113. Wu, Y.; Xu, W.; Wang, F.; Zhao, S.; Yang, J. Increasing Cytosine Base Editing Scope and Efficiency With Engineered Cas9-PmCDA1 Fusions and the Modified sgRNA in RiceData_Sheet_1.docx. *Front. Genet.* **2019**, *10*, 379. [CrossRef]
114. Hou, Z.; Zhang, Y.; Propson, N.E.; Howden, S.E.; Chu, L.F.; Sontheimer, E.J.; Thomson, J.A. Efficient genome engineering in human pluripotent stem cells using Cas9 from Neisseria meningitidis. *Proc. Natl. Acad. Sci. USA* **2013**, *110*, 15644–15649. [CrossRef] [PubMed]
115. Ran, F.A.; Cong, L.; Yan, W.X.; Scott, D.A.; Gootenberg, J.S.; Kriz, A.J.; Zetsche, B.; Shalem, O.; Wu, X.; Makarova, K.S.; et al. In vivo genome editing using Staphylococcus aureus Cas9. *Nature* **2015**, *520*, 186–191. [CrossRef] [PubMed]
116. Müller, M.; Lee, C.M.; Gasiunas, G.; Davis, T.H.; Cradick, T.J.; Siksnys, V.; Bao, G.; Cathomen, T.; Mussolino, C. Streptococcus thermophilus CRISPR-Cas9 Systems Enable Specific Editing of the Human Genome. *Mol. Ther.* **2016**, *24*, 636–644. [CrossRef]
117. Hirano, H.; Gootenberg, J.S.; Horii, T.; Abudayyeh, O.O.; Kimura, M.; Hsu, P.D.; Nakane, T.; Ishitani, R.; Hatada, I.; Zhang, F. Structure and Engineering of Francisella novicida Cas9. *Cell* **2016**, *164*, 950–961. [CrossRef]
118. Kim, E.; Koo, T.; Park, S.W.; Kim, D.; Kim, K.; Cho, H.Y.; Song, D.W.; Lee, K.J.; Jung, M.H.; Kim, S.; et al. In vivo genome editing with a small Cas9 orthologue derived from Campylobacter jejuni. *Nat. Commun.* **2017**, *8*, 14500. [CrossRef]
119. Steinert, J.; Schiml, S.; Fauser, F.; Puchta, H. Highly efficient heritable plant genome engineering using Cas9 orthologues from Streptococcus thermophilus and Staphylococcus aureus. *Plant J.* **2015**, *84*, 1295–1305. [CrossRef]
120. Abudayyeh, O.O.; Gootenberg, J.S.; Konermann, S.; Joung, J.; Slaymaker, I.M.; Cox, D.B.; Shmakov, S.; Makarova, K.S.; Semenova, E.; Minakhin, L.; et al. C2c2 is a single-component programmable RNA-guided RNA-targeting CRISPR effector. *Science* **2016**, *353*, aaf5573. [CrossRef]
121. Cox, D.B.T.; Gootenberg, J.S.; Abudayyeh, O.O.; Franklin, B.; Kellner, M.J.; Joung, J.; Zhang, F. RNA editing with CRISPR-Cas13. *Science* **2017**, *358*, 1019–1027. [CrossRef] [PubMed]
122. Aman, R.; Ali, Z.; Butt, H.; Mahas, A.; Aljedaani, F.; Khan, M.Z.; Ding, S.; Mahfouz, M. RNA virus interference via CRISPR/Cas13a system in plants. *Genome Biol.* **2018**, *19*, 1–9. [CrossRef] [PubMed]
123. Harrington, L.B.; Burstein, D.; Chen, J.S.; Paez-Espino, D.; Ma, E.; Witte, I.P.; Cofsky, J.C.; Kyrpides, N.C.; Banfield, J.F.; Doudna, J.A. Programmed DNA destruction by miniature CRISPR-Cas14 enzymes. *Science* **2018**, *362*, 839–842. [CrossRef] [PubMed]
124. Yan, W.X.; Hunnewell, P.; Alfonse, L.E.; Carte, J.M.; Keston-Smith, E.; Sothiselvam, S.; Garrity, A.J.; Chong, S.; Makarova, K.S.; Koonin, E.V.; et al. Functionally diverse type V CRISPR-Cas systems. *Science* **2019**, *363*, 88. [CrossRef]
125. Kim, H.; Kim, J.S. A guide to genome engineering with programmable nucleases. *Nat. Rev. Genet.* **2014**, *15*, 321–334. [CrossRef] [PubMed]
126. Globus, R.; Qimron, U. A technological and regulatory outlook on CRISPR crop editing. *J. Cell Biochem.* **2018**, *119*, 1291–1298. [CrossRef]
127. USDA Re: Confirmation of Regulatory Status of Waxy Com Developed by CRISPR-Cas Technology. Available online: https://www.aphis.usda.gov/biotechnology/downloads/reg_loi/15-352-01_air_response_signed.pdf (accessed on 2 October 2020).
128. USDA Re: Request for Confirmation that Transgene-Free, CRISPR-Edited Mushroom Is Not a Regulated Article. Available online: https://www.aphis.usda.gov/biotechnology/downloads/reg_loi/15-321-01_air_response_signed.pdf (accessed on 2 October 2020).
129. Callaway, E. CRISPR plants now subject to tough GM laws in European Union. *Nature* **2018**, *560*, 16. [CrossRef]
130. Mallapaty, S. Australian gene-editing rules adopt 'middle ground'. *Nature* **2019**. [CrossRef]
131. Metje-Sprink, J.; Sprink, T.; Hartung, F. Genome-edited plants in the field. *Curr. Opin. Biotechnol.* **2020**, *61*, 1–6. [CrossRef]
132. Zhu, H.; Li, C.; Gao, C. Applications of CRISPR-Cas in agriculture and plant biotechnology. *Nat. Rev. Mol. Cell Biol.* **2020**, *21*, 661–677. [CrossRef]
133. Li, Q.; Sapkota, M.; van der Knaap, E. Perspectives of CRISPR/Cas-mediated cis-engineering in horticulture: Unlocking the neglected potential for crop improvement. *Hortic. Res.* **2020**, *7*, 36. [CrossRef] [PubMed]
134. Schmitz, D.J.; Ali, Z.; Wang, C.; Aljedaani, F.; Hooykaas, P.J.J.; Mahfouz, M.; de Pater, S. CRISPR/Cas9 Mutagenesis by Translocation of Cas9 Protein Into Plant Cells via the Agrobacterium Type IV Secretion System. *Front. Genome Ed.* **2020**, *2*, 6. [CrossRef]
135. Schmidt, C.; Pacher, M.; Puchta, H. Efficient induction of heritable inversions in plant genomes using the CRISPR/Cas system. *Plant J.* **2019**, *98*, 577–589. [CrossRef]
136. Beying, N.; Schmidt, C.; Pacher, M.; Houben, A.; Puchta, H. CRISPR–Cas9-mediated induction of heritable chromosomal translocations in Arabidopsis. *Nat. Plants* **2020**, *6*, 638–645. [CrossRef]
137. Chuang, Y.F.; Phipps, A.J.; Lin, F.L.; Hecht, V.; Hewitt, A.W.; Wang, P.Y.; Liu, G.S. Approach for in vivo delivery of CRISPR/Cas system: A recent update and future prospect. *Cell Mol. Life Sci.* **2021**, *78*, 2683–2708. [CrossRef]
138. Lowe, K.; Wu, E.; Wang, N.; Hoerster, G.; Hastings, C.; Cho, M.J.; Scelonge, C.; Lenderts, B.; Chamberlin, M.; Cushatt, J.; et al. Morphogenic Regulators Baby boom and Wuschel Improve Monocot Transformation. *Plant Cell* **2016**, *28*, 1998–2015. [CrossRef]
139. Ali, Z.; Abul-faraj, A.; Li, L.; Ghosh, N.; Piatek, M.; Mahjoub, A.; Aouida, M.; Piatek, A.; Baltes, N.J.; Voytas, D.F.; et al. Efficient Virus-Mediated Genome Editing in Plants Using the CRISPR/Cas9 System. *Mol. Plant* **2015**, *8*, 1288–1291. [CrossRef] [PubMed]

horticulturae

Article

Identification of Major Loci and Candidate Genes for Anthocyanin Biosynthesis in Broccoli Using QTL-Seq

Chunqing Liu [1,†], Xueqin Yao [1,†], Guangqing Li [1], Lei Huang [1], Xinyan Wu [2] and Zhujie Xie [1,*]

[1] Shanghai Key Lab of Protected Horticultural Technology, Horticulture Research Institute, Shanghai Academy of Agricultural Sciences, Shanghai 201403, China; liuchunqing@saas.sh.cn (C.L.); yaoxueqin@saas.sh.cn (X.Y.); liguangqing0212@163.com (G.L.); hl646532431@163.com (L.H.)

[2] School of Ecological Technology and Engineering, Shanghai Institute of Technology, Shanghai 201418, China; 18355093278@163.com

* Correspondence: xiezj8@163.com; Tel.: +86-189-1816-2046

† These authors contributed equally to this work.

Abstract: Anthcyanins determine the colors of flowers, fruits, and purple vegetables and act as important health-promoting antioxidants. BT 126 represents a broccoli variety with a high content of anthocyanins (5.72 mg/g FW). Through QTL-seq bulk segregant analysis, the present study aimed to determine the quantitative trait loci (QTLs) involved in anthocyanin biosynthesis in the F2 population (n = 302), which was obtained by crossing BT 126 with a non-anthocyanin-containing SN 60. The whole-genome resequencing of purple (n = 30) and green (n = 30) bulk segregates detected ~1,117,709 single nucleotide polymorphisms (SNPs) in the *B. oleracea* genome. Two QTLs, tightly correlated with anthocyanin biosynthesis ($p < 0.05$), were detected on chromosomes 7 (*BoPur7.1*) and 9 (*BoPur9.1*). The subsequent high-resolution mapping of *BoPur9.1* in the F2 population (n = 280) and F3 population (n = 580), with high-throughput genotyping of SNPs technology, narrowed the major anthocyanin biosynthesis QTL region to a physical distance of 73 kb, containing 14 genes. Among these genes, *Bo9g174880*, *Bo9g174890*, and *Bo9g174900* showed high homology with *AT5G07990* (gene encoding flavonoid 3′ hydroxylase), which was identified as a candidate gene for *BoPur9.1*. The expression of *BoF3′H* in BT 126 was significantly higher than that in SN60. Multiple biomarkers, related to these QTLs, represented potential targets of marker-assisted selection (MAS) foranthocyanin biosynthesis in broccoli. The present study provided genetic insights into the development of novel crop varieties with augmented health-promoting features and improved appearance.

Keywords: *Brassica oleracea*; broccoli; QTL; candidate gene; anthocyanin

Citation: Liu, C.; Yao, X.; Li, G.; Huang, L.; Wu, X.; Xie, Z. Identification of Major Loci and Candidate Genes for Anthocyanin Biosynthesis in Broccoli Using QTL-Seq. *Horticulturae* **2021**, *7*, 246. https://doi.org/10.3390/horticulturae7080246

Academic Editor: Yuyang Zhang

Received: 6 June 2021
Accepted: 9 August 2021
Published: 13 August 2021

Publisher's Note: MDPI stays neutral with regard to jurisdictional claims in published maps and institutional affiliations.

Copyright: © 2021 by the authors. Licensee MDPI, Basel, Switzerland. This article is an open access article distributed under the terms and conditions of the Creative Commons Attribution (CC BY) license (https://creativecommons.org/licenses/by/4.0/).

1. Introduction

Broccoli (*Brassica oleracea* var. *italic*) is a popular vegetable of *B. oleracea* that differs from most Brassica species, including Chinese cabbage, turnip, cabbage, broccoli, cauliflower, and oilseed rape. Most varieties of broccoli are domesticated from crop wild relatives in the Mediterranean Basin and grow as annuals, producing a large head with florets, buds, leaves, stalks, and stems for consumption. Both of broccoli and cauliflower cultivar groups are members of the CC genome *B. oleracea* (2n = 18) coenospecies. High-quality reference genomes of cauliflower have been reported, and the assembled cauliflower genome was 584.60 Mb in size [1]. As a great food source of essential vitamins and minerals, broccoli contains antioxidant phytochemicals, such as glucoraphanin, which may help prevent cancer [2]. Purple broccoli attracts increasing attention as a functional food, owing to its pleasing appearance and high level of health-promoting effects [3]. The purple coloration has been identified as one of the signs of anthocyanin accumulation [4].

Anthocyanins belong to a class of flavonoid compounds that impart color to plants and play an important role in plant protection against a variety of biotic and abiotic stresses [5–7]. Anthocyanin biosynthetic pathway genes have been extensively character-ized in Arabidopsis (*Arabidopsis thaliana*), maize (*Zea mays*), tomato (*Solanum lycopersicum*), eggplant (*Solanum melongena*), and other plant species [8–11]. The induction of structural genes and transcription factors is considered to be an important mechanism for the reg-ulation of anthocyanin biosynthesis in the Brassica species [12–14]. A cauliflower purple mutation exhibited a tissue-specific pattern of anthocyanin overproduction [15]. Due to the insertion of Harbinger transposon, the upregulation of *BoMYB2* specifically activates BobHLHs and some downstream anthocyanin structural genes to generate the ectopic ac-cumulation of anthocyanin. Similarly, the accumulation of anthocyanin is caused by the ac-tivation of *TT8* and *MYB2* genes in red cabbages [14]; of *BrMYB2* and downstream genes, such as *BrTT8*, *BrF3'H*, *BrDFR1*, *BrANS1*, *BrUGTs*, *BrATs*, and *BrGSTs*, under the control of *BrMYB2* in a purple head of Chinese cabbage cultivar 11S91 [16]. In purple cabbages, deleting or replacing nucleotides in the exon of *BoMYBL2-1* is solely responsible for the pur-ple coloration [17]. In addition, temperature and light are the major environmental factors that affect anthocyanin accumulation. In purple head Chinese cabbage, *BrMYB2* and *BrTT8* activated anthocyanin structural genes after low temperature induction [18]. Elevated temperature could suppress anthocyanin accumulation via COP1-HY5 signaling, and *MYBL2* down-regulation partially modulated the high-temperature-associated suppression of anthocyanin production [19]. Some efforts have also led to the identification of candidate genes that regulate the coloration of the Brassica species. In broccoli, three QTLs have been mapped to the purple sepal trait of the flower head on chromosome C01 [20]. In orna-mental kale, the genes that individually conferred pink and purple leaf colorations have been mapped to chromosomes C3 [21] and C9 [22], respectively. In Zicaitai, an important locus on chromosome 7 highly controlled the stalk color trait, which was significantly correlated with *bHLH49* expression in the F2 population [23]. In purple-heading Chinese cabbage, the purple inner leaf trait was solely regulated by the dominant gene *BrPur*, which was mapped to A07, between SSR markers A710 and A714, with a genetic distance of 3.1 and 3.5 cM, respectively [24].

The majority of agronomic traits are controlled by QTLs, which are critical for improve-ment in crop breeding, through marker-assisted selection (MAS). The classical method of QTL mapping is linkage mapping, which is laborious and requires a great amount of time. Next-generation sequencing (NGS) has become the new strategy for establishing associations between agronomic traits and biomarkers or genes. It has been demonstrated that the QTL-seq method, which combines bulk segregant analysis (BSA) with NGS, is an effective tool for mapping and isolating QTLs [25]. QTL-seq is performed with two groups of individual plants, with a contrasting phenotype on a trait of interest, from segregating population-either F2 re-combinant inbred lines, double haploid, or backcross populations. Through high-throughput SNP (Hi-SNP) technology, the genotype analysis of two mixed pools identifies the genomic position of the polymorphic molecular markers, and the major QTL region with significant segregation of genotypes is identified. Hi-SNP is a technique for large-scale SNP genotyping, based on multiplex-PCR, combined with the next generation sequencing and bioinformatics tools. Amplicon sequencing, combined multiplex-PCR and NGS with higher depth and low false discovery rate (and was more accurate than the whole-genome resequencing (WGS)), has been used for known SNPs genotyping in diploid but was not reported in allopolyploid crops. QTL-seq used for SNP genotyping had many advantages, such as simple primer design, single short reads sequences, high-throughput, high-depth sequencing, and easy to automate and process, e.g., recently, the QTL-seq method has been utilized for mapping QTLs related to resistance to rice blast disease and seedling vigor [26], leaf spot resistance in peanuts [27,28], heat tolerance in broccoli [29], resistance to Fusarium oxysporum f. sp. niveum race 1 in watermelon [30], cucumber early flowering [31], and tomato fruit weight and lobule amounts [32].

Mutant analyses, leading to purple or red organs, has been extensively studied in flowers, fruits, and model plants [15]. Although there is much research on the underlying mechanisms of anthocyanin biosynthesis in cauliflower [15], cabbage [17], kohlrabi [33], kale [22], and

Chinese kale [34], only few have focused on broccoli [20]. The purple broccoli mutant represents an interesting mutation, in which the accumulation of anthocyanins in flower buds causes the mutant heads to exhibit purple coloration. The broccoli purple mutation was found to be controlled by QTLs. We designate the symbol *Pur* for the Purple allele.

Through QTL-seq, this work aimed to identify QTLs involved in anthocyanin biosynthesis in the F2 population, which was obtained by crossing the purple broccoli line BT 126 with SN 60 (a cultivar with green heads).

2. Materials and Methods

2.1. Plant Materials

Two broccoli inbred lines, including BT 126 (with purple heads) and SN 60 (with green heads), were used as parental lines in this study. To develop segregating populations for anthocyanin biosynthesis, the purple head plant BT 126 was crossed with the green head plant SN 60 (Figure 1). The F1 plants were self-pollinated to produce the two F2 populations with 302 and 280 individuals, respectively.

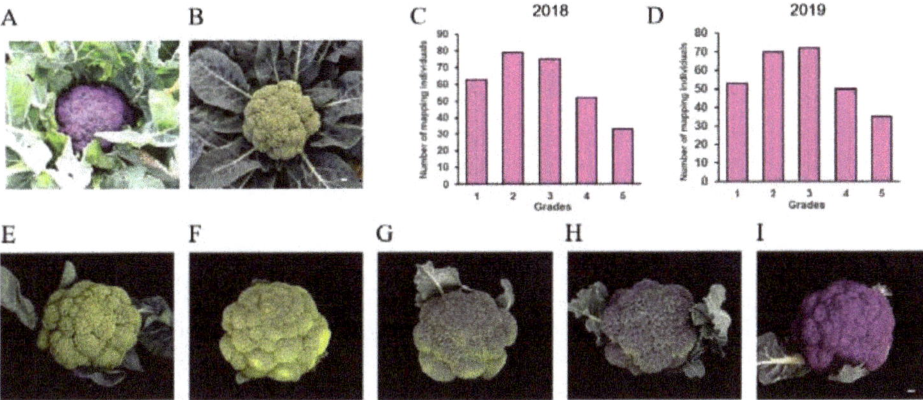

Figure 1. Phenotypes of the parents and F2 individuals and their frequency distributions. (**A**) Maternal line BT 126; (**B**) paternal line, SN 60; (**C,D**) the frequency distribution of the purple head trait in the F2 populations with 302 and 280 individuals planted in 2018 and 2019, respectively; (**E–I**) phenotype of two F2 populations with 1st to 5th grades of head color. The DNAs of 30 F2 individuals with extreme phenotypes (1st and 5th grade) were used to develop high anthocyanin and low anthocyanin bulks; scale bar = 1 cm.

Initially, a population of 302 F2 individuals, along with 10 plants from each parental line, were grown at Zhuanghang Experimental Base of Shanghai Academy of Agricultural Sciences, Shanghai in 2018. The phenotyping of each F2 individual was carried out on at least three separate days, after the flower head reached maturity in the field. Curd color, which showed a color distribution from green to purple, was visually scored from 1 to 5: 1–green, 2–slight purple, 3–light purple, 4–slightly darker purple, and 5–purple (Figure 1E–I). The total anthocyanin content has been determined by the high-performance liquid chromatography (HPLC) method at six biological replicates. The BT 126 is a purple broccoli variety, with an average content of anthocyanin 5.72 mg/g FW. It was crossed to SN 60, with very low anthocyanin content (0.64 mg/g FW), to generate F1 progeny. The average contents of anthocyanin, for the 5 phenotypic categories, were 5.38, 3.26, 2.29, 1.27, and 0.78 mg/g FW. All the above information was increased in the revised manuscript. Subsequently, two populations of 280 F2 and 580 F3 individuals were grown under routine management at Zhuanghang Experimental Base of Shanghai Academy of Agricultural Sciences, Shanghai in 2019 and 2020, respectively. We started the broccoli seeds in the plastic tunnel in August. Four weeks after germination, we transplanted them to field plots. The head color of each F2 and F3 plants was visually phenotyped in 60 days, when grown from transplants, when the heads usually had a diameter larger than $2\frac{1}{4}$ inches. To avoid the sunlight and temperature effects, all the plants were grown in the same plastic tunnel. Leaf tissues were harvested and stored at $-20\,°C$ for DNA extraction. The head tissues of both parents (BT 126 and SN 60), at the full-size, mature stage, were collected from the same site of the top head, at three biological replicates, and used for RT-qPCR analysis. The anthocyanins in the broccoli heads were isolated and assessed, according to the method proposed by Liu C. et al. [4].

All materials were obtained from the Institute of Horticulture, Shanghai Academy of Agricultural Sciences.

2.2. DNA Purification, Library Generation, and Whole-Genome ReSequencing

Total DNA was extracted from fresh leaves using the CTAB method [35]. A total of 1.5 µg DNA, per specimen, was utilized for DNA sample preparations. The DNA quality was assessed by 1% agarose gel electrophoresis. Among the F2 segregating population, composed of 302 plants, 30 with purple heads (high anthocyanin biosynthesis, HAB) and 30 with green heads (low anthocyanin biosynthesis, LAB) were selected to extract equal amount of DNA, pooled to construct the DNA bulks. The DNA bulks, together with the two parental DNA, were used for whole-genome resequencing.

By ultrasonication, 350-bp fragments were obtained from the tested DNA sample. A sequencing library was generated using the Truseq Nano DNA HT Sample Preparation Kit (Illumina, San Diego, CA, USA), according to the manufacturer's instructions. The constructed library was sequenced on the Illumina HiSeq4000 platform (Illumina, CA, USA), and 150-bp paired-end reads were produced with approximately 350-bp inserts. Stringent quality control (QC) steps were applied to ensure the reliability and accuracy of the reads. Then, the filtered, high-quality sequences from the DNA bulks and parental genotypes were aligned and mapped to the public *B. oleracea* genome database (TO1000) (http://plants.ensembl.org/Brassica_oleracea) (accessed on 5 May 2021) with the Burrows-Wheeler alignment (BWA) tool [36,37]. Alignment files were converted into BAM files with SAMtools (Wellcome Trust Genome Campus, Cambridge, UK) [38]. SNP detection was carried out for each specimen, utilizing the UnifiedGenotyper function in the GATK3.8 software (The Broad Institute of Harvard and MIT, Cambridge, MA, USA) [39]. ANNOVAR (Children's Hospital of Philadelphia, Philadelphia, PA, USA) was employed for SNP annotation, based on the GFF3 file of the reference genome [40].

2.3. QTL-Seq Analysis

By applying an established QTL-seq method, based on SNP-index and Δ(SNP-index) estimates, candidate genomic regions harboring the main QTL(s) involved in anthocyanin biosynthesis in broccoli were identified. Only the SNPs homozygous in either parent and polymorphic between the parents were prepared for the further analysis. Then 1,117,709 SNPs, between both parents, were selected. The read depth for the above-mentioned homozygous SNPs, in LAB and HAB bulks, was obtained to evaluate the SNP-index. The SNP-index, calculated according to the reads order-checking depth information, utilized short reads covering the given nucleotide position, calculated the differs reads bar number, and accounted for the ratio of the total number [26]. The genotype of one parent was utilized as a reference to determine the number of reads for the parental genotype, or other genotypes, in the LAB and HAB libraries. Then, the number of the various reads was divided by the total read number, and the ratio constituted the SNP-index of the base sites. The SNP-index points below 0.3, in both libraries, were removed. The sliding window method was utilized to present the SNP-indexes for the entire genome. All SNP-indexes in a given window were averaged to obtain the SNP-index for that particular window. The window size of 1Mb and the step size of 1Kb were routinely utilized. The Δ (SNP-index) was then calculated using the following formula: [SNP index (HAB bulk)—SNP index (LAB bulk)].

2.4. High-Throughput Genotyping of SNPs in Intra-Specific Mapping Individuals

In order to validate the *BoPur* QTL, obtained by QTL-seq, classical QTL mapping was performed, via the selection of the SNP, with a polymorphism between the genotypes of the parents (BT 126 and SN 60). Through the Hi-SNP method, a total of 33 SNPs, differentiating BT 126 and SN60, were utilized for validation and high-throughput genotyping in the F2 segregating population with 280 individuals and parents. The HI-SNP technology was developed by Shanghai BioWing Applied Biotechnology Company (http://www.biowing.com.cn/) (accessed on 9 June 2021) the genotyping of 33 SNPs was carried out by multiplex PCR with NGS on Illumina X-10 (Illumina, CA, USA) [41]. The genotyping primers were shown in the Supplementary Table S1. Based on the phenotypic and genotypic data, linkage maps were constructed by QTL IciMapMaker (Chinese Academy of Agricultural Sciences, Beijing, China) with a logarithm of the odds (LOD) value [42]. The LOD score is a measure of the strength of the evidence for the presence of a QTL at a particular location (the LOD score = log10 likelihood ratio, comparing single-QTL model to the "no QTL anywhere" mode).

2.5. Analysis of the Candidate Gene

To infer the gene regulation patterns of *BoF3'H* during anthocyanin biosynthesis, a qRT-PCR assay was performed. The RNA was isolated from BT 126 and SN60 at three head developmental stages, with both low and high anthocyanin contents. Reverse transcription was conducted using oligo-dT primers and PrimeScript™ RT Master Mix (Takara BIO, Inc., Shiga, Japan). The primer pair sequences for *BoF3'H*, *BoANS2*, and *BoTTG1* used in this study were previously designed for the RT-qPCR analysis. *BoLBD38.3*-specific primers were designed based on the sequences in NCBI (accession number: XM_013739916). The qPCR primer pair sequences for *BoLBD38.3* were 5′-GCCCAAACGGAGACGATTAG-3′ (forward) and 5′-AACCGTTCACTGGCGATGTG-3′ (reverse), gene-specific primers that were confirmed to produce specific gene products by sequencing [15]. Real-time PCR was performed using TB Green$^®$ Premix Ex Taq™ (Tli RNaseH Plus) (Takara, Dalian, China) on an ABI QuantStudio 5 real-time PCR system(Thermo Fisher Scientific, MA, USA) according to the manufacturer's instructions. Actin was used as a reference gene [43]. The data were processed using the $2^{-\Delta\Delta Ct}$ method [44]. Sequence data of *BoF3'H* can be found in the GenBank data libraries, under accession number XM_013751545.

3. Results

3.1. Inheritance of Head Color of BT 126 with SN 60

Significant differences in head color were observed among the BT 126 (parent 1, P1, purple head), SN 60 (parent 2, P2, green head), and F1 hybrids planted in 2018 and 2019 (Figure 1). The BT 126 purple line exhibited a high anthocyanin accumulation in the heads (5.72 mg/g FW) and developed normally when compared to green broccoli, under normal conditions, in the field and greenhouse [4]. It was crossed to SN 60, with very low anthocyanin content (0.241 mg/g FW), to generate the F1 progeny. F1 individuals displayed third grade color. As expected, the F2 population was divided into purple and green classes, with varying degrees of continuous distribution in 1st–5th color grades (Figure 1E–I). No transgressive segregation was observed in either direction of the parental genotype in the mapping population. We could infer that the purple trait was controlled by QTL.

3.2. QTL-Seq Analysis

The high-throughput, whole-genome resequencing of both parents, as well as the LAB and HAB libraries, yielded 158.49 to 261.57 million reads per sample. Based on the *B. oleracea* reference genome (estimated genome size~630 Mb), the read mapping ratio was >93% (Table 1). The mean coverage of the reference genome, across all samples, was 33×. As depicted in Table 1, the 4× coverage of each specimen ranged from 86.85% to 92.43%. Meanwhile, the value of Q20 was above 93% in all specimens (data not shown). The comparative genome sequence analysis of the parental DNA and DNA bulks with the reference genome revealed 1,117,709 polymorphic SNPs.

Table 1. Whole-genome mapping statistics for parents, as well as LAB and HAB bulks.

Sample	Mapped Reads	Total Reads	Mapping Rate (%)	Average Depth (X)	Coverage 1× (%)	Coverage 4× (%)
SN 60	245,092,342	261,565,560	93.7	41.81	92.33	89.41
BT 126	148,780,556	158,487,152	93.88	26.2	91.51	86.85
LAB	218,978,598	232,601,048	94.14	34.9	95.03	92.43
HAB	192,459,282	203,800,728	94.44	31.13	94.77	91.78

The SNP-index of each SNP, that distinguished LAB from HAB, was determined. The mean SNP-index, covering 1 Mb genomic sequence, was assessed separately in LAB and HAB, by the 1 kb sliding window method, and mapped against all *B. oleracea* reference chromosomes (9 in total). The Δ (SNP-index) was determined by combining the SNP-index data of LAB and HAB and was plotted against the positions (Mb) on the *B. oleracea* genome. The QTL-seq analysis detected two major genomic regions on chromosomes 7 (36,784,249–44,791,849) (*BoPur7.1*) and 9 (46,217,406–52,600,419) (*BoPur9.1*), with significant associations with anthocyanin biosynthesis in broccoli (Figure 2). Moreover, LAB and HAB mapping plants with reduced or elevated anthocyanin biosynthesis had the majority of SNP alleles from SNP 1 to SNP 26 (Table S1). An important genomic region [BoSNP5 (46,217,406 bp) to BoSNP 26 (52,600,419 bp)] harboring the QTL *Pur* on chromosome 9 showed a Δ (SNP-index) that was markedly different from 0 ($p < 0.05$) (Figures 2A and 3B). The QTL-seq data confirmed an important QTL (*BoPur9.1*) located at the 6.38 Mb genomic interval [46,217,406 (BoSNP 5) to 52,600,419 (BoSNP 26) bp] on chromosome 9, which was associated with the regulation of anthocyanin biosynthesis in broccoli.

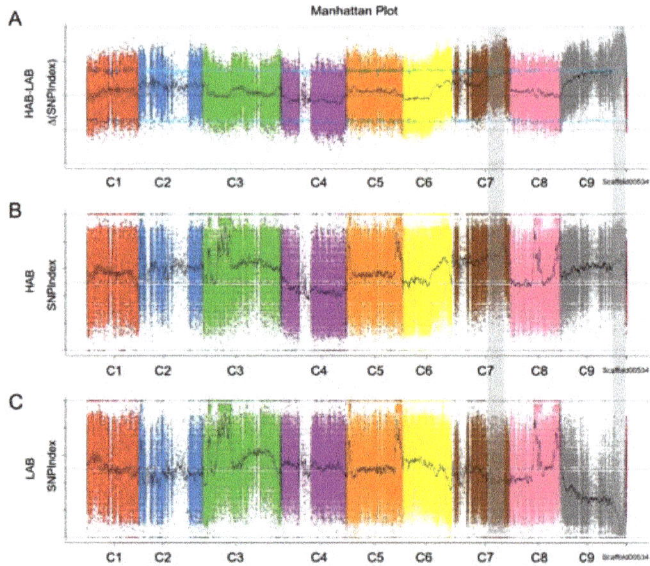

Figure 2. Distribution of two progeny SNP-index on the chromosome length, represented by the horizontal axis (Mb); vertical axis: SNP-index. (**A**) Δ (SNP-index) graphs generated from QTL-seq study; (**B**) SNP-index graphs depicting the HAB (high anthocyanin biosynthesis bulk); (**C**) SNP-index graphs depicting the LAB (low anthocyanin biosynthesis bulk). Gray shaded boxes indicate significant QTLs.

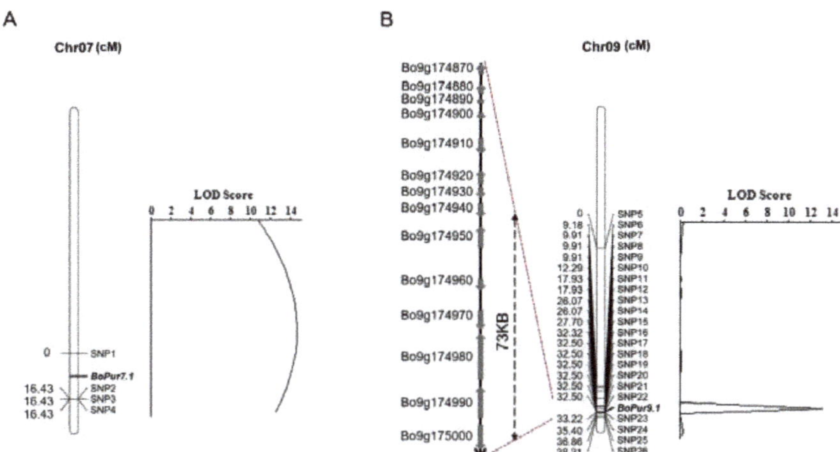

Figure 3. Maps of the *Pur* loci on C07 (**A**) and C09 (**B**) constructed using 26 SNP markers.

3.3. Validation of QTL-Seq-Derived Anthocyanin Biosynthesis QTL through High-Throughput SNP

To assess *BoPur7.1* and *BoPur9.1*, detected by QTL-seq, a high-throughput SNP was carried out. The genotyping data of 26 SNP markers were mapped to a highly dense intraspecific genetic region of chromosomes 7 and 9, depicting polymorphisms between parental genotypes and between LAB and HAB, and were combined with phenotypic features of a second F2 mapping population with 280 individuals. Classical QTL analysis, based on interval mapping and composite interval mapping, revealed two highly important genomic regions: [BoSNP37 (32.5 cM) to BoSNP38 (33.22 cM)], harboring a potent (LOD: 13.1) *Pur* QTL (*BoPur9.1*) on broccoli chromosome 9, and [BoSNP1 (0 cM) to BoSNP2 (16.43 cM)], harboring a potent (LOD: 14.6) *Pur* QTL (*BoPur7.1*) on broccoli chromosome 7. The detected QTLs had an interval of 0.72 cM with 72,683 bp [BoSNP22 (51,716,244 bp) to BoSNP23 (51,788,927 bp)] on chromosome 9 and an interval of 16.43 cM with 6,920,903 bp [BoSNP1 (36,784,249 bp) to BoSNP2 (43,705,152 bp)] on chromosome 7 (Figure 3). The proportions of phenotypic variants caused by the *BoPur9.1* and *BoPur7.1* QTLs were 28.19% and 38.12%, respectively (Table 2).

Table 2. Quantitative trait loci ($p < 0.05$) associated with anthocyanin biosynthesis in broccoli.

Chromosome	Start	End	LOD	PVE (%)	Add	Dom
7	SNP1	SNP2	14.6193	38.1207	−0.8350	−0.4366
9	SNP22	SNP23	13.1231	28.1935	0.0150	1.0399

LOD-logarithm of odds; PVE-phenotypic variance explained; Add-additive effect; Dom-dominance effect.

To further delineate the *BoPur9.1* QTL, the genotyping data of 17 selected SNP markers, the most tightly linked to *BoPur9.1* (SNP10–SNP26), were analyzed in 580 F3 mapping individuals. Loci mapping analysis identified a major QTL for anthocyanin biosynthesis, designed by two SNP markers, including SNP22 and SNP23. This result was consistent with the QTL analysis supporting a major anthocyanin biosynthesis QTL *BoPur9.1*, which was located in the genomic interval of 51.71–51.79 Mb on chromosome 9 (Table S2).

3.4. Prediction and Analysis of the Candidate Gene

Based on the *B. oleracea* genome databases (TO1000) (http://plants.ensembl.org/Brassica_oleracea) (accessed on 5 May 2021), a total of 3 anthocyanin-related and 14 predicted protein-coding genes were found in the interval of *BoPur7.1* and *BoPur9.1*, respectively (Table 3). According to domain annotations from InterPro and BLASTX (best hit) analyses, two of these fourteen genes have not been annotated (Table 3). The other fifteen candidate genes were as follows: *Bo7g096780* (homologous gene AT5G24520), encoding the transparent testa glabra 1; *Bo7g099150* (homologous gene AT3G49940), encoding the LOB domain-containing protein 38; *Bo7g108300* (homologous gene AT4G22880), encoding the anthocyanin synthase; *Bo9g174870* (homologous gene AT5G08000), encoding the X8-GPI family of proteins; *Bo9g174880*, *Bo9g174890*, and *Bo9g174900* (homologous gene AT5G07990), encoding flavonoid 3′ hydroxylase (F3'H), which catalyzed the conversion of dihydrokaempferol to dihydroquercetin in anthocyanin biosynthesis [45]; *Bo9g174920* (homologous gene AT5G07960), encoding the Asterix-like protein; *Bo9g174940* (homologous gene AT5G07890), encoding the myosin heavy chain-like protein; *Bo9g174950* (homologous gene AT5G07830), encoding the glucuronidase 2; *Bo9g174960* (homologous gene AT5G07820), encoding the calmodulin-binding, protein-like protein; *Bo9g174970* (homologous gene AT5G07800), encoding the flavin-containing monooxygenase; and *Bo9g174980*, *Bo9g174990*, and *Bo9g175000* (homologous gene AT5G07740), encoding the actin-binding protein. *BoF3'H* was annotated as a homologue of the anthocyanin biosynthesis gene.

Table 3. Annotation of *B. oleracea* genes in the candidate region.

Bo Genes	Chromosome	Gene Position (bp)	AT ID [a]	E-Value	AT GO [b] Annotation
Bo7g096780	C7	37466680–37466680	AT5G24520	0.0	TRANSPARENT TESTA GLABRA 1
Bo7g099150	C7	38873553–38874347	AT3G49940	0.0	LOB domain-containing protein 38
Bo7g108300	C7	42567442–42568605	AT4G22880	0.0	anthocyanin synthase
Bo9g174870	C9	51722173–51723584	AT5G08000	1×10^{-127}	X8-GPI family of proteins
Bo9g174880	C9	51725909–51727132	AT5G07990	0.0	flavonoid 3′ hydroxylase activity
Bo9g174890	C9	51728802–51729245	AT5G07990	0.0	flavonoid 3′ hydroxylase activity
Bo9g174900	C9	51731919–51732761	AT5G07990	0.0	flavonoid 3′ hydroxylase activity
Bo9g174910	C9	51736550–51738488	-	-	-
Bo9g174920	C9	51742405–51744195	AT5G07960	1×10^{-128}	Asterix-like protein
Bo9g174930	C9	51744445–51745784	-	-	-
Bo9g174940	C9	51747328–51749122	AT5G07890	0.0	myosin heavy, chain-like protein
Bo9g174950	C9	51751745–51754512	AT5G07830	0.0	glucuronidase 2
Bo9g174960	C9	51758770–51760263	AT5G07820	0.0	calmodulin-binding, protein-like protein
Bo9g174970	C9	51764406–51766923	AT5G07800	0.0	Flavin-containing monooxygenase
Bo9g174980	C9	51767719–51777672	AT5G07740	0.0	actin-binding protein
Bo9g174990	C9	51779465–51787686	AT5G07740	0.0	actin-binding protein
Bo9g175000	C9	51788944–51794354	AT5G07740	0.0	actin-binding protein

[a] The best hits of the seven *B. oleracea* genes compared to *A. thaliana* (AT). [b] GO annotations for seven Bo to AT best-hit genes obtained from TAIR.

A primer pair, spanning the full-length CDS of *BoF3'H* was designed, and PCR was performed using the cDNA of BT 126 and SN 60 as a template. DNA sequencing revealed that the full-length of *BoF3'H* in purple-head SN 60 is 1669 bp, whereas it was 1600 bp in green-head SN 60. Compared with *BoF3'H* in BT 126, a deletion of 68 bp and 1bp was found at nucleotide 1080 and 1501, respectively, and nine SNPs were present in SN60 (sequences of BT 126 and SN60 were illustrated in Table S3). The polymorphism of the candidate gene BoF3'H was further confirmed in the segregating population. To delineate the potential candidate genes regulating the anthocyanin biosynthesis in broccoli, differ-ential expression profiling of *BoF3'H* was performed in two parental lines, at three head developmental stages. The expression of *BoF3'H*, *BoANS2*, and *BoLED38* in BT 126 were up-regulated at three head developmental stages of anthocyanin accumulation, and the expression levels were significantly higher than those in SN60 (Figure 4B–F).

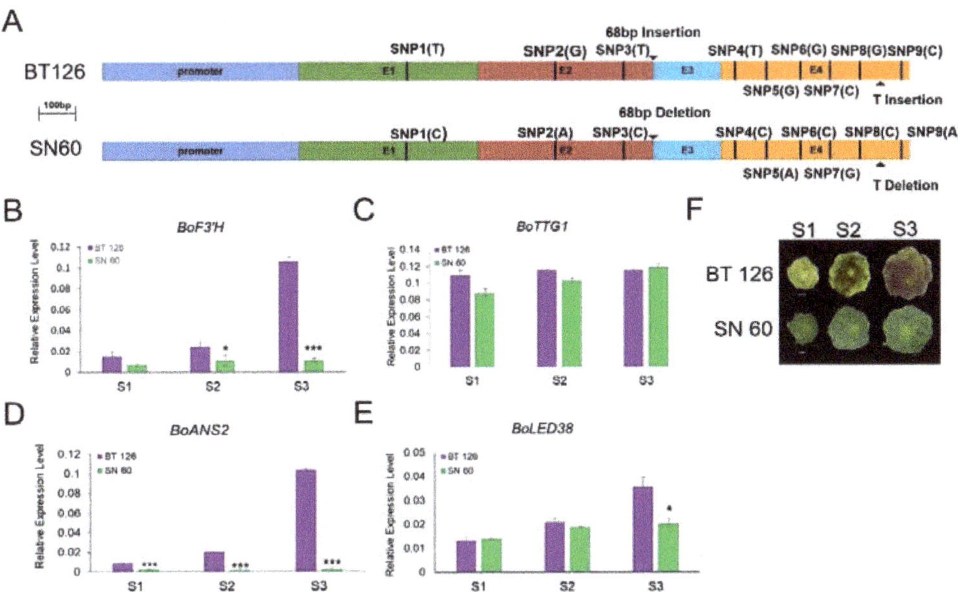

Figure 4. Structure identity and expression of broccoli gene *BoF3'H*. (**A**) Comparison of *BoF3'H* promoter and cDNA sequence between parental lines BT 126 and SN 60; (**B–E**) relative expression of *BoF3'H*, *BoTTG1*, *BoANS2*, and *BoLED38* at three head developmental stages. Significant differences between BT 126 and SN 60 are indicated by * based on Students *t* test (* $p < 0.05$, ** $p < 0.01$; *** $p < 0.001$); (**F**) the phenotype of the head during three developmental stages. The upper part depicts the head of BT 126, while the lower part depicts the head of SN 60. Scale bar = 1 cm.

4. Discussion

Increasing the anthocyanin content in Brassica vegetables represents an important goal for nutriment breeding. This work discovered two important genomic regions harboring anthocyanin biosynthesis QTLs *BoPur7.1* and *Bopur9.1* on chromosomes 7 and 9, based on intra-specific broccoli mapping individuals, through whole-genome NGS-based, high-throughput QTL-seq. The two QTL-seq-derived *Pur* QTLs were subsequently verified by Hi-SNP analysis. In previous studies, a major locus was reported to control anthocyanin pigmentation. In purple cauliflower (*B. oleracea* var *botrytis*), a *Pur* gene, encoding the transcription factor R2R3 MYB, was isolated [43]. Broccoli flower head's purple sepal trait is regulated by a major loci and two minor loci on chromosome C01 [20]. The purple gene of non-heading Chinese cabbage is subject to a single dominant inheritance mode but does not follow the Mendel law [33]. Due to the considerable variations in anthocyanin content, a novel locus was mapped in the linkage group R07 in purple turnip plants [46]. Moreover, a single dominant gene on C09, named *BoPr*, was reported to control anthocyanin pigmentation in leaves. The physical region of the C09 QTL was from 19,018,694 to 24,359,626 (5.3 Mb in internal length), which was different from our major QTLs for head anthocyanin contents. The major locus on C09 from 51,716,244 to 51,788,927 was included in this study. This indicates a population-specific inheritance modality for certain QTLs controlling anthocyanin biosynthesis in the Brassica species. Hence, the integrated approach developed here could be used for rapidly identifying the target QTLs, important genes, and/or alleles involved in the qualitative and quantitative traits of various crops. Diverse models of purple traits, related to anthocyanins, may come from different genetic backgrounds. The purple gene in ornamental kale (*B. oleracea* L. var. *acephala*) shows a single dominant inheritance pattern on chromosome C09, and *Bo9g058630* encoding dihydroflavonol 4-reductase (DFR) was considered to be a candidate gene [22].

Combining QTL-seq with Hi-SNP analysis, the 0.73 Mb QTL *Bopur9.1* region [BoSNP37 (51,716,244 bp) to BoSNP38 (51,788,927 bp)], encompassing 14 genes, was detected on chromosome 9, which accounted for ~28.19% of all phenotypic variations in anthocyanin biosynthesis. Among these genes, according to their respective annotations (Table 3), three genes, *Bo9g174880*, *Bo9g174890*, and *Bo9g174900* were homologues of Arabidopsis *F3′H*, encoding flavanone 3′-hydroxylase. The *BoF3′H* gene has been suggested to have played a critical role in modifying plant coloration [45], which contained a coding sequence length of 1536 bp and encoded a protein of 511 amino acids with four exons. All of the 3 candidate genes became aligned to the different regions of *BoF3′H*. An increased expression of *F3′H* was found to be responsible for anthocyanin production in a number of anthocyanin-accumulating mutants. For example, the ectopic expression of apple *MdF3′H* can increase the production of flavonols and cyanidin-based pigments in the Arabidopsis *tt7* mutant, under nitrogen pressure [47]. The *VvF3′H* gene was identified from grapevine, and its ectopic expression results in high accumulation levels of anthocyanin and flavonols in the petunia *ht1* mutant line [48]. *OsF3′H* editing in the Heugseonchal or Sinmyungheugchal variety, using the CRISPR/Cas9 system, might be responsible for ocher seeds with lower anthocyanin contents than wild-type black rice plants [49]. Two copies of the *F3′H* gene were isolated in the barley genome and exhibited a tissue-specific expression pattern of anthocyanin synthesis [50]. Furthermore, nine SNPs and a 68-bp insertion, between the 3rd and 4rd exon of *BoF3′H* CDS sequences, were found, which may cause a gain-of-function mutation (Figure 4A). The 68 bp InDel was in the open reading frame (between E3 and E4), as well as the T InDel in the 3′ end. Therefore, a frame shift is expected, resulting in a putative truncated or extended protein. In addition, since most enzymes had conserved regions, anthocyanin discoloration might occur, as a result of *BoF3′H* enzyme inactivity, due to one of the two isoforms. In addition, the 6.92 Mb QTL *Bopur7.1* region [BoSNP1 (36,784,249 bp) to BoSNP2 (43,705,152 bp)], encompassing 3 anthocyanin-related genes, was detected on chromosome 7, which accounted for ~38.12% of all phenotypic variations in anthocyanin biosynthesis. Among these genes, *BoANS2* and *BoLED38* were activated at three head developmental stages of anthocyanin accumulation [12,13]. However, further experiments involving transformation are needed to verify whether the function of this gene is responsible for the purple head in broccoli.

Most of the agronomic traits are controlled by multiple QTLs. The traditional QTL fine-mapping method usually requires several generations of backcrossing, screening of a large population for recombinants and exhaustive field phenotyping, which is both time-consuming and labor-intensive [51]. Although the genotyping-by-sequencing (GBS) approach has been shown to be an efficient way to develop high-resolution genetic mapping [52], the genotyping of a large population with genome-wide markers is expensive. It is important to note that the QTL-seq approach has been successfully used to conduct a BSA analysis, as well as deployed for the mapping of a segregated population of homozygous parental lines with opposite phenotypes. Using QTL-seq, only two samples need to be sequenced. The sequence coverage is decided by genome size and complexity, but as an example for broccoli, sufficient data could be generated by sequencing each bulk to 30× coverage. Further, it's an effective method to identify markers most tightly linked to the trait.

In this study, two identified QTLs were distributed on chromosome 7 and 9, respectively. The purple intensity and size of the heading leaves had great variations among F2 individuals. The anthocyanin content of purple head BT 126 was also affected by the temperature condition. Therefore, environmental factors might have an important effect on anthocyanin synthesis. The genetic mechanism of anthocyanin biosynthesis is complex.

Candidate genes that control the purple trait may be different in diverse genetic backgrounds and may have different spatial and temporal expression patterns.

In summary, the above findings indicated that combining QTL-seq, Hi-SNP analysis and differential gene expression profiling might help identify candidate genes manipulating anthocyanin biosynthesis at major QTL intervals in broccoli. *Bo9g174880*, *Bo9g174890*, and *Bo9g174900* were first selected as the strong candidate genes at the *BoPur* locus. The functional assessment of the candidate gene homologs, identified within the confidence intervals of both QTLs, would provide novel insights into the detailed mechanisms of anthocyanin biosynthesis in broccoli.

Supplementary Materials: The following are available online at https://www.mdpi.com/article/10.3390/horticulturae7080246/s1. Table S1: SNPs genetically mapped on chromosome 7 and 9, used for anthocyanin biosynthesis targeted QTL-seq in broccoli; Table S2: quantitative trait loci ($p < 0.05$), associated with anthocyanin biosynthesis in F3 population; Table S3: sequences of BT 126 and SN60.

Author Contributions: Conceptualization, Z.X.; methodology, C.L. and X.Y.; investigation, C.L., L.H., and X.W.; resources, X.Y. and G.L.; data curation, C.L.; writing—original draft preparation, C.L.; writing—review and editing, Z.X.; supervision, Z.X.; funding acquisition, Z.X. and X.Y. All authors have read and agreed to the published version of the manuscript.

Funding: This research was funded by the Shanghai Agriculture Applied Technology Development Program, grant number G2016060105″ and the Youth Talent Development Plan of Shanghai Municipal Agricultural System, grant number 20180116.

Institutional Review Board Statement: Not applicable.

Informed Consent Statement: Not applicable.

Data Availability Statement: Not applicable.

Acknowledgments: We acknowledge Bin Wang, Shanghai BioWing Applied Biotechnology Company, China for the kind help in data analysis.

Conflicts of Interest: The authors declare no conflict of interest.

References

1. Sun, D.; Wang, C.; Zhang, X.; Zhang, W.; Jiang, H.; Yao, X.; Liu, L.; Wen, Z.; Niu, G.; Shan, X. Draft genome sequence of cauliflower (*Brassica oleracea* L. var. botrytis) provides new insights into the C genome in Brassica species. *Hortic. Res.* **2019**, *6*, 1–11. [CrossRef] [PubMed]
2. Fahey, J.W.; Holtzclaw, W.D.; Wehage, S.L.; Wade, K.; Stephenson, K.K.; Talalay, P. Sulforaphane Bioavailability from Glucoraphanin-Rich Broccoli: Control by Active Endogenous Myrosinase. *PLoS ONE* **2015**, *10*, e0140963. [CrossRef]
3. Rodríguez-Hernández, M.d.C.; Moreno, D.A.; Carvajal, M.; García-Viguera, C.; Martínez-Ballesta, M.d.C. Natural antioxidants in purple sprouting broccoli under Mediterranean climate. *J. Food Sci.* **2012**, *77*, C1058–C1063. [CrossRef] [PubMed]
4. Liu, C.; Yao, X.; Li, G.; Huang, L.; Xie, Z. Transcriptomic profiling of purple broccoli reveals light-induced anthocyanin biosynthetic signaling and structural genes. *PeerJ* **2020**, *8*, e8870. [CrossRef]
5. Steyn, W.J.; Wand, S.J.E.; Holcroft, D.M.; Jacobs, G. Anthocyanins in vegetative tissues: A proposed unified function in photoprotection. *New Phytol.* **2002**, *155*, 349–361. [CrossRef] [PubMed]
6. Pourcel, L.; Routaboul, J.-M.; Cheynier, V.; Lepiniec, L.; Debeaujon, I. Flavonoid oxidation in plants: From biochemical properties to physiological functions. *Trends Plant Sci.* **2007**, *12*, 29–36. [CrossRef]
7. Lev-Yadun, S.; Gould, K.S. Role of Anthocyanins in Plant Defence. In *Anthocyanins*; Springer: Berlin, Germany, 2008; pp. 22–28. [CrossRef]
8. Shi, M.-Z.; Xie, D.-Y. Biosynthesis and metabolic engineering of anthocyanins in Arabidopsis thaliana. *Recent Pat. Biotechnol.* **2014**, *8*, 47–60. [CrossRef]
9. Sun, C.; Deng, L.; Du, M.; Zhao, J.; Chen, Q.; Huang, T.; Jiang, H.; Li, C.-B.; Li, C. A Transcriptional Network Promotes Anthocyanin Biosynthesis in Tomato Flesh. *Mol. Plant* **2020**, *13*, 42–58. [CrossRef]
10. Liu, X.; Li, S.; Yang, W.; Mu, B.; Jiao, Y.; Zhou, X.; Zhang, C.; Fan, Y.; Chen, R. Synthesis of Seed-Specific Bidirectional Promoters for Metabolic Engineering of Anthocyanin-Rich Maize. *Plant Cell Physiol.* **2018**, *59*, 1942–1955. [CrossRef]
11. Li, J.; Ren, L.; Gao, Z.; Jiang, M.; Liu, Y.; Zhou, L.; He, Y.; Chen, H. Combined transcriptomic and proteomic analysis constructs a new model for light-induced anthocyanin biosynthesis in eggplant (*Solanum melongena* L.). *Plant Cell Environ.* **2017**, *40*, 3069–3087. [CrossRef]

12. Guo, N.; Cheng, F.; Wu, J.; Liu, B.; Zheng, S.; Liang, J.; Wang, X. Anthocyanin biosynthetic genes in Brassica rapa. *BMC Genom.* **2014**, *15*, 426. [CrossRef] [PubMed]

13. Goswami, G.; Nath, U.K.; Park, J.-I.; Hossain, M.R.; Biswas, M.K.; Kim, H.-T.; Kim, H.R.; Nou, I.-S. Transcriptional regulation of anthocyanin biosynthesis in a high-anthocyanin resynthesized Brassica napus cultivar. *J. Biol. Res. Thessalon.* **2018**, *25*, 19. [CrossRef] [PubMed]

14. Yuan, Y.; Chiu, L.-W.; Li, L. Transcriptional regulation of anthocyanin biosynthesis in red cabbage. *Planta* **2009**, *230*, 1141–1153. [CrossRef]

15. Chiu, L.-W.; Zhou, X.; Burke, S.; Wu, X.; Prior, R.L.; Li, L. The Purple Cauliflower Arises from Activation of a MYB Transcription Factor. *Plant Physiol.* **2010**, *154*, 1470–1480. [CrossRef]

16. He, Q.; Wu, J.; Xue, Y.; Zhao, W.; Li, R.; Zhang, L. The novel gene BrMYB2, located on chromosome A07, with a short intron 1 controls the purple-head trait of Chinese cabbage (*Brassica rapa* L.). *Hortic. Res.* **2020**, *7*, 1–19. [CrossRef] [PubMed]

17. Song, H.; Yi, H.; Lee, M.; Han, C.-T.; Lee, J.; Kim, H.; Park, J.-I.; Nou, I.-S.; Kim, S.-J.; Hur, Y. Purple Brassica oleracea var. capitata F. rubra is due to the loss of BoMYBL2–1 expression. *BMC Plant Biol.* **2018**, *18*, 82. [CrossRef]

18. He, Q.; Ren, Y.; Zhao, W.; Li, R.; Zhang, L. Low Temperature Promotes Anthocyanin Biosynthesis and Related Gene Expression in the Seedlings of Purple Head Chinese Cabbage (*Brassica rapa* L.). *Genes* **2020**, *11*, 81. [CrossRef]

19. Kim, S.; Hwang, G.; Lee, S.; Zhu, J.-Y.; Paik, I.; Nguyen, T.T.; Kim, J.; Oh, E. High Ambient Temperature Represses Anthocyanin Biosynthesis through Degradation of HY5. *Front. Plant Sci.* **2017**, *8*, 1787. [CrossRef]

20. Yu, H.; Wang, J.; Sheng, X.; Zhao, Z.; Shen, Y.; Branca, F.; Gu, H. Construction of a high-density genetic map and identification of loci controlling purple sepal trait of flower head in *Brassica oleracea* L. italica. *BMC Plant Biol.* **2019**, *19*, 228. [CrossRef]

21. Zhu, P.; Cheng, M.; Feng, X.; Xiong, Y.; Liu, C.; Kang, Y. Mapping of Pi, a gene conferring pink leaf in ornamental kale (*Brassica oleracea* L. var. acephala DC). *Euphytica* **2015**, *207*, 377–385. [CrossRef]

22. Liu, X.-P.; Gao, B.-Z.; Han, F.-Q.; Fang, Z.-Y.; Yang, L.-M.; Zhuang, M.; Lv, H.-H.; Liu, Y.-M.; Li, Z.-S.; Cai, C.-C.; et al. Genetics and fine mapping of a purple leaf gene, BoPr, in ornamental kale (*Brassica oleracea* L. var. acephala). *BMC Genom.* **2017**, *18*, 230. [CrossRef] [PubMed]

23. Li, G.-H.; Chen, H.-C.; Liu, J.-L.; Luo, W.-L.; Xie, D.-S.; Luo, S.-B.; Wu, T.-Q.; Akram, W.; Zhong, Y.-J. A high-density genetic map developed by specific-locus amplified fragment (SLAF) sequencing and identification of a locus controlling anthocyanin pigmentation in stalk of Zicaitai (*Brassica rapa* L. ssp. chinensis var. purpurea). *BMC Genom.* **2019**, *20*, 343. [CrossRef]

24. Wu, J.; Zhao, J.; Qin, M.; Ren, Y.; Zhang, H.; Dai, Z.; Hao, L.; Zhang, L. Genetic Analysis and Mapping of the Purple Gene in Purple Heading Chinese Cabbage. *Hortic. Plant J.* **2016**, *2*, 351–356. [CrossRef]

25. Michelmore, R.W.; Paran, I.; Kesseli, R.V. Identification of markers linked to disease-resistance genes by bulked segregant analysis: A rapid method to detect markers in specific genomic regions by using segregating populations. *Proc. Natl. Acad. Sci. USA* **1991**, *88*, 9828–9832. [CrossRef]

26. Takagi, H.; Abe, A.; Yoshida, K.; Kosugi, S.; Natsume, S.; Mitsuoka, C.; Uemura, A.; Utsushi, H.; Tamiru, M.; Takuno, S.; et al. QTL-seq: Rapid mapping of quantitative trait loci in rice by whole genome resequencing of DNA from two bulked populations. *Plant J.* **2013**, *74*, 174–183. [CrossRef] [PubMed]

27. Pandey, M.K.; Khan, A.W.; Singh, V.K.; Vishwakarma, M.K.; Shasidhar, Y.; Kumar, V.; Garg, V.; Bhat, R.S.; Chitikineni, A.; Janila, P. QTL-seq approach identified genomic regions and diagnostic markers for rust and late leaf spot resistance in groundnut (*A rachis hypogaea* L.). *Plant Biotechnol. J.* **2017**, *15*, 927–941. [CrossRef]

28. Clevenger, J.; Chu, Y.; Chavarro, C.; Botton, S.; Culbreath, A.; Isleib, T.G.; Holbrook, C.C.; Ozias-Akins, P. Mapping Late Leaf Spot Resistance in Peanut (*Arachis hypogaea*) Using QTL-seq Reveals Markers for Marker-Assisted Selection. *Front. Plant Sci.* **2018**, *9*, 83. [CrossRef]

29. Branham, S.E.; Farnham, M.W. Identification of heat tolerance loci in broccoli through bulked segregant analysis using whole genome resequencing. *Euphytica* **2019**, *215*, 34. [CrossRef]

30. Fall, L.A.; Clevenger, J.; McGregor, C. Assay development and marker validation for marker assisted selection of Fusarium oxysporum f. sp. niveum race 1 in watermelon. *Mol. Breed.* **2018**, *38*, 130. [CrossRef]

31. Lu, H.; Lin, T.; Klein, J.; Wang, S.; Qi, J.; Zhou, Q.; Sun, J.; Zhang, Z.; Weng, Y.; Huang, S. QTL-seq identifies an early flowering QTL located near Flowering Locus T in cucumber. *Theor. Appl. Genet.* **2014**, *127*, 1491–1499. [CrossRef] [PubMed]

32. Illa-Berenguer, E.; Van Houten, J.; Huang, Z.; Van Der Knaap, E. Rapid and reliable identification of tomato fruit weight and locule number loci by QTL-seq. *Theor. Appl. Genet.* **2015**, *128*, 1329–1342. [CrossRef] [PubMed]

33. Zhang, Y.; Hu, Z.; Zhu, M.; Zhu, Z.; Wang, Z.; Tian, S.; Chen, G. Anthocyanin accumulation and molecular analysis of correlated genes in purple kohlrabi (*Brassica oleracea* var. gongylodes L.). *J. Agric. Food Chem.* **2015**, *63*, 4160–4169. [CrossRef] [PubMed]

34. Tang, Q.; Tian, M.; An, G.; Zhang, W.; Chen, J.; Yan, C. Rapid identification of the purple stem (Ps) gene of Chinese kale (*Brassica oleracea* var. alboglabra) in a segregation distortion population by bulked segregant analysis and RNA sequencing. *Mol. Breed.* **2017**, *37*, 153. [CrossRef]

35. Porebski, S.; Bailey, L.G.; Baum, B.R. Modification of a CTAB DNA extraction protocol for plants containing high polysaccharide and polyphenol components. *Plant Mol. Biol. Rep.* **1997**, *15*, 8–15. [CrossRef]

36. Li, H.; Durbin, R. Fast and accurate short read alignment with Burrows-Wheeler transform. *Bioinformatics* **2009**, *25*, 1754–1760. [CrossRef]

37. Yu, J.; Zhao, M.; Wang, X.; Tong, C.; Huang, S.; Tehrim, S.; Liu, Y.; Hua, W.; Liu, S. Bolbase: A comprehensive genomics database for Brassica oleracea. *BMC Genom.* **2013**, *14*, 664. [CrossRef]
38. Li, H.; Handsaker, R.; Wysoker, A.; Fennell, T.; Ruan, J.; Homer, N.; Marth, G.; Abecasis, G.; Durbin, R. The Sequence Alignment/Map format and SAMtools. *Bioinformatics* **2009**, *25*, 2078–2079. [CrossRef]
39. McKenna, A.; Hanna, M.; Banks, E.; Sivachenko, A.; Cibulskis, K.; Kernytsky, A.; Garimella, K.; Altshuler, D.; Gabriel, S.B.; Daly, M.J.; et al. The Genome Analysis Toolkit: A MapReduce framework for analyzing next-generation DNA sequencing data. *Genome Res.* **2010**, *20*, 1297–1303. [CrossRef]
40. Wang, K.; Li, M.; Hakonarson, H. ANNOVAR: Functional annotation of genetic variants from high-throughput sequencing data. *Nucleic Acids Res.* **2010**, *38*, e164. [CrossRef] [PubMed]
41. Chen, K.; Zhou, Y.-X.; Li, K.; Qi, L.-X.; Zhang, Q.-F.; Wang, M.-C.; Xiao, J.-H. A novel three-round multiplex PCR for SNP genotyping with next generation sequencing. *Anal. Bioanal. Chem.* **2016**, *408*, 4371–4377. [CrossRef] [PubMed]
42. Meng, L.; Li, H.; Zhang, L.; Wang, J. QTL IciMapping: Integrated software for genetic linkage map construction and quantitative trait locus mapping in biparental populations. *Crop. J.* **2015**, *3*, 269–283. [CrossRef]
43. Chiu, L.-W.; Li, L. Characterization of the regulatory network of BoMYB2 in controlling anthocyanin biosynthesis in purple cauliflower. *Planta* **2012**, *236*, 1153–1164. [CrossRef] [PubMed]
44. Livak, K.J.; Schmittgen, T.D. Analysis of relative gene expression data using real-time quantitative PCR and the $2-\Delta\Delta CT$ method. *Methods* **2001**, *25*, 402–408. [CrossRef] [PubMed]
45. Schoenbohm, C.; Martens, S.; Eder, C.; Forkmann, G.; Weisshaar, B. Identification of the Arabidopsis thaliana Flavonoid 3′-Hydroxylase Gene and Functional Expression of the Encoded P450 Enzyme. *Biol. Chem.* **2000**, *381*, 749–753. [CrossRef] [PubMed]
46. Hayashi, K.; Matsumoto, S.; Tsukazaki, H.; Kondo, T.; Kubo, N.; Hirai, M. Mapping of a novel locus regulating anthocyanin pigmentation in *Brassica rapa. Breed. Sci.* **2010**, *60*, 76–80. [CrossRef]
47. Han, Y.; Vimolmangkang, S.; Soria-Guerra, R.E.; Rosales-Mendoza, S.; Zheng, D.; Lygin, A.V.; Korban, S.S. Ectopic Expression of Apple F3′H Genes Contributes to Anthocyanin Accumulation in the Arabidopsis tt7 Mutant Grown Under Nitrogen Stress. *Plant Physiol.* **2010**, *153*, 806–820. [CrossRef]
48. Bogs, J.; Ebadi, A.; McDavid, D.; Robinson, S.P. Identification of the Flavonoid Hydroxylases from Grapevine and Their Regulation during Fruit Development. *Plant Physiol.* **2005**, *140*, 279–291. [CrossRef]
49. Jung, Y.J.; Lee, H.J.; Kim, J.H.; Kim, D.H.; Kim, H.K.; Cho, Y.-G.; Bae, S.; Kang, K.K. CRISPR/Cas9-targeted mutagenesis of F3′H, DFR and LDOX, genes related to anthocyanin biosynthesis in black rice (*Oryza sativa* L.). *Plant Biotechnol. Rep.* **2019**, *13*, 521–531. [CrossRef]
50. Vikhorev, A.V.; Strygina, K.V.; Khlestkina, E.K. Duplicated flavonoid 3′-hydroxylase and flavonoid 3′, 5′-hydroxylase genes in barley genome. *PeerJ* **2019**, *7*, e6266. [CrossRef]
51. Zhang, H.; Wang, X.; Pan, Q.; Li, P.; Liu, Y.; Lu, X.; Zhong, W.; Li, M.; Han, L.; Li, J. QTG-Seq accelerates QTL fine mapping through QTL partitioning and whole-genome sequencing of bulked segregant samples. *Mol. Plant* **2019**, *12*, 426–437. [CrossRef]
52. Zhou, X.; Xia, Y.; Ren, X.; Chen, Y.; Huang, L.; Huang, S.; Liao, B.; Lei, Y.; Yan, L.; Jiang, H. Construction of a SNP-based genetic linkage map in cultivated peanut based on large scale marker development using next-generation double-digest restriction-site-associated DNA sequencing (ddRADseq). *BMC Genom.* **2014**, *15*, 351. [CrossRef] [PubMed]

horticulturae

Article

Development of Molecular Markers Associated with Resistance to Gray Mold Disease in Onion (*Allium cepa* L.) through RAPD-PCR and Transcriptome Analysis

So-Jeong Kim [1,†], Jee-Soo Park [1,†], TaeHoon Park [2], Hyun-Min Lee [1], Ju-Ri Choi [1] and Young-Doo Park [1,*]

[1] Department of Horticultural Biotechnology, Kyung Hee University, 1732 Deogyoung-daero, Giheung-gu, Yongin-si 17104, Gyeonggi-do, Korea; coalla110@naver.com (S.-J.K.); jeesoo_92@naver.com (J.-S.P.); ahawhgek@naver.com (H.-M.L.); bmokh@naver.com (J.-R.C.)

[2] Seeds & People Co., 483 Bongdeok-ro, Yeomsan-myeon, Yeonggwang-gun 57066, Jeollanam-do, Korea; onion888@naver.com

[*] Correspondence: ydpark@khu.ac.kr; Tel.: +82-10-3338-9344

[†] These authors contributed equally to this work.

Abstract: Onions (*Allium cepa* L.) are one of the most consumed vegetable crops worldwide and are damaged by several fungal diseases in the field or during storage. Gray mold disease caused by the necrotrophic pathogens *Botrytis cinerea* and *Botrytis squamosa* is a disease that reduces the productivity and storage life in onions. However, it is difficult to control gray mold disease in onions by using physical and chemical methods. Breeding resistant onions against gray mold disease can reduce the damage caused by pathogens, reduce the labor required for control, and reduce environmental pollution caused by fungicides. However, onions have a large genome size (16Gb), making them difficult to analyze, and have a biennial cycle, resulting in a very long breeding period. Therefore, in this study, markers were developed to shorten the onion breeding period. First, random amplified polymorphic DNA (RAPD) was performed to confirm the genetic relationship between the gray mold disease-resistant and -susceptible lines through a dendrogram. In addition, the sequence characterized amplified region (SCAR)-OPAN1 marker to select resistant lines was developed using a polymorphic RAPD fragment. Second, the RNA-seq of the gray mold-resistant and -susceptible onion lines were analyzed using NGS technology. Using the RNA-seq results and DEG and GO analyses were performed, and the variants, such as SNPs and indels, were analyzed to develop a selectable marker for the resistant line. This study developed the SNP-3 HRM marker for selecting gray mold disease-resistant lines by using the SNPs present in the aldo-keto reductase (AKR) gene with high expression levels in these lines. The SCAR-OPAN1 and SNP-3 HRM markers developed in this study could be used to select gray mold disease-resistant onions in breeding programs to reduce the damage caused by gray mold disease.

Keywords: HRM; molecular marker; phylogenetic analysis; RNA sequencing; SCAR; SNP

Citation: Kim, S.-J.; Park, J.-S.; Park, T.; Lee, H.-M.; Choi, J.-R.; Park, Y.-D. Development of Molecular Markers Associated with Resistance to Gray Mold Disease in Onion (*Allium cepa* L.) through RAPD-PCR and Transcriptome Analysis. *Horticulturae* **2021**, *7*, 436. https://doi.org/10.3390/horticulturae7110436

Academic Editor: Yuyang Zhang

Received: 25 September 2021
Accepted: 20 October 2021
Published: 25 October 2021

Publisher's Note: MDPI stays neutral with regard to jurisdictional claims in published maps and institutional affiliations.

Copyright: © 2021 by the authors. Licensee MDPI, Basel, Switzerland. This article is an open access article distributed under the terms and conditions of the Creative Commons Attribution (CC BY) license (https://creativecommons.org/licenses/by/4.0/).

1. Introduction

Onions (*Allium cepa* L.) are one of the most economically and nutritionally important crops worldwide. They are also one of the oldest cultivated crops and are used as an ingredient in various foods and sauces to enhance flavor and promote health, such as for lowering cholesterol levels [1–3]. Therefore, it is important to breed and produce higher-quality onions to improve their competitive advantage in the market. Onion breeding is performed for various purposes, such as to improve the onion yield; for qualities like size, taste, or color; for male sterility; and for a resistance against biotic and abiotic stresses [3–5].

Onions are susceptible to many pathogens and insects [3]; therefore, breeding for resistant onions has been extensively studied to reduce the damage caused by various diseases, many of which are caused by the genus *Botrytis*. Onion botrytis leaf blight is caused by *B. squamosa*, and onion neck rot is caused by *B. aclada*, *B. alli*, *B. squamosa*, and *B. porri* [6]. In particular, *B. cinerea* is a necrotrophic pathogen with more than 200 host crop species, causing severe damage to onions [7]. Gray mold, caused by *B. cinerea* and *B. squamosa*, reduces the yield and storage capacity. Gray mold disease affecting the onion bulb is caused by *B. squamosa* during bulb formation and bulb filling and by *B. cinerea* during the later cultivation and storage periods [8,9]. Previous studies have attempted chemical and biological controls to prevent damage to onions by gray mold. To prevent this disease, onions should be kept dry, infected onions should be rapidly removed, and crops must be rotated every 3 to 4 years. In addition, many fungicides are used to control gray mold, with approximately 10% of the global fungicide market focused on controlling *B. cinerea* [10]. However, despite these efforts, it is difficult to control gray mold disease. Furthermore, synthetic fungicides can cause problems, such as residue concerns and a negative impact on human health, the emergence and increase of resistant pathogen populations, and environmental pollution [11–14]. Therefore, breeding disease-resistant onions can reduce the damage caused by diseases, such as gray mold disease, increase production, and reduce labor and environmental pollution.

Disease-resistant onions have a long breeding period, as onions are a biennial plant; therefore, a complete generation of onions requires two years. Marker assistant selection (MAS) can be used to shorten the breeding period of onions. In addition, a genome analysis of onions is difficult, because onions have a large genome size (16Gb), which is 100 times larger than that of Arabidopsis genomes [3,4]. Molecular markers such as random amplified polymorphic DNA (RAPD) and sequence characterized amplified region (SCAR) have been developed for various purposes. These markers are used for MAS, facilitating selection in breeding and shortening the breeding periods. RAPD is a PCR-based marker using short random primers; therefore, a RAPD analysis can reveal small genetic polymorphisms between large genomes, such as that of onions [15,16]. The disadvantage of RAPD markers is their low reproducibility. Therefore, in this study, we developed a SCAR marker by using a more specific primer than RAPD from the results of polymorphism studies between resistant and susceptible lines of onion.

In addition, NGS technologies such as RNA sequencing have enabled large-scale transcriptome data analysis, which has improved the efficiency of gene discovery despite no prior knowledge of reference genome sequences [4,17,18]. In this study, RNA-seq was performed to develop molecular markers for breeding gray mold-resistant onions. RNA-seq was used for a DEG analysis and the analysis of variants such as SNPs and insertions and deletions (InDels) between the resistant and susceptible groups. The HRM markers confirm the fluorescence of the PCR product's melting curve when the double-stranded DNA becomes single-stranded DNA. The sequence region of interest was amplified with a fluorescent dsDNA-binding dye, and the PCR melting curve was measured when the product was gradually melted. The melting curve varies depending on the sequence, such as the presence of SNPs, GC content, length, and heterozygosity [19,20]. In this study, the HRM marker was also developed through selected transcripts from gene ontology (GO) and variant analysis using RNA-seq data to breed gray mold-resistant onions.

2. Materials and Methods

2.1. Plant Materials and Genomic DNA Extraction

To develop molecular markers, four (S&P 7522, S&P 7521, S&P 7129, and S&P 7168) gray mold-resistant and three (S&P 7130, S&P 7175, and S&P 7483) susceptible onion (*Allium cepa* L.) lines grown in 'Seeds & People' Co., Ltd. (Yeonggwang, Korea) were used in this study. The characteristics of the seven onion lines of 'Seeds & People' Co. are shown in Table S1. The pedigree of the seven onion lines used in this study is shown in Figure S1. In addition, to confirm the versatility of the developed molecular markers, three (Asia-12, Asia-42, and Asia-53) resistant and four (Asia-30, Asia-35, Asia-45, and Asia-50) susceptible onion lines against gray mold were provided by 'Asia Seed' Co., Ltd. (Seoul, Korea) and analyzed. Genomic DNA was isolated from the leaf tissues of each line by using cetyltrimethylammonium bromide (CTAB) [21]. The concentration and purity of the gDNA were measured using a Nanodrop® ND-1000 spectrophotometer (NanoDrop Technologies, Wilmington, NC, USA).

2.2. RAPD and Phylogenetic Analysis

The gDNA of each line was used for a RAPD analysis to identify the genetic relationship between the resistant and susceptible lines. For a RAPD analysis, OPERON random primers OPAN-1~OPAN-20 and OPL-1~OPL-20 were used. The sequences of the 40 OPERON random primers are shown in Table S2. The RAPD PCR was performed in 20 μL by using the Maxime PCR Premix (iNtRON Biotechnology, Seongnam, Korea) containing 2.5-U i-Taq™ *Taq* polymerase, 2.5-mM dNTPs, 1X reaction buffer with 10 pmol primer, and 50 ng of gDNA. The amplification program was as follows: denaturation at 95 °C for 10 min, 40 cycles of 1 min at 95 °C, 1 min at 37 °C, 1 min at 72 °C, and a final extension at 72 °C for 10 min. The amplified RAPD PCR product was electrophoresed at 100 mA for 3 h in a 1% agarose gel. The polymorphic bands obtained from the RAPD-PCR of each line were converted into binary data, depending on the presence of a band; the presence of a polymorphic band was scored as 1, while its absence was scored as 0. The dendrogram was obtained using the unweighted pair group method (UPGMA) with an arithmetic mean [22] by using the Jaccard coefficient [23] through the XLSTAT program (Addinsoft, New York, NY, USA).

2.3. Development of the SCAR Marker

To develop the SCAR marker, the 2-Kb polymorphic bands obtained between four resistant and three susceptible lines by using the OPAN-1 primer (5′-ACT CCA CGT C-3′) were sequenced and identified. The amplified polymorphic products were eluted from a 1% agarose gel and purified using the NucleoSpin® Gel and PCR Clean-up Kit (Macherey-Nagel, Duren, Germany). The eluted products were ligated into the pGEM-T easy vector (Promega, Madison, WI, USA) for sequencing. The plasmid was purified using the Fast DNA-spin™ Plasmid DNA Purification Kit (iNtRON Biotechnology, Seongnam, Korea), and the fragments were sequenced by Macrogen® (Seoul, Korea). The SCAR primer set was designed from the sequence of the 2-Kb polymorphic band between the resistant and susceptible lines. PCR using the SCAR primer set was performed under the following conditions: initial denaturation for 10 min at 95 °C, 35 cycles of 30 s at 95 °C, 30 s at 61 °C, 1 min at 72 °C, and a final extension for 10 min at 72 °C. Thereafter, the amplified products of the resistant onion lines were confirmed by electrophoresis in a 1% agarose gel. In addition, to confirm the versatility of the SCAR marker, three gray mold-resistant and four susceptible onion lines from the 'Asia Seed' Co. were also analyzed.

2.4. RNA Sequencing and Variant Analysis

For RNA sequencing, the leaves of four resistant and three susceptible lines were ground, and the total RNA was extracted using the RNease® kit (QIAGEN, Hilden, Germany). Before analyzing the RNA sequence, the OD values were measured using Dropsense96 (Trinean, Gentbrugge, Belgium), and the total RNA quality was checked using a Bioanalyzer RNA Chip (Agilent Technologies, Santa Clara, CA, USA). The library was constructed using the TruSeq Stranded mRNA kit (Illumina, San Diego, CA, USA), and RNA-seq was performed using HiseqX (Illumina, San Diego, CA, USA) by the DNAcare Company (Seoul, Korea). To remove the low-quality base and Illumina adapters, the Trimmomatic program (USADELLAB, Aachen, Germany) was used on the RNA-seq raw data of each line. After trimming, only paired reads containing at least 50 nucleotides were used for the analysis. In addition, quality trimming was performed by applying options such as a sliding window, average quality, and minimum read size. Thereafter, the generated trimmed data were used for the de novo assembly of resistant and susceptible lines by using the Trinity program. As a reference for mapping, the onion reference transcript (National Agricultural Biotechnology Information Center (NABIC), Rural Development

Administration (RDA), JeonJu, Korea) and the data obtained through the de novo assembly were used. Read mapping was performed using the BWA-Mem algorithm. To remove the duplicated PCR reads from the produced BAM file, the Picard program (Broad Institute, Cambridge, MA, USA) was used. Thereafter, the variant information of seven lines was produced using haplotypeCaller of the Genome Analysis Tool Kit (GATK, Broad Institute, Cambridge, MA, USA), and the final raw variant call file (vcf) was generated by integrating the variant files of each line. Variant filtering was performed using Vcftools (VCFtools, 1000 Genomes Project Analysis Group, http://vcftools.sourceforge.net/ accessed on 1 October 2021) to remove the low-quality genotypes, target missing levels of depth coverage (DP), genotype quality (GQ), and genotype data. The filtering conditions were set to min DP = 5, max DP = 100, and min GQ = 220, and the missing was set to 20%. Various information was used to select the variants showing polymorphisms between the resistant and susceptible lines. For the selection condition for the variants, those showing the same genotype in each resistant and susceptible group and showing polymorphisms between these groups were selected.

2.5. Differentially Expressed Gene (DEG) and Gene Ontology (GO) Analysis for Selection of Transcript Related to Disease Resistance

Among the sequences generated from Trinity, sequences with a length of more than 100 amino acids were selected using TransDecoder. Based on these data and onion reference transcript data (NABIC, RDA, JeonJu, Korea, https://nabic.rda.go.kr, accessed on 7 February 2021), the sequences of each onion line were aligned with HISAT2. The read count of the transcript expression level was then calculated using the StringTie program. The transcripts obtained through StringTie were calculated at the transcript level, and a comparative analysis was performed between each onion line based on the read count of each transcript. After dividing into resistant and susceptible groups, a DEG analysis was performed using DEGseq (Bioconductor, http://www.bioconductor.org/packages/ release/bioc/html/DEGseq.html, accessed on 2 October 2021) [24]. First, after normalizing the raw read count data, a correlation analysis was performed between each onion line based on the normalized data. The analysis was conducted using Pearson's correlation coefficient and the average linkage method. The DEGseq of the R package was used to confirm the statistical significance of the expression differences between resistant and susceptible groups. After comparing the average expression levels between the two groups, the conditions were set as follows to select genes using significantly different expressions. The test was conducted using the equation $\log_2\left(\frac{\text{Base Mean of R}}{\text{Base Mean of S}}\right)$. A negative value was set for the transcripts more expressed in the susceptible group than in the resistant group, and a positive value was set for the transcripts more expressed in the resistant group than in the susceptible groups. Thereafter, the transcripts with DEGs satisfying the conditions of $|\log_2$ fold change$| \geq 2$ and PADJ < 0.05 were selected. From the results of variants analysis and DEG analysis, transcripts that commonly satisfied each condition were selected. A gene function analysis was performed to identify whether the selected variants were related to the disease-resistant mechanisms. To identify the functions related to disease resistance, the selected transcripts were analyzed using The Arabidopsis Information Resource (TAIR) ID derived from Arabidopsis, a model plant. Thereafter, GO annotation was performed using the transcripts' confirmed gene functions by TAIR ID, and the transcripts with functions related to disease resistance were selected.

2.6. HRM Primer Designs from Selected Transcripts and HRM Analysis

HRM primers were designed from 14 selected transcripts with SNPs that exist between the resistant and susceptible lines. The conditions of the designed primers were as follows: the amplified product size was between 80 and 200 bp, the variant region was inside the PCR product, and the annealing temperature was approximately 60 ± 1 °C. The sequences and annealing temperature information of the HRM primers are listed in Table 1. HRM was performed using a total 10-μL reaction mixture containing the BioFact^{TM} 2X Real-Time PCR

Master Mix (BIOFACT, Daejeon, Korea) with a 10-pmol primer and 50 ng of gDNA. The reaction conditions were: pre-denaturation at 95 °C for 15 min, 40 cycles of denaturation at 95 °C for 20 s, annealing at 60 °C for 30 s, and extension at 72 °C for 20 s. Thereafter, the temperature was sequentially increased from 65 °C to 95 °C, and the melt curve and peak value were measured. At least six repetitions were performed for each line. The HRM results were statistically grouped using the ANOVA of the XLSTAT program. In addition, to confirm the versatility of the selected HRM marker, which was able to distinguish between resistant and susceptible onion lines, an additional analysis was performed using 'Asia Seed' Co. onion lines.

Table 1. HRM primers designed from 14 selected transcripts with SNPs that exist between the resistant and susceptible lines.

No.	SNP Name	Primer Name	Sequence (5′→3′)	Tm	PCR Product Size
1	SNP-1	SNP 1 F	CTTTGAACTTCGGGCAATACCCG	60.5	199 bp
		SNP 1 R	CCTCATCAGGCGAGTGAGTGGAC	59.6	
2	SNP-2	SNP 2 F	AACGTCCGCCGAAGAAGCTGA	60.7	204 bp
		SNP 2 R	TTTGCTGGAGGAGGTGGTGGTG	60.1	
3	SNP-3	SNP 3 F	CGTTAGCTCAAGTGGGTTTGAGGTG	59.9	134 bp
		SNP 3 R	TTCTTCCAGCTCTTCTTCGCT	59.2	
4	SNP-4	SNP 4 F	AGGGTTCAGAACCAAAACAGCATCA	59.8	163 bp
		SNP 4 R	CGATGCTTTTTTGGTAACTGGGAAG	59.1	
5	SNP-5	SNP 5 F	TCGATGGCATTAAGGATGCTAAGGA	59.8	163 bp
		SNP 5 R	ATTGCCTTTGCTAGGGAGCCATAA	59.1	
6	SNP-6	SNP 6 F	TGGTGACAAGAAATTCTTCAACGGC	60.2	158 bp
		SNP 6 R	TCTCCATGCATCTCTTTCCCCACT	59.8	
7	SNP-7	SNP 7 F	TGAGCTCCTTTCAGACTCCTTTCCC	60.1	170 bp
		SNP 7 R	CGACCACCTTAACAGCTTGATCGTC	59.9	
8	SNP-8	SNP 8 F	CTTTCTCAGGGTTAATAGAGGCGGG	60.1	171 bp
		SNP 8 R	GCCAAACTGGCTGAAAACCTTTTCT	59.9	
9	SNP-9	SNP 9 F	TTCATGGTCACAGAAACGCCAAGA	60.1	181 bp
		SNP 9 R	GGCAGAACTTCTTTGTTCATCCGCT	59.9	
10	SNP-10	SNP 10 F	AATCTCACAATCGAACCTCACTGCC	59.4	175 bp
		SNP 10 R	TGCGAGGTGAATTCCAGTCAAAGAG	60.4	
11	SNP-11	SNP 11 F	GCAACAAGGGCTGCAAATTACAGTT	60.1	169 bp
		SNP 11 R	GTTTGTGTGCATGAATCTGTGCAGG	59.9	
12	SNP-12	SNP 12 F	CGACTATGGCTGGGACACTGCA	59.6	164 bp
		SNP 12 R	TCCCCGAACTTGACCCCGTTAC	60.7	
13	SNP-13	SNP 13 F	AGCAATGTTGTCCGGTACTCCAAAG	59.4	185 bp
		SNP 13 R	CGCTCAAAAACCCAGCTCGTACA	60.1	
14	SNP-14	SNP 14 F	TATCGTACCTTCCTACCCTGAGCGA	60.1	182 bp
		SNP 14 R	TCCGAACATGGGCAGCTTCC	59.9	

2.7. Quantitative Real-Time PCR (qPCR) to Identify Expression Level of Transcripts

The total RNA was extracted from the leaves of the resistant and susceptible lines by using the Plant RNA Extraction Kit (Takara, Shiga, Japan). The concentration and purity of the RNA were measured using a Nanodrop® ND-1000 spectrophotometer (NanoDrop Technologies, Wilmington, NC, USA). Thereafter, cDNA was synthesized from 500 ng of RNA extracted from each line by using the HiSenScript™ RH[-] RT PreMix Kit (iNtRON, Seongnam, Korea). PCR for cDNA synthesis was performed using the following cycle: reverse transcription step at 42 °C for 1 h and RTase inactivation extension step at 85 °C for 10 min. The synthesized cDNA was stored at −20 °C. Quantitative real-time PCR (qPCR) was performed to compare the expression levels of the SNP-3 transcript related to the aldo-keto reductase (AKR) gene with in silico data, which showed significant HRM results in 14 transcripts. The qPCR primer conditions were as follows: the amplified product size was less than 200 bp, and the annealing temperature was approximately 60 ± 1 °C. The forward primer was 5′-CGT TAG CTC AAG TGG GTT TGA GGT G-3′, and the reverse primer was 5′-CTC CAG CAC ACG CCC TCC A-3′. Before the qPCR analysis, it was confirmed that the 171-bp band targeted by the qPCR primer set was amplified by RT-PCR using the Maxime™ PCR PreMix Kit (iNtRON, Seongnam, Korea). The reaction cycle was as follows: 10 min of initial denaturation at 95 °C, followed by 40 cycles of 95 °C for 30 s, 60 °C for 30 s, 72 °C for 40 s, and a final extension at 72 °C for 10 min. The amplified product was electrophoresed using 1% agarose gel.

The qPCR analysis was performed with the TransStart Top Green qPCR Super Mix (TransGen, Beijing, China) containing 1 pmol of qPCR primer and 500 ng of cDNA by using Roter-Gene™6000 (Corbett, Melbourne, Vic, Australia), and the qPCR program proceeded at 95 °C for 10 s, followed by 40 cycles consisting of 2 steps: 95 °C for 10 s and 60 °C for 30 s. After amplification, a fluorescence melting curve was obtained by heating the samples from 60 °C to 95 °C. AKR gene expression was identified using the Ct value. To compare the differences in the gene expression levels between the resistant and susceptible lines, the actin gene, the housekeeping gene, was used as a control. After obtaining the Ct value using Roter-Gene Q Series Software, the ΔCt value ((Ct value of target gene)$-$(Ct value of actin gene)) was calculated. To compare the expression levels based on the susceptible S&P 7483 lines, the $\Delta\Delta$Ct value ((ΔCt of each onion line)$-$(ΔCt of S&P 7483)) was obtained. The difference in the expression levels was confirmed by calculating the 2$\Delta\Delta$Ct. The expression level of each line was calculated from the results of six repetitions. Data were presented as the means with standard errors, and the means were compared using Duncan's multiple comparison test ($p \leq 0.05$).

3. Results

3.1. RAPD and Phylogenetic Analysis

Four (S&P 7522, S&P 7521, S&P 7129, and S&P 7168) gray mold-resistant and three (S&P 7130, S&P 7175, and S&P 7483) susceptible onion lines were amplified using 40 OPERON random primers, and the polymorphic bands were confirmed by electrophoresis. Most of the amplified band sizes ranged from 200 bp to 2500 bp, and 124 bands (38%) of the 330 bands showed polymorphisms. The RAPD analysis was repeated at least three times to obtain reproducibility, and the presence or absence of amplified bands was transformed into binary data. The binary data were converted into dendrograms by using the UPGMA methods of the XLSTAT program using Jaccard coefficients (Figure 1). The genetic similarity within the group was 78.27%, and the genetic similarity between the groups was 21.73%. Based on the phylogenetic analysis, a genetic relationship between the seven lines was confirmed. The results of the phylogenetic analysis were compared with the pedigrees of the seven breeding lines to confirm the reliability of the RAPD (Figure S1). The phylogenic analysis results were consistent with the genetic relationship of the pedigree of the breeding lines.

3.2. Development of a SCAR Marker for the Selection of Resistant Lines

From the RAPD results, a SCAR marker was developed using a specific RAPD product showing polymorphisms between the resistant and susceptible lines (Figure S2). The 2-Kb bands were amplified in the resistant lines S&P 7522, S&P 7521, S&P 7129, and S&P 7168 by using the OPAN-1 primer. The amplified polymorphic bands were eluted, purified, and sequenced. From the sequencing results, 5′-ACT CCA CGT C-3′, which is a sequence of random primer OPAN-1, was identified at both ends of the fragment sequence (Figure 2). In addition, the sequences were analyzed using BLAST at the National Center for Biotechnology Information (NCBI), and insignificant results were obtained. Therefore, further studies are required to analyze this unknown DNA sequence. The SCAR marker primer was designed to include the 5′-ACT CCA CGT C-3′ sequence of the OPAN-1 random primer. The forward primer was 5′-ACT CCA CGT CAT CGA TTC GAA-3′, and the reverse primer was 5′-ACT CCA CGT CCG AAC TAC AGA A-3′. The primer set for the SCAR marker was designed for considering the annealing temperature, GC content, the possibility of dimer formation, and hairpin loops. The developed SCAR marker was designated as SCAR-OPAN1.

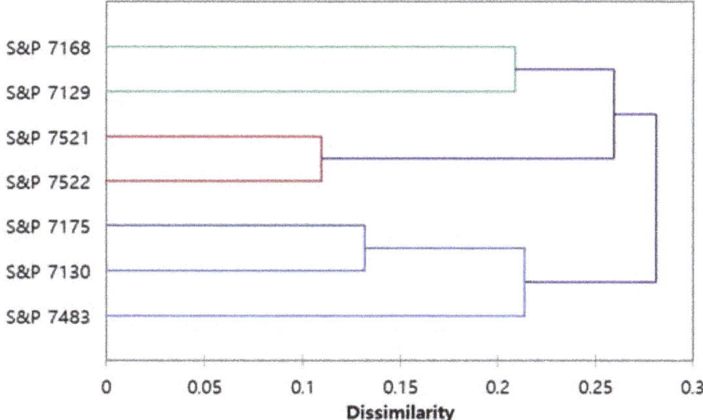

Figure 1. Phylogenetic tree of the resistant and susceptible lines provided by 'Seeds & People' Co. by a RAPD analysis. The resistant lines S&P 7168 and S&P 7129 showed a close genetic relationship, and also, resistant lines S&P 7522 and S&P 7521 showed a close genetic relationship. The susceptible onion lines S&P 7175, S&P 7130, and S&P 7483 appeared to have a closer genetic relationship.

```
ACTCCACGTC ATCGATTCGA AGGAGCATCG ATACGTATGT ATCGATATGC AGCCATCAGT CCACTCGCAA AATCAACTTT
GATCGATCAT AACTCGAAAA CAACAAGTTT AAATGCTATG AAACTTCCAT AGATGCTTGC TTATTCGCTT AAGAATGTCC
ATGATCTACC GCAGCTCAAA TGGAGCTAAA CCTGATATAT ACGAAGATGA TAGAGATCGA TCCTAAACGA CCCTAGATCC
ATCTGGTGGG CCCATCCACG TATGAGAAGA GGTATTTCGT GGTTTGAGAG AGATATTTTT AGATCTACAC CTATAAATAG
CACTTCTACG CCTGTAGAAA AAAAGACTTT TTGGCTTTTC TTCGTCGTTA CTTTGCTGAA ACTCCATTGT AACTTGGATG
TCATCGATTG TACTATTGTA AATTAATTGA GAATCATTAA TAAACATTGA ACTCTTCAAA GTGGACGTAG GCTACAACGC
CGAACCACTA TAAATCTCGT ATCTTTACTG CTTTCTTTTT AATGTTTGTC GTTCTTCATA TTTATATTTT CTTCTCATAT
TCTATCTGTG ATCTGTGACC GTGAATTGGT GTTTTCATTC GCAATTGGGT ATTTGTGATC AATTTTGACG CAAACAGTGT
GTATCGCTGA AGACATAATT AGGTTTATCT TCATCTTCAT CTATCATAGG TTTTTTGTCT AGACATTGAT ATGCAAAACT
CGTAAAACCT ACCCCACCCC AATATTTTAA CTGTCGCTAG AACTTTATTG GTATTTCAGG CGTTATGCAT ATGGTTACGC
ATATAAAGTG TGGCTTATCT TCACAAAATA AACTTCGCCA TCGACCATTT TATAGATTGA CTTGTGAGTT AGGTAACTAA
GCTTGTGGTT AAAGTTTTTT TTTTTGCCCG GGGGAAAAAA TTTTTTAATA AAAAAAAAAA TTCCCGGAAA CGGGGGGCCT
TGTTTTTTTT GAAATTAAAA ATTTCCAGGG AAAATTTTCA GTTTTCCGAC TTCGATGAGA TTTCAGAAAT TATTGGTTTG
TACTTTACAT CCAGCTGAGA TACATTTTTA TGTCGGTCCT TTTAGCCTGC TAGTAGATTC AATCAAATTA AAAAGGATTT
TCGTACATAG ATTTGGCATA GGTTAAAATT GTAGCTGAAT GTAACTACAG CTATTAGGCG AAATATACCT ACAGTTACAG
GGTGCATCAA ATGAGAAGAA AAAAAAATTG TCTTAACCAT TTTTTTCTCT ATAGCTTCTT TTTACATCAC TGTTTCTATC
AAATCTTAAA CTTTAGTTCT GTAATAGTGT AATTAACTTG CACAATTGGT ATCAAGATAA AATTATGTTT TAGCTTCCAT
CAGCTGCAGC GTCGCAAGTT CTGATGATCC ATTTAAATAA ATATTTACTT CCGGTGAATT AATAGTTTTC AATGTCGACC
CTTTGATTGT GCATTGCATG GATTGTATGC ACTGGGTTCC ACATGTTTAC CATACTAGTT GTCCGCTCCT TTGTACGTCA
GAAATTAATG GATTACTAAT TACAACATTT CACTTTACCG TTTTTTGATT TGTTACATTT TCACTTTCCT TAGAATTTTG
TTCTGCAAAA ATTAATGTTA ATGACATTAT TAAAATTCTT GCTTTTTTGC CTAGTGAATG AATCATTGGC GGGCAAACAT
AGTTTATCAT GTTCGATTTC ATTTAGAAAA TGGTGGAGGT ATCATCATAA TTTGGTGTAT ATATTTTGTA TGACATAAAA
GCCCATTTTA TCTTTCAAAG AATATGTGTA GTAGTTGTAA ATCGGTTATA ATGTAAAATC ATTAATCTAG GAAACCCATA
ATCTTTCGTT TAAATTTTTT TTACTGCACA CTATAAATGA GCAAAAATAC AGGCGCCAAT AAATAGAATT ATAATCCGTG
TTGTAAACTT ATGTATCTTT GTTAGTACAC AATATATTGA AGTTTTTGTG TGATATACAG ACCCGTTGCA GGCAAATAAT
TCTGTAGTTC GGACGTGGAG T
```

Figure 2. The SCAR-OPAN-1 primer and sequence of a 2-Kb product amplified in four resistant lines by an OPAN-1 random primer. The SCAR-OPAN-1 primer contained the OPAN-1 random primer sequence. The sequence of the designed SCAR-OPAN-1 primer is shown in bold. The underlined sequence is the sequence of OPAN-1 random primer, 5′-ACT CCA CGT C-3′.

3.3. Validation of the Developed SCAR Marker Using the Resistant and Susceptible Onion Lines

The SCAR-OPAN1 marker was used to screen the resistant and susceptible lines. The 2-Kb-sized polymorphic band was amplified in the resistant lines S&P 7522, S&P 7521, S&P 7129, and S&P 7168 and was not amplified in the susceptible lines S&P 7483, S&P 7130, and S&P 7175 (Figure 3a). The developed SCAR-OPAN1 marker amplified the specific band only from the resistant lines, demonstrating its potential as a molecular marker for the selection of resistant lines. To validate the SCAR-OPAN1 marker, three resistant lines and four susceptible lines of 'Asia Seed' Co. were screened using the SCAR-OPAN1 marker (Figure 3b). Similar to the screening results of the 'Seeds & People' Co. onion lines, a product size of 2 Kb was amplified only in the resistant lines (Asia-12, Asia-42, and

Asia-53). These polymorphic bands between the resistant and susceptible lines allow the selection of gray mold-resistant onion lines for breeding programs.

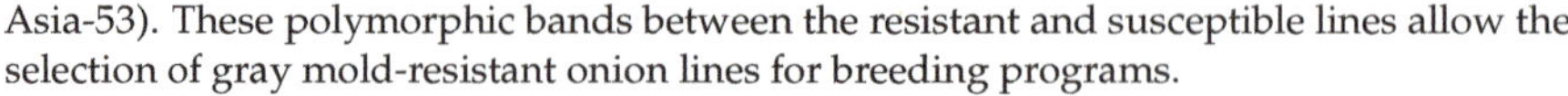
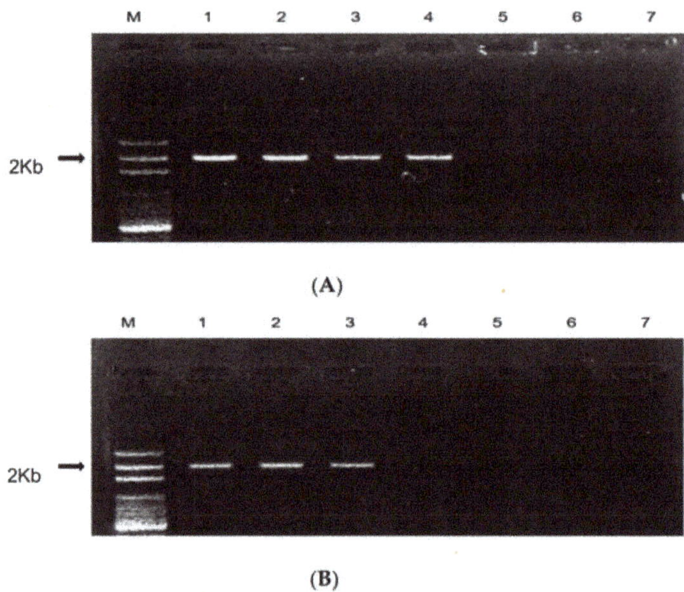

Figure 3. Identification of resistant onion lines of 'Seeds & People' Co. and 'Asia Seed' Co. using the SCAR-OPAN1 marker. (**A**) The 2-Kb-sized polymorphic products were amplified in the resistant lines (S&P 7522, S&P 7521, S&P 7129, and S&P 7168) and were not amplified in the susceptible lines (S&P 7483, S&P 7130, and S&P 7175). M: 100-bp DNA ladder marker, Lane 1: S&P 7522, Lane 2: S&P 7521, Lane 3: S&P 7129, Lane 4: S&P 7168, Lane 5: S&P 7483, Lane 6: S&P 7130, and Lane 7: S&P 7175. (**B**) The 2-Kb products were amplified only in the resistant lines (Asia-12, Asia-42, and Asia-53) and were not amplified in the susceptible lines (Asia-30, Asia-35, Asia-45, and Asia-50). M: 100-bp DNA ladder marker, Lane 1: Asia-12, Lane 2: Asia-42, Lane 3: Asia-53, Lane 4: Asia-30, Lane 5: Asia-35, Lane 6: Asia-45, and Lane 7: Asia-50.

3.4. Preprocessing of Raw Data of the RNA Sequence

After the quality check of the extracted RNA was completed, a library was constructed, and the quality of distribution by cDNA size and concentration of the constructed library were measured using a Bioanalyzer DNA Chip (Agilent Technologies, Santa Clara, CA, USA). Thereafter, RNA sequencing was performed using the constructed library information and HiseqX (Illumina, San Diego, CA, USA), and the raw RNA sequencing data of each onion line were obtained. The raw RNA-seq data are shown in Table S3. The raw reads of the seven onion lines ranged from 28,899,894 to 41,664,542, the total reads of the trimmed reads ranged from 20,744,300 to 27,932,656, and the mapped reads ranged from 16,852,911 to 21,958,205. Trimming was performed by removing low-quality bases and Illumina adapters from the raw data. Mapping was performed on the reference data of NABIC (RDA, JeonJu, Korea, https://nabic.rda.go.kr, accessed on 7 February 2021), and the number of paired reads containing at least 50 nucleotides was confirmed. As a result of mapping, the lengths of the mapping reads ranged from 267 nucleotides to 5478 nucleotides, the average length of the reads was 2236 nucleotides, and the median was 1988 nucleotides. The mapping results determined that the length of the mapped reads was sufficient for the next analysis, and the following analysis was performed.

3.5. Variant Analysis and Filtering for Selection of Transcripts

A variant analysis was conducted to confirm the genetic variations between the gray mold disease-resistant and -susceptible lines. The total raw variants was 359,288, which was generated using haplotypeCaller of the Genome Analysis Tool Kit (GATK, Broad Institute, Cambridge, MA, USA). These variants were confirmed to have occurred in 28,602 transcripts out of a total of 87,427 acquired transcripts. To remove the low-quality genotypes, the minimum depth coverage was set to five, and the maximum depth coverage was set to 100. The minimum genotype quality was filtered to 20%. As a result, the total filtered-out variants were 28,602, including 27,303 SNPs and 1299 indels. Among them, the variants that matched the following conditions were selected: four resistant lines showed the same genotype within a group, and three susceptible lines also showed the same genotype within a group but showed differences between the resistant and susceptible groups. A total of 233 variants were obtained that matched these conditions, and 118 corresponding transcripts were selected.

3.6. Selection of Transcripts Related to Disease Resistance through DEG Analysis and GO Annotation

A DEG analysis between the resistant and susceptible lines was performed to develop molecular markers related to disease resistance. TransDecoder was used to select 109,521 sequences with a length of 100 or more amino acids among the sequences generated using Trinity de novo assembly. As a result of calculating the read count of the transcript expression level using the StringTie program, we confirmed that the overall average mapping rate was 85%, and a total of 87,427 transcript read counts were obtained. Based on the read count of the transcript of each line, a comparative analysis between each onion line was performed using DESeq2. First, the raw read count data were normalized through size factor and dispersion. Thereafter, a correlation analysis between each line was performed based on the standardized value. In addition, from the RNA-seq results, the DEGs were analyzed for the resistant and susceptible groups. Using the DEGseq of the R package, it was confirmed that the difference in resistance versus susceptibility was statistically significant. Significantly expressed genes were verified using the MA plot results (Figure S3). In the MA plot, a false discovery rate (FDR) of less than 0.05 was indicated in red. Through the identification of the significantly expressed genes shown in red, it was confirmed that it was sufficient for use in the subsequent resistance-related gene analysis. In addition, a heatmap was constructed using z-scores to analyze the differences in expression for each line by using a group of significantly expressed genes (Figure 4). The heatmap was analyzed by comparing and analyzing the Pearson correlation coefficients for each line and gene after hierarchical clustering using the average linkage method. In addition, a volcano plot of the DEGs was obtained (Figure S4). The results of the volcano plot are shown in different colors according to the following conditions: FDR < 0.05 and $|\log2$ fold change$| \geq 2$. In the volcano plot, it was expressed in different colors according to the following conditions: red: FDR < 0.01 and $|\log2$ fold change$| \leq 2$, green: FDR < 0.01 and $|\log2$ fold change$| > 2$, and orange: FDR ≥ 0.01 and $|\log2$ fold change$| > 2$. Through this visualized DEG result, it was confirmed that there was a transcript showing a significant DEG difference between the gray mold-resistant and -susceptible groups.

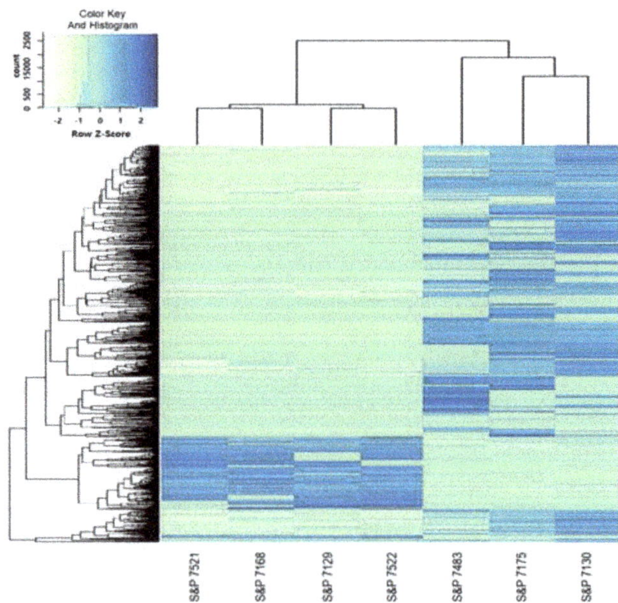

Figure 4. The heatmap constructed using z-scores to analyze the differences of expression in each line by using a group of significantly expressed genes.

The differences between resistant and susceptible lines of 'Seeds & People' Co. were significant, and it could be statistically confirmed that a resistance-related mechanism analysis was possible. Among the obtained 87,427 transcripts, the transcripts with significantly different expression levels between the resistant and susceptible lines were selected to satisfy the following conditions: PADJ < 0.05 and $|$ log2 fold change $| \geq 2$. A total of 1636 transcripts were selected. Among the 1636 transcripts, 320 transcripts showed higher expression levels in the resistant group, and 1316 showed higher expression in the susceptible group. Among the 320 transcripts that showed a higher level in the resistant group, only 182 transcripts matched to TAIR ID, while 138 did not, suggesting an unknown onion gene. In addition, among the 1316 transcripts that showed higher expression levels in the susceptible group, only 897 transcripts matched, while 419 did not. The matched transcripts were analyzed by GO annotations of the cellular components, molecular functions, and biological processes using TAIR ID.

To select genes with increased expression in relation to resistance, among the 182 transcripts with increased expression levels, 22 transcripts related to 'response to stress' and seven transcripts related to 'response to biotic stimuli' were analyzed by GO annotation (Figure S5). Finally, 29 resistance-related transcripts were confirmed to be related to disease resistance, and variants were also observed between the resistant and susceptible groups. In addition, the gene functions of the identified transcripts were analyzed using TAIR ID

3.7. Transcripts Selection to Develop SNP Markers and a HRM Analysis

Based on variant analyses, including the DEG and GO analyses, the transcripts were selected to develop SNP markers for the screening of gray mold-resistant onions. A total of 118 transcripts with 233 variants, such as SNPs and indels, between the resistant and susceptible groups were selected by variant analyses. Thereafter, 29 transcripts that were found to be related to disease resistance through the GO analysis and showed higher gene expression levels in the resistant group than in the susceptible group in the DEG analysis were selected. Finally, a total of 14 transcripts with SNPs were selected to develop SNP markers, and HRM primers were designed from these 14 selected transcripts with SNPs that existed between the resistant and susceptible lines (Table 1). These selected transcripts were associated with genes with functions related to 'response to biotic stimulus' and 'response to stress'. Plants are known to show resistance to pathogens in various ways. Although it prevents infection by pathogens structurally, such as through cell walls, it has several defense systems that show resistance to invading pathogens. The gene functions of the selected transcripts were 'lipoxygenase 3', 'Glutathione S-transferase', and 'systemic acquired resistance', which were genes related to plant resistance (Table 2).

Table 2. Gene functions of 14 selected transcripts with SNPs that exist between the resistant and susceptible lines.

No	Transcripts No	TAIR ID [z]	Gene Function
1	SNP-1 transcript	AT1G17420	Lipoxygenase 3
2	SNP-2 transcript	AT1G42970	Glyceraldehyde-3-phosphate dehydrogenase B subunit
3	SNP-3 transcript	AT1G59960	Aldo/keto reductase
4	SNP-4 transcript	AT1G67090	Ribulose bisphosphate carboxylase small chain 1A
5	SNP-5 transcript	AT1G68090	Annexin Arabidopsis thaliana 5
6	SNP-6 transcript	AT1G78380	Glutathione S-transferase
7	SNP-7 transcript	AT3G16640	Translationally controlled tumor protein
8	SNP-8 transcript	AT4G39260	Glycine-rich protein
9	SNP-9 transcript	AT5G13750	Zinc induced facilitator-like 1
10	SNP-10 transcript	AT5G52810	Systemic acquired resistance
11	SNP-11 transcript	AT1G14290	Sphingoid base hydroxylase2
12	SNP-12 transcript	AT1G29930	Chlorophyll A/B protein
13	SNP-13 transcript	AT5G25220	KNAT3
14	SNP-14 transcript	AT5G38430	Rubisco small subunit 1B

[z] The Arabidopsis Information Resource ID.

Each HRM amplicon obtained from the 14 HRM primers was confirmed to match the expected size by electrophoresis. In addition, the sequence analysis confirmed that SNPs existed between the resistant and susceptible lines and revealed the amino acid sequences that were altered by the SNP (Figure 5). The SNPs targeted by the HRM primers are shown in green, and the nonsynonymous mutations are shown in red. Synonymous mutations are shown in blue and were present in SNP-1, SNP-2, SNP-4, SNP-12, and SNP 14. Nonsynonymous mutations were identified in SNP-3, SNP-5, SNP-6, SNP-7, SNP-8, SNP-9, SNP-10, SNP-11, and SNP-13. The melting peak values obtained from the HRM analysis were confirmed by ANOVA grouping. As a result, SNP-3 primers targeting the AKR gene transcript capable of distinguishing resistance and susceptibility were selected. The expected amplicon size of the SNP-3 primer was 134 bp. The HRM amplicon sizes of all seven onion lines were identified as the same. In addition, the sequences of the four resistant and three susceptible lines targeted by the SNP-3 primer were analyzed (Figure 6). Two SNPs were identified within the sequence. One SNP showed the 'T' allele in the resistant lines and the 'C' allele in the susceptible lines, while the other SNP showed the 'C' allele in the resistant lines and the 'G' allele in the susceptible ones.

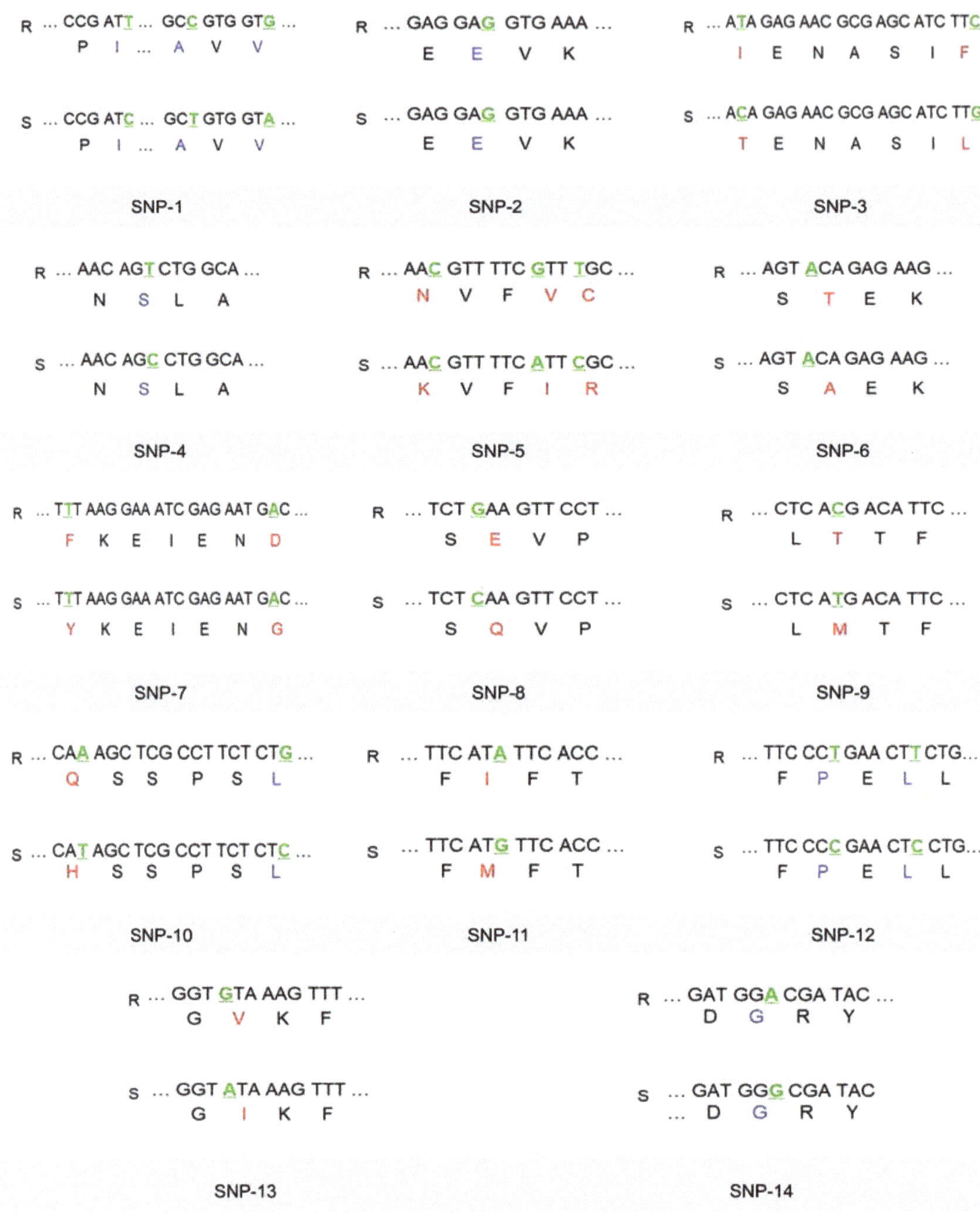

Figure 5. SNPs and amino acid sequences targeted by the HRM primers in the resistant and susceptible lines. The SNPs targeted by HRM primers are shown in green. The synonymous mutations are shown in blue, and the nonsynonymous mutations between the resistant and susceptible groups are shown in red.

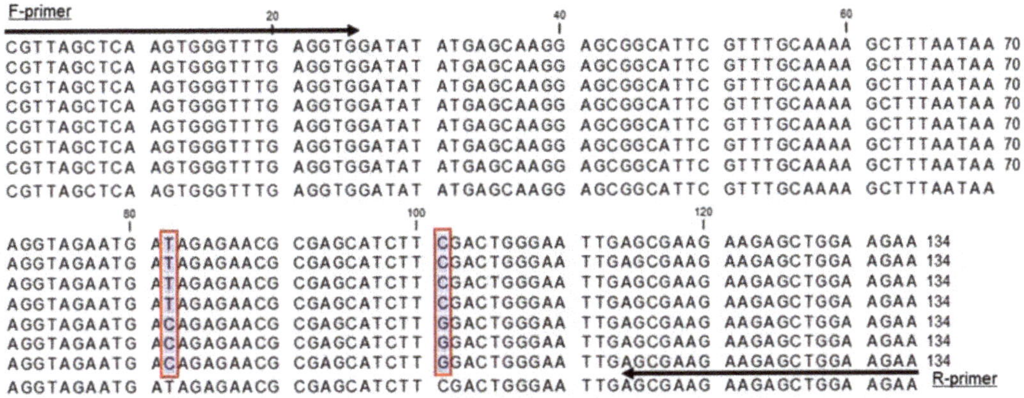

```
F-primer                    20                            40                            60
CGTTAGCTCA AGTGGGTTTG AGGTGGATAT ATGAGCAAGG AGCGGCATTC GTTTGCAAAA GCTTTAATAA 70
CGTTAGCTCA AGTGGGTTTG AGGTGGATAT ATGAGCAAGG AGCGGCATTC GTTTGCAAAA GCTTTAATAA 70
CGTTAGCTCA AGTGGGTTTG AGGTGGATAT ATGAGCAAGG AGCGGCATTC GTTTGCAAAA GCTTTAATAA 70
CGTTAGCTCA AGTGGGTTTG AGGTGGATAT ATGAGCAAGG AGCGGCATTC GTTTGCAAAA GCTTTAATAA 70
CGTTAGCTCA AGTGGGTTTG AGGTGGATAT ATGAGCAAGG AGCGGCATTC GTTTGCAAAA GCTTTAATAA 70
CGTTAGCTCA AGTGGGTTTG AGGTGGATAT ATGAGCAAGG AGCGGCATTC GTTTGCAAAA GCTTTAATAA 70
CGTTAGCTCA AGTGGGTTTG AGGTGGATAT ATGAGCAAGG AGCGGCATTC GTTTGCAAAA GCTTTAATAA 70
CGTTAGCTCA AGTGGGTTTG AGGTGGATAT ATGAGCAAGG AGCGGCATTC GTTTGCAAAA GCTTTAATAA

          80                           100                           120
AGGTAGAATG ATAGAGAACG CGAGCATCTT CGACTGGGAA TTGAGCGAAG AAGAGCTGGA AGAA 134
AGGTAGAATG ATAGAGAACG CGAGCATCTT CGACTGGGAA TTGAGCGAAG AAGAGCTGGA AGAA 134
AGGTAGAATG ATAGAGAACG CGAGCATCTT CGACTGGGAA TTGAGCGAAG AAGAGCTGGA AGAA 134
AGGTAGAATG ATAGAGAACG CGAGCATCTT CGACTGGGAA TTGAGCGAAG AAGAGCTGGA AGAA 134
AGGTAGAATG ACAGAGAACG CGAGCATCTT GGACTGGGAA TTGAGCGAAG AAGAGCTGGA AGAA 134
AGGTAGAATG ACAGAGAACG CGAGCATCTT GGACTGGGAA TTGAGCGAAG AAGAGCTGGA AGAA 134
AGGTAGAATG ACAGAGAACG CGAGCATCTT GGACTGGGAA TTGAGCGAAG AAGAGCTGGA AGAA 134
AGGTAGAATG ATAGAGAACG CGAGCATCTT CGACTGGGAA TTGAGCGAAG AAGAGCTGGA AGAA R-primer
```

Figure 6. Sequences of the products amplified with the SNP-3 HRM primer in four resistant and three susceptible onion lines. Two SNPs were identified within the sequence. One SNP showed the 'T' allele in the resistant lines and the 'C' allele in the susceptible lines, while the other SNP showed the 'C' allele in the resistant lines and the 'G' allele in the susceptible ones. The forward (F) and reverse (R) primers for the SNP-3 HRM marker are underlined with the arrow.

In the results of the HRM analysis, which was repeated six times using the SNP-3 primer, the melting values of the resistant strains were 79.642–79.838 and those of the susceptible strains were 80.353–80.472. Upon statistically grouping these values by the Duncan method using ANOVA, significant results were obtained by grouping four resistant lines into one group (A) and three susceptibility lines into another group (B) (Table 3a). In the HRM result graph, the resistant lines (red) show a distinct curve from the susceptible lines (blue) (Figure 7a). Compared with the sequence results, it was confirmed that the resistant group that showed 'T' and 'C' at the SNP position had a lower melting temperature than the susceptible group that had 'C' and 'G'. In addition, for validation, the 'Asia Seed' Co. onion lines were analyzed using the SNP-3 HRM marker. As a result of ANOVA grouping, the resistant lines (Asia-12, Asia-48, and Asia-53) were grouped into one group (A), and the susceptible lines (Asia-30, Asia-35, Asia-45, and Asia-50) were also grouped together (B) (Table 3b). Similar to the HRM results of the 'Seeds & People' Co. onion lines, the HRM curve graph of the 'Asia Seed' Co. onion lines showed that the resistant lines were melted at a lower temperature than the susceptible lines (Figure 7b). Based on this result, it was confirmed that the SNP-3 HRM marker is versatile and can be used to select resistant onions.

Table 3. HRM analysis with the SNP-3 primers and statistically grouped by using ANOVA.

(A) HRM analysis results of the 'Seeds & People' Co. onion lines.

HRM Marker	Line	LS Means	Group *	
	S&P 7522	79.642	A	
	S&P 7521	79.768	A	
	S&P 7129	79.782	A	
SNP-3	S&P 7168	79.838	A	
	S&P 7483	80.388		B
	S&P 7130	80.353		B
	S&P 7175	80.472		B

(B) HRM analysis results of the 'Asia seed' Co. onion lines.

HRM Marker	Line	LS Means	Group *	
	Asia-12	82.183	A	
	Asia-53	82.217	A	
	Asia-48	82.253	A	
SNP-3	Asia-50	82.473		B
	Asia-35	82.500		B
	Asia-45	82.527		B
	Asia-30	82.550		B

* In each column, means with the same letter are not significantly different.

(A)

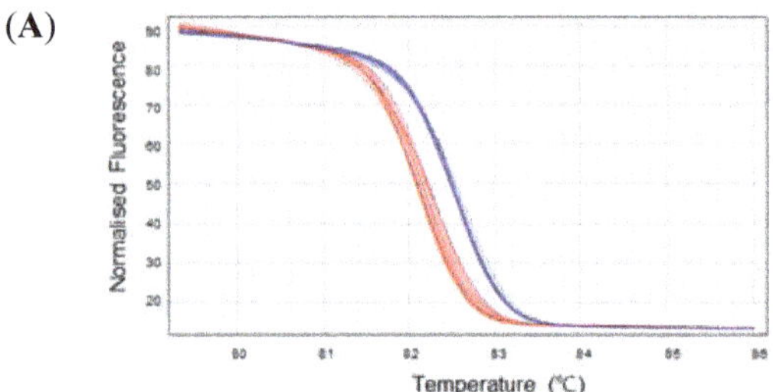

(B)

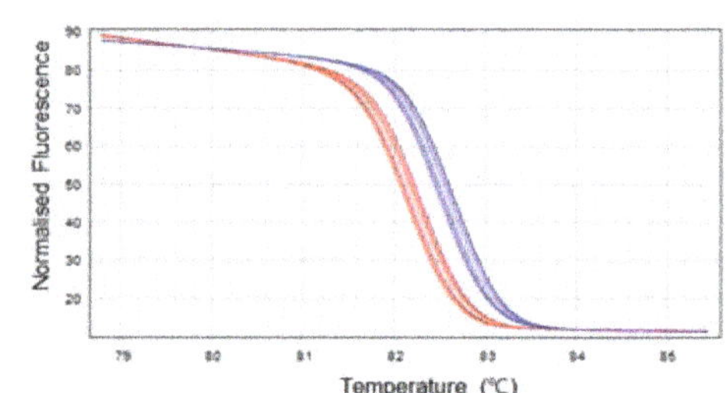

Figure 7. HRM analysis of the 'Seeds & People' Co. and 'Asia Seed' Co. onion lines. (**A**) HRM curve of the 'Seeds & People' Co. onion lines. Red: resistant lines (S&P 7522, S&P 7521, S&P 7129, and S&P 7168) and blue: susceptible lines (S&P 7130, S&P 7175, and S&P 7483). (**B**) HRM curve of the 'Asia Seed' Co. onion lines. Red: resistant lines (Asia-12, Asia-42, and Asia-53) and, blue: susceptible lines (Asia-30, Asia-35, Asia-45, and Asia-50).

3.8. Confirmation of AKR Gene Expression Level in the Onion Lines through qPCR Analysis

The SNP-3 HRM marker was selected from the AKR transcript and showed higher expression levels in the resistant line group in the in silico data. qPCR was performed to confirm the gene expression levels of AKR. Before performing qPCR, RT-PCR was performed to confirm the amplification using the primer set designed for qPCR. RT-PCR showed that all seven onion lines were amplified into one band of 171 bp. qPCR was performed in six replicates by the SNP-3 primer, and the results were calculated using the ΔΔCt method. The gene expression level was normalized to the actin gene expression level and compared with the susceptible line S&P 7483 with the highest susceptibility (Figure 8). In the qPCR analysis, the expression level of the AKR gene was higher in the resistant group than in the susceptible group, with S&P 7522 showing the highest expression level and S&P 7130 showing the lowest expression level.

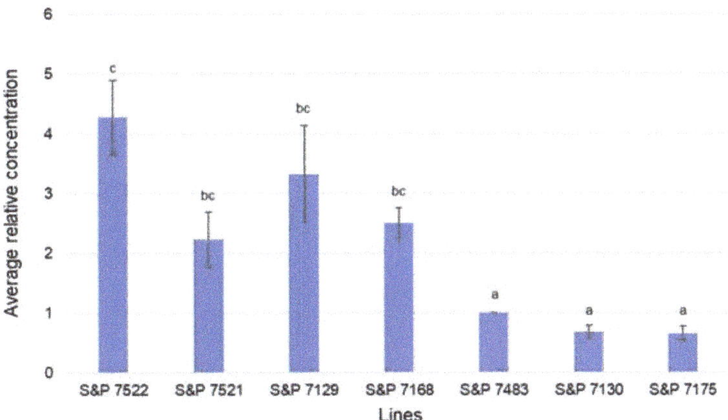

Figure 8. The gene expression level of the aldo-keto reductase (ARK) gene in the seven onion lines using the SNP-3 primer by the qPCR analysis. Four resistant lines (S&P 7522, S&P 7521, S&P 7129, and S&P 7168) and three susceptible lines (S&P 7483, S&P 7130, and S&P 7175) were analyzed. The qPCR analysis was performed in six replicates, and the standard errors are indicated as bars. Means with the same letter on the bar were not significantly different according to Duncan's multiple comparison test.

4. Discussion

Using molecular markers, genetic differences can be quickly and easily identified at the DNA level without phenotyping. RAPD markers can easily identify polymorphisms by using short random primers, commonly 10 bp in length. In addition, it does not rely on knowing the target DNA sequence information, and RAPD is inexpensive, simple, quick, and easy to use [25–27]. Based on the RAPD results, small genetic differences between large genomes can be identified. Genetic relationships between the onion lines can also be confirmed through the unweighted pair group method with an arithmetic mean cluster dendrogram [15,16].

In this study, a phylogenetic analysis of gray mold disease-resistant and -susceptible lines was conducted using RAPD. The phylogenic analysis obtained from the RAPD showed that the resistant lines were closely related to other resistant lines, and the susceptible lines were closely related to other susceptible lines. S&P 7522, S&P 7521, S&P 7129, and S&P 7168 belong to the gray mold disease-resistant line group, and S&P 7483, S&P 7130, and S&P 7175 belong to the gray mold disease-susceptible line group. Additionally, a close genetic relationship appeared between the resistant S&P 7522 and S&P 7521 lines and between the resistant S&P 7129 and S&P 7168 lines. Similarly, susceptible lines S&P 7483, S&P 7130, and S&P 7175 showed close genetic relationships in the phylogenic analysis. These results were similar to the pedigree of the 'Seeds & People' Co. onion lines. According to the pedigrees, two gray mold disease-resistant lines, S&P 7129 and S&P 7168, were bred by crossing the resistant lines S&P 7522 and S&P 7521. The gray mold disease-susceptible lines S&P 7130 and S&P 7175 were bred through the crossing of the susceptible lines S&P 7406 and S&P 7405. A comparison of the results of the phylogenetic analysis obtained from the RAPD with the pedigree of the breeding lines revealed similar genetic relationships, thereby demonstrating the reliability of the RAPD results.

The SCAR-OPAN1 marker was developed using the polymorphic fragment in the RAPD analysis. The SCAR-OPAN1 marker sequence was designed to extend longer, including the 5′-ACT CCA CGT C-3′ sequence of the OPAN-1 random primer. The primer set for the SCAR- OPAN1 marker was designed considering the annealing temperature, GC content, the possibility of the dimer formation, and the hairpin loops. A BLAST analysis was performed to analyze the products amplified by the SCAR marker. In the BLAST analysis of the DNA sequence of the fragment amplified from the SCAR marker, there was no significant percentage of sequences matched to the other sequences from onions and other species; therefore, additional analysis is required. SCAR markers have been used to distinguish between various cultivars and species [28,29]. SCAR markers can be used to select individuals with specific traits and to identify disease-resistant individuals. Using SCAR markers, the selection of male infertility-dominant *Ms* and -recessive *ms* in onions [30], anthrax resistance in grapes [31], and Acokita blight resistance in lentils [32] have been reported. It was confirmed that the selection of gray mold disease-resistant onions was possible using this SCAR- OPAN1 marker in the onion breeding lines of 'Seeds & People' Co.

In addition, to validate the developed SCAR-OPAN1 marker, additional gray mold disease-resistant and -susceptible lines from the 'Asia Seed' Co. were analyzed. In the analysis of the gray mold disease-resistant and -susceptible 'Asia Seed' Co. lines, it was confirmed that a product of 2Kb was amplified only in the resistant lines. Therefore, it was confirmed that the SCAR-OPAN1 marker developed in this study not only facilitates the selection of gray mold-resistant onions but also facilitates easy selection in a considerably short time.

The HRM marker was developed to select gray mold disease-resistant onion lines by using the selected transcripts through a DEG analysis from RNA sequencing. After analyzing the RNA-seq, the selected 14 transcripts were analyzed using the Arabidopsis Information Resource (TAIR) ID derived from Arabidopsis. The matched transcripts were analyzed using GO annotation and were largely divided into cellular components, molecular functions, and biological processes. Among them, the transcripts related to disease resistance, classified as 'response to biotic stimulus' and 'response to stress', were selected.

Plant cells synthesized reactive oxygen species (ROS) and pathogenesis-related proteins (PR proteins) after the detection of the presence of pathogens, along with a hypersensitive response (HR) to prevent the growth of pathogens. These resistance responses can induce a systemic acquired resistance (SAR) response, which is a resistance response of the whole plant. When SAR is activated by pathogens such as fungi, bacteria, and viruses, salicylic acid (SA) accumulates, and the accumulation of SA is essential for SAR expression [33,34].

In this study, the SAR gene was targeted by the SNP-10 transcript and showed DEGs and variants between the gray mold disease-resistant and -susceptible line groups (Table 2). After analyzing the relationship between SA accumulation and SAR in the early 1990s, SARs have been identified in *Arabidopsis thaliana*, tobacco, and cucumbers. The immune response by SAR is induced by PR proteins and has been studied to recognize SA as a signal [35–37]. When tobacco began to show resistance to *B. cinerea* and *Psequdomonas syringae*, it was confirmed that the resistance mechanism was initiated by the involvement of SAR, PR genes (PR-1 and PR-5), and SA before preparation in whole plants [38].

In contrast to SAR, the induced systemic resistance (ISR) with other signaling pathways induced by the plant hormones jasmonic acid (JA) and ethylene is also one of the resistance responses of plants. JA is a plant hormone that is related to disease resistance. The transcript targeted by SNP-1 was found to be the lipoxygenase (LOX) 3 gene (Table 2). The LOX gene is involved in JA biosynthesis. In particular, LOX3 was rapidly upregulated when the pathogen *B. cinerea* was inoculated in *Arabidopsis thaliana*. This LOX-upregulated response was shown in LOX3 and LOX4, which was presumably related to the early JA response of oligogalacturonides acting as damage-associated molecular patterns (DAMPs) [39].

In addition, the glutathione S-transferase (GST) gene, the transcript targeted by SNP-6, has been identified in many studies on disease resistance in plants (Table 2) [40–42]. GST plays a major role as an antioxidant and is resistant to plants in relation to a hypersensitive response to cell death. The transformed *Nicotiana benthamiana* was resistant to *Colletotrichum destructivum* and *C. orbiculare* by the GST gene [43].

In other selected transcripts, it was also confirmed that target genes have various functions related to plant resistance. Among them, the SNP-3 transcript developed as an HRM marker was confirmed to be an aldo-keto reductase gene (AKR) (Table 2). It was confirmed that the SNP-3 HRM marker related to the AKR gene can also select gray mold disease-resistant lines from 'Seeds & People' Co. and 'Asia Seed' Co. AKR gene expression was higher in the gray mold disease-resistant line group than in the -susceptible line group. AKR has been mentioned in previous studies and is known to increase the resistance at high expression levels [44,45].

Since the SNP-3 HRM marker was a gene-based marker, a qPCR analysis was conducted to confirm the expression level of the AKR gene. The expression level of AKR was higher in the resistant group than in the susceptible group. The qPCR results were compared with the in silico data. In the RNA-seq results, the expression level of AKR was higher in the resistant group than in the susceptible group, and S&P7522 showed the highest expression level. The similarity of the AKR gene expression levels between the qPCR analysis and in silico RNA-seq results showed that the gene was associated with resistance to gray mold disease, increasing the reliability of the in silico data used to develop the molecular markers.

It has also been identified that the AKR gene provides multiple stress tolerances in plants. In particular, there have been studies related to abiotic stresses such as herbicide resistance, heat stress tolerance, and biotic stress, such as mildew [41,46,47]. For biotic stress from pathogens such as microorganisms, the AKR gene group is mainly involved in plant secondary metabolic pathways, including flavonoid biosynthesis in plant-microbial interactions. Therefore, these AKRs primarily function as plant defense systems against biological stresses, such as a pathogen attack. However, despite reports identifying plant AKR as a potential target for developing abiotic and biotic stress-tolerant plant species, the importance of plant AKR has not yet been emphasized [46,48].

Various studies have revealed that AKR plays a stress-related role. This study also showed differences in the resistance and gene expression levels according to SNPs between the gray mold-resistant and -susceptible line groups. It was considered that the AKR gene affected the resistant group to show a resistance to gray mold disease. Therefore, further studies of AKR will be needed, and the SNP-3 HRM marker developed in this study could be used to select gray mold disease-resistant onion lines.

5. Conclusions

Onions have a large genome size, and the information on the existing DNA sequences is insufficient. In this study, a molecular marker was developed to breed resistant onions against gray mold disease. A phylogenetic analysis between the gray mold disease-resistant and -susceptible lines was performed using a dendrogram derived from the RAPD analysis. The constructed dendrogram was compared with the pedigree of the gray mold disease-resistant and -susceptible breeding lines provided by 'Seeds & People' Co. The results of the phylogenetic analysis were consistent with the genetic relationship of the pedigree of the breeding lines and showed the reliability of the RAPD results. Thereafter, a SCAR-OPAN1 marker was developed based on the RAPD results showing polymorphic fragments between the resistant and susceptible lines. The SCAR-OPAN1 marker only amplified the resistant lines of the specific 2-Kb product, and no product was amplified in the susceptible lines.

In addition, the RNA-seq of the gray mold disease-resistant and -susceptible onion lines was analyzed using NGS technology. Based on the RNA-seq results, DEG and GO analyses were performed to identify the variants, such as SNPs and indels. As a result, a selectable marker, SNP-3 HRM, was developed to select gray mold-resistant lines. The SNP-3 HRM marker for the selection of the resistant lines includes SNPs present in the AKR gene exhibiting high expression levels in these lines. Consequently, in this study, the SCAR-OPAN1 and SNP-3 HRM markers were developed for the selection of resistant onion lines in breeding programs to reduce the damage caused by gray mold disease. Thus, using these molecular markers, the breeding period of biennial onions can be shortened by selecting an onion line that is resistant to gray mold disease, thereby alleviating the economic loss of onions.

Supplementary Materials: The following are available online at https://www.mdpi.com/article/10.3390/horticulturae7110436/s1: Figure S1: The pedigree of the onion lines provided by 'Seeds & People' Co. Figure S2: RAPD analysis of the resistant and susceptible lines provided by 'Seeds & People' Co. by using the OPAN-1 random primer. Figure S3: MA plot to verify the significantly expressed genes in the groups of resistant and susceptible lines. Figure S4: Volcano plot of the transcripts that satisfied the following conditions: FDR < 0.05 and |log2 fold change| $\geq$ 2. Figure S5: GO annotation of 182 transcripts with increased expression levels. Table S1: Characteristics of the four resistant lines (S&P 7129, S&P 7168, S&P 7521, and S&P 7522) and three susceptible lines (S&P 7130, S&P 7175, and S&P 7483) provided by 'Seeds & People' Co., Ltd. (Yeonggwang, Korea). Table S2: Sequence information of 40 OPERON random primers used for the RAPD analysis. Table S3: Results of the raw data obtained by RNA-seq, trimming, and mapping.

Author Contributions: Conceptualization, S.-J.K., J.-S.P., and Y.-D.P.; investigation and writing—original draft preparation, S.-J.K. and J.-S.P.; resources, T.P.; data curation, H.-M.L.; investigation, J.-R.C.; and conceptualization, supervision, writing—review and editing, and funding acquisition, Y.-D.P. All authors have read and agreed to the published version of the manuscript.

Funding: This research was funded by the Ministry of Agriculture, Food and Rural Affairs (MAFRA).

Data Availability Statement: The datasets generated during and/or analyzed during the current study are available from the corresponding author upon reasonable request.

Acknowledgments: This work was carried out with the support of the Korea Institute of Planning and Evaluation for Technology in Food, Agriculture, and Forestry (IPET) through the Golden Seed Project (213007-05-5-SBB20).

Conflicts of Interest: The authors declare that they have no conflicts of interest. The funders had no role in the design of the study; in the collection, analyses, or interpretation of the data; in the writing of the manuscript; or in the decision to publish the results.

References

1. Nam, K.H.; Baik, H.W.; Choi, T.Y.; Yoon, S.G.; Park, S.W.; Joung, H.J. Effects of ethanol extract of onion on the lipid profiles in patients with hypercholesterolemia. *J. Nutr. Health* **2007**, *40*, 242–248.
2. Kim, J.Y.; Seo, Y.J.; Noh, S.K.; Cha, Y.J. Concentrated onion extract lowers serum lipid levels in rats fed a high-fat diet. *Korean J. Food Preserv.* **2010**, *17*, 398–404.
3. Khosa, J.S.; McCallum, J.; Dhatt, A.S.; Macknight, R.C. Enhancing onion breeding using molecular tools. *Plant. Breed.* **2016**, *135*, 9–20. [CrossRef]
4. Jo, J.; Purushotham, P.M.; Han, K.; Lee, H.R.; Nah, G.; Kang, B.C. Development of a genetic map for onion (*Allium cepa* L.) using reference-free genotyping-by-sequencing and SNP assays. *Front. Plant. Sci.* **2017**, *8*, 1606.
5. Havey, M.J. Onion Breeding. In *Plant Breeding Review*; Wiley: New Jersey, NJ, USA, 2018; Volume 42, pp. 39–85.
6. Chilvers, M.I.; du Toit, L.J. Detection and identification of *Botrytis* species associated with neck rot, scape blight, and umbel blight of onion. *Plant. Health Prog.* **2006**, *7*, 38. [CrossRef]
7. Williamson, B.; Tudzynski, B.; Tudzynski, P.; van Kan, J.A.L. *Botrytis cinerea*: The cause of grey mould disease. *Mol. Plant. Pathol.* **2007**, *8*, 561–580. [CrossRef] [PubMed]
8. Choquer, M.; Fournier, E.; Kunz, C.; Levis, C.; Pradier, J.M.; Simon, A.; Viaud, M. Botrytis cinerea virulence factors: New insights into a necrotrophic and polyphageous pathogen. *FEMS Microbiol. Lett.* **2007**, *277*, 1–10. [CrossRef]
9. Carisse, O.; Tremblay, D.M.; McDonald, M.R.; Brodeur, L.; McRoberts, N. Management of botrytis leaf blight of onion: The Québec experience of 20 years of continual improvement. *Plant. Dis.* **2001**, *95*, 504–514. [CrossRef] [PubMed]
10. Abbey, J.A.; Percival, D.; Abbey, L.; Asiedu, S.K.; Prithiviraj, B.; Schilder, A. Biofungicides as alternative to synthetic fungicide control of grey mould (*Botrytis cinerea*)—prospects and challenges. *Biocontrol. Sci. Technol.* **2018**, *29*, 207–228. [CrossRef]
11. Howell, G.S. Sustainable grape productivity and the growth-yield relationship: A review. *Am. J. Enol. Vitic.* **2001**, *52*, 165–174.
12. Tripathi, P.; Dubey, N.K.; Shukla, A.K. Use of some essential oils as post-harvest botanical fungicides in the management of grey mould of grapes caused by *Botrytis cinerea*. *World J. Microbiol. Biotechnol.* **2008**, *24*, 39–46. [CrossRef]
13. Fan, F.; Hamada, M.S.; Li, N.; Li, G.Q.; Luo, C.X. Multiple fungicide resistance in *Botrytis cinerea* from greenhouse strawberries in Hubei province, China. *Plant. Dis.* **2017**, *101*, 601–606. [CrossRef]
14. Rupp, S.; Weber, R.W.S.; Rieger, D.; Detzel, P.; Hahn, M. Spread of *Botrytis cinerea* strains with multiple fungicide resistance in German horticulture. *Front. Microbiol.* **2017**, *7*, 2075. [CrossRef]

15. Maniruzzaman, M.; Haque, M.E.; Haque, M.M.; Sayem, M.A.; Al-Amin, M. Molecular characterization of onion (*Allium cepa*) using RAPD markers. *Bangladesh J. Agric. Res.* **2010**, *35*, 313–322.

16. Sudha, G.S.; Ramesh, P.; Sekhar, A.C.; Krishna, T.S.; Bramhachari, P.V.; Riazunnisa, K. Genetic diversity analysis of selected onion (*Allium cepa* L.) germplasm using specific RAPD and ISSR polymorphism markers. *Biocatal. Agric. Biotechnol.* **2019**, *17*, 110–118. [CrossRef]

17. Wang, Z.; Gerstein, M.; Snyder, M. RNA-Seq: A revolutionary tool for transcriptomics. *Nat. Rev. Genet.* **2009**, *10*, 57–63. [CrossRef] [PubMed]

18. Han, J.; Thamllarasan, S.K.; Natarajan, S.; Park, J.I.; Chung, M.Y.; Nou, I.S. De novo assembly and transcriptome analysis of bulb onion (*Allium cepa* L.) during cold acclimation using contrasting genotypes. *PLoS ONE* **2016**, *11*, e0161987. [CrossRef]

19. Graham, R.; Liew, M.; Meadows, C.; Lyon, E.; Wittwer, C.T. Distinguishing different DNA Heterozygotes by high-resolution melting. *Clin. Chem.* **2005**, *51*, 1295–1298. [CrossRef] [PubMed]

20. Reed, G.H.; Kent, J.O.; Wittwer, C.T. High-resolution DNA melting analysis for simple and efficient molecular diagnostics. *Pharmacogenomics* **2007**, *8*, 597–608. [CrossRef]

21. Doyle, J.J.; Doyle, J.L. Isolation of plant DNA from fresh tissue. *Focus* **1990**, *12*, 13–15.

22. Sokal, R.R.; Michener, C.D. A statistical method for evaluation systematic relationships. *Univ. Kans. Sci. Bull.* **1958**, *28*, 1409–1438.

23. Jaccard, P. Étude comparative de la distribution florale dans une portion des Alpes et des Jura. *Bull. Soc. Vaud. Sci. Nat.* **1901**, *37*, 547–579.

24. Wang, L.; Feng, Z.; Wang, X.; Wang, X.; Zhang, X. DEGseq: An R package for identifying differentially expressed genes from RNA-seq data. *Bioinformatics* **2010**, *26*, 136–138. [CrossRef]

25. Welsh, J.; McClelland, M. Fingerprinting genomes using PCR with arbitrary primers. *Nucleic. Acids Res.* **1990**, *18*, 7213–7218. [CrossRef]

26. Williams, J.G.K.; Kubelik, A.R.; Livak, K.J.; Rafalski, J.A.; Tingey, S.V. DNA polymorphisms amplified by arbitrary primers are useful as genetic markers. *Nucleic. Acids Res.* **1990**, *18*, 6531–6535. [CrossRef]

27. Sidiq, Y.; Subiastuti, A.S.; Wibowo, W.A.; Daryono, B.S. Development of SCAR marker linked to *begomovirus* resistance in melon (*Cucumis melo* L.). *Jordan J. Biol. Sci.* **2020**, *13*, 145–151.

28. Baite, M.S.; Upadhyay, B.K.; Dube, S.C. Development of a sequence-characterized amplified region marker for detection of Ascochyta rabiei causing Ascochyta blight in chickpea. *Folia Microbiol.* **2020**, *65*, 103–108. [CrossRef] [PubMed]

29. Mei, Z.; Khan, M.A. Genetic authentication of *Eclipta prostrate* (Asteraceae) from *Penthorum chinense* (Penthoraceae) by sequence characterized amplified region (SCAR) markers. *Rev. Biol. Trop.* **2020**, *68*, 180–188. [CrossRef]

30. Yang, Y.Y.; Huo, Y.M.; Miao, J.; Liu, B.J.; Kong, S.P.; Gao, L.M.; Liu, C.; Wang, Z.B.; Tahara, Y.; Kitano, H.; et al. Identification of two SCAR markers co-segregated with the dominant *Ms* and recessive *ms* alleles in onion (*Allium cepa* L.). *Euphytica* **2013**, *190*, 267–277. [CrossRef]

31. Kim, G.H.; Yun, H.K.; Choi, C.S.; Park, J.H.; Jung, Y.J.; Park, K.S.; Dane, F.; Kang, K.K. Identification of AFLP and RAPD markers linked to anthracnose resistance in grapes and their conversion to SCAR markers. *Plant. Breed.* **2008**, *127*, 418–423. [CrossRef]

32. Chowdhury, M.A.; Andrahennadi, C.P.; Slinkard, A.E.; Vandenberg, A. RAPD and SCAR markers for resistance to acochyta blight in lentil. *Euphytica* **2001**, *118*, 331–337. [CrossRef]

33. Hunt, M.D.; Ryals, J.A.; Reinhardt, D. Systemic acquired resistance signal transduction. *Crit. Rev. Plant. Sci.* **1996**, *15*, 583–606. [CrossRef]

34. Durrant, W.E.; Dong, X. Systemic acquired resistance. *Annu. Rev. Phytopathol.* **2004**, *42*, 185–209. [CrossRef] [PubMed]

35. Ward, E.R.; Uknes, S.J.; Williams, S.C.; Dincher, S.S.; Wiederhold, D.L.; Alexander, D.C.; Aphl-Goy, P.; Metraux, J.P.; Ryals, J.A. Coordinate gene activity in response to agents that induce systemic acquired resistance. *Plant. Cell* **1991**, *3*, 1085–1094. [CrossRef] [PubMed]

36. Sticher, L.; Mauch-Mani, B.; Métraux, A.J. Systemic acquired resistance. *Annu. Rev. Phytopathol.* **1997**, *35*, 235–270. [CrossRef] [PubMed]

37. Mauch-Mani, B.; Métraux, J.P. Salicylic acid and systemic acquired resistance to pathogen attack. *Ann. Bot.* **1998**, *82*, 535–540. [CrossRef]

38. Frías, M.; Brito, N.; González, C. The *Botrytis cinerea* cerato-platanin BcSpl1 is a potent inducer of systemic acquired resistance (SAR) in tobacco and generates a wave of salicylic acid expanding from the site of application. *Mol. Plant. Pathol.* **2013**, *14*, 191–196. [CrossRef] [PubMed]

39. Windram, O.; Madhou, P.; McHattie, S.; Hill, C.; Hickman, R.; Cooke, E.; Jenkins, D.J.; Penfold, C.A.; Baxter, L.; Breeze, E.; et al. *Arabidopsis* defense against *Botrytis cinerea*: Chronology and regulation deciphered by high-resolution temporal transcriptomic analysis. *Plant. Cell* **2012**, *24*, 3530–3557. [CrossRef] [PubMed]

40. Poland, C.A.; Duffin, R.; Kinloch, I.; Maynard, A.; Wallace, W.A.; Seaton, A.; Stone, V.; Brown, S.; MacNee, W.; Donaldson, K. Carbon nanotubes introduced into the abdominal cavity of mice show asbestos-like pathogenicity in a pilot study. *Nat. Nanotechnol.* **2008**, *3*, 423–428. [CrossRef] [PubMed]

41. Turóczy, Z.; Kis, P.; Török, K.; Cserháti, M.; Lendvai, Á.; Dudits, D.; Horváth, G.V. Overproduction of a rice aldo-keto reductase increases oxidative and heat stress tolerance by malondialdehyde and methylglyoxal detoxification. *Plant. Mol. Biol.* **2011**, *75*, 399–412. [PubMed]

42. Gullner, G.; Komives, T.; Király, L.; Schröder, P. Glutathione S-transferase enzymes in plant-pathogen interactions. *Front. Plant Sci.* **2018**, *9*, 1836. [CrossRef]

43. Dean, J.D.; Goodwin, P.H.; Hsiang, T. Induction of glutathione S-transferase genes of *Nicotiana benthamiana* following infection by *Colletotrichum destructivum* and *C. orbiculare* and involvement of one in resistance. *J. Exp. Bot.* **2005**, *56*, 1525–1533. [CrossRef] [PubMed]

44. Simpson, P.J.; Tantitadapitak, C.; Reed, A.M.; Mather, O.C.; Bunce, C.M.; White, S.A.; Ride, J.P. Characterization of two novel aldo—keto reductases from *Arabidopsis*: Expression patterns, broad substrate specificity, and an open active-site structure suggest a role in toxicant metabolism following stress. *J. Mol. Biol.* **2009**, *392*, 465–480. [CrossRef]

45. Kanayama, Y.; Mizutani, R.; Yaguchi, S.; Hojo, A.; Ikeda, H.; Nishiyama, M.; Kanahama, K. Characterization of an uncharacterized aldo-keto reductase gene from peach and its role in abiotic stress tolerance. *Phytochemistry* **2014**, *104*, 30–36. [CrossRef] [PubMed]

46. Sengupta, D.; Naik, D.; Reddy, A.R. Plant aldo-keto reductases (AKRs) as multi-tasking soldiers involved in diverse plant metabolic processes and stress defense: A structure-function update. *J. Plant. Physiol.* **2015**, *179*, 40–55. [CrossRef] [PubMed]

47. Vemanna, R.S.; Vennapusa, A.R.; Easwaran, M.; Chandrashekar, B.K.; Rao, H.; Ghanti, K.; Sudhakar, C.; Mysore, K.S.; Makarla, U. Aldo-keto reductase enzymes detoxify glyphosate and improve herbicide resistance in plants. *Plant. Biotechnol. J.* **2017**, *15*, 794–804. [CrossRef]

48. Wang, H.; He, Z.; Luo, L.; Zhao, X.; Lu, Z.; Luo, T.; Li, M.; Zhang, Y. An aldo-keto reductase, Bbakr1, is involved in stress response and detoxification of heavy metal chromium but not required for virulence in the insect fungal pathogen, *Beauveria bassiana*. *Fungal Genet. Biol.* **2018**, *111*, 7–15. [CrossRef]

Article

Generation of a High-Density Genetic Map of Pepper (*Capsicum annuum* L.) by SLAF-seq and QTL Analysis of *Phytophthora capsici* Resistance

Yi-Fei Li [1,†], Shi-Cai Zhang [1,†], Xiao-Miao Yang [1], Chun-Ping Wang [2,*], Qi-Zhong Huang [1,*] and Ren-Zhong Huang [1,*]

1 The Institute of Vegetable and Flower Research, Chongqing Academy of Agricultural Sciences, Chongqing 401329, China; liyifei201@163.com (Y.-F.L.); shicaiz@126.com (S.-C.Z.); xiaomiaoyangy@sina.com (X.-M.Y.)
2 Chongqing Key Laboratory of Adversity Agriculture, Biotechnology Research Institute, Chongqing Academy of Agricultural Sciences, Chongqing 401329, China
* Correspondence: wcp49930000@sina.com (C.-P.W.); hjshqz@126.com (Q.-Z.H.); HRZ2323@163.com (R.-Z.H.); Tel.: +86-023-68527162 (Q.-Z.H.); +86-023-68527061 (R.-Z.H.)
† Yi-Fei Li and Shi-Cai Zhang contributed equally to this article.

Abstract: Pepper (*Capsicum annuum* L.) is an economically significant global crop and condiment. Its yield can be severely reduced by the oomycete plant pathogen, *Phytophthora capsici* (*P. capsici*). Here, a high-density genetic map was created with a mapping panel of F_2 populations obtained from 150 individuals of parental lines PI201234 and 1287 and specific-locus amplified fragment sequencing (SLAF) that was then utilized to identify loci that are related to resistance to *P. capsici*. The sequencing depth of the genetic map was 108.74-fold for the male parent, 126.25-fold for the female parent, and 22.73-fold for the offspring. A high-resolution genetic map consisting of 5565 markers and 12 linkage groups was generated for pepper, covering 1535.69 cM and an average marker distance of 0.28 cM. One major quantitative trait locus (QTL) for the *P. capsici* resistance (*CQPc5.1*) was identified on Chr05 that explained the observed 11.758% phenotypic variance. A total of 23 candidate genes located within the QTL *CQPc5.1* interval were identified, which included the candidate gene *Capana05g000595* that encodes the RPP8-like protein as well as two candidate genes *Capana05g000596* and *Capana05g000597* that encodes a RPP13-like protein. Quantitative reverse-transcription PCR (qRT-PCR) revealed higher expression levels of *Capana05g000595*, *Capana05g000596*, and *Capana05g000597* in *P. capsici* resistance accessions, suggesting their association with *P. capsici* resistance in pepper.

Keywords: pepper; *Capsicum annuum*; *Phytophthora capsici*; high-density genetic map; QTL

Citation: Li, Y.-F.; Zhang, S.-C.; Yang, X.-M.; Wang, C.-P.; Huang, Q.-Z.; Huang, R.-Z. Generation of a High-Density Genetic Map of Pepper (*Capsicum annuum* L.) by SLAF-seq and QTL Analysis of *Phytophthora capsici* Resistance. *Horticulturae* **2021**, *7*, 92. https://doi.org/10.3390/horticulturae7050092

Academic Editor: Yuyang Zhang

Received: 5 March 2021
Accepted: 16 April 2021
Published: 1 May 2021

Publisher's Note: MDPI stays neutral with regard to jurisdictional claims in published maps and institutional affiliations.

Copyright: © 2021 by the authors. Licensee MDPI, Basel, Switzerland. This article is an open access article distributed under the terms and conditions of the Creative Commons Attribution (CC BY) license (https://creativecommons.org/licenses/by/4.0/).

1. Introduction

Pepper (*Capsicum annuum* L.) is a common condiment and an economically significant vegetable crop. It is not only used in many cuisines but also found to have many medicinal properties. In 2019, approximately 212.04 million tons of chilies and peppers were grown on about 49.31 Mha around the world (http://www.fao.org/faostat/zh/#data/QC). However, pepper is susceptible to a variety of pathogens such as CMV, TMV, *Colletotrichum* spp., and *Phytophthora capsici* (*P. capsici*) [1–4]. *Phytophthora* blight can significantly decrease pepper yield and quality [5]. The disease is caused by the oomycete plant pathogen *P. capsici* that initially infects the roots and crown roots, then subsequently spread to every plant part, including the roots, stems, fruits, and leaves [6]. *Phytophthora* blight is a severe disease that commonly occurs under warm (25–28 °C) and highly humid conditions [7–9]. No effective and safe measures to control *Phytophthora* blight have been established to date, except for chemical control [10–13]. Therefore, the utilization of resistant varieties has become a simple, effective, and safe way of resolving *Phytophthora* blight occurrence in pepper. Plant breeders have also focused on selecting varieties with high levels of resistance.

The three physiological races of *P. capsici*, named "races 1–3," have been determined by their virulence on four pepper varieties: early calwonder (sensitive), PI201234 (resistant),

PBC137 (partially resistant), and PBC602 (partial resistance) [14]. Previous studies have reported several pepper accessions that are resistant to *P. capsici*, including PI123469, PI201232, PI201234, AC2258, and CM334 (Criollo de Morelos 334) [4,14–18]. Resistance to *P. capsici* is mainly regulated by a single dominant gene in PI201234 or by one dominant gene in the presence of modifiers [9,19–21], and AC2258, which has been derived from PI201234, is resistant to *P. capsici* [17,18]. Studies have shown that resistance to *P. capsici* in CM334 is controlled by a minimum of two genes [22,23]. In addition, these reports revealed that the regulatory mechanism underlying *P. capsici* resistance in pepper is highly complex. Numerous reports have investigated the effect of a pepper QTLs on chromosomes that are associated with resistance against *P. capsici* [18,23–31]. *Pc5.1* is a homologous QTL on chromosome 5 of CM334, PI201234, and Perennial that has been associated with resistance to *P. capsici* [23,29,31]. Mallard et al. (2013) have identified resistance QTLs among three meta-QTLs (*MetaPc5.1*, *MetaPc5.2*, and *MetaPc5.3*) by meta-analysis [31]. Siddique et al. (2019) identified three QTLs on chromosome P5, including *QTL5.1*, *QTL5.2*, and *QTL5.3*, which were associated with resistance to three *P. capsici* isolates (race 1, race 2, and race 3) by traditional QTL mapping combined with GWAS strategy [30]. In addition, a few minor-effect QTLs has been identified on different chromosomes [23,27,28,32].

Large-scale SNP markers have recently been discovered by next-generation sequencing (NGS) that have expedited the construction of the pepper genetic map. SLAF-seq is a novel high-throughput sequencing technique that is less expensive and complex than high-quality reference genome libraries [33]. In addition, the SLAF-seq strategy has been generally utilized in constructing high-density genetic maps of different species and in QTL mapping [34–42]. This strategy had also been successfully used in the creating high-density pepper genetic maps [40,43,44]. For instance, Guo et al. (2017) determined two candidate CMV resistance genes on pepper chromosomes 2 and 11 using SLAF-seq along with BSA technologies [43]. In addition, Zhang et al. (2019) utilized SLAF-seq in detecting two major QTLs that were strongly associated with FFN [40].

In this work, we developed a high-density pepper linkage map with SLAF-seq as well as identified QTLs that are related to *P. capsici* resistance using F_2 populations that were obtained from a cross between parental lines 1287 (*P. capsici* susceptible, female) and PI201234 (*P. capsici* resistant, male). Finally, we investigated the main effect of QTLs as well as select candidate genes. Our results could potentially facilitate the elucidation of the genetic mechanism underlying *P. capsici* resistance in pepper and lay the foundation for breeding highly resistance pepper cultivars.

2. Materials and Methods

2.1. Mapping Population

The *P. capsici*-susceptible sweet pepper line "1287" was obtained from Zhongjiao808, whereas the *P. capsici*-resistant "PI201234" was collected from Central America. The present study used an F_2 mapping population, comprising 150 individuals that were obtained by crossing female parent 1287 and male parent PI201234, which was then used as mapping population. The parental lines and the F_2 population were grown at the Chongqing experimental station of the Chongqing Academy of Agricultural Sciences (Chongqing, China). Sowing of pepper seeds was performed using 50-cell trays containing a mixture of peat and vermiculite that was autoclave sterilized for 30 min in 2018.

2.2. Pathogen Preparation and Plant Inoculation with P. capsici

P. capsici isolate HT1 was used for *P. capsici* resistance identification in pepper. HT1 has been identified as physiological race 3 and was isolated from infected pepper fruit at the experimental station in Jiulongpo District, Chongqing, China. The isolate was cultured on V8 juice-agar medium at 28 °C in an incubator. To prepare the inoculums for disease screening, the cultures were soaked in 5 mL ddH$_2$O and cultivated at 4 °C for 1 h and then set at room temperature for 1 h to promote sporulation. Spore density was determined using a hemocytometer and adjusted to 1×10^5 spores/mL in distilled water. Before inoculation, pepper plants were soaked in water. Then, 5 mL of suspension was injected into the root of each six- to seven-true-leaf stage pepper plant. The inoculated plants were then grown at 28 °C for 16 h/day and at 80% relative humidity.

2.3. Disease Evaluation

Seven days post inoculation, the plants were assessed for disease symptoms using the 0–5 scale of the Chinese standard NY/T 2060.1-2011 (Ministry of Agriculture of the People's Republic of China 2011), which consisted of the following: 0 = no disease symptoms; 1 = emergence of brown lesions in the roots and stems with no to slight wilting of leaves; 2 = extension of root and stem lesions by 1–2 cm, the leaves wilted and had fallen off; 3 = root and stem lesions exceed 2 cm and leaves clearly show wilting or defoliation; 4 = large brown lesions on stems are extended and dehydrated, with the exception of the uppermost leaves which have been lost; and 5 = plant death. According to the disease grade of each plant, the disease index (DI) of each identification material was calculated. The DI was calculated using the equation below:

$$DI = \frac{\Sigma(s \times n)}{N \times S} \times 100,$$

where s is the disease level ranging between 0 and 5; n is the number of plants with corresponding disease level; N is the number of plants investigated in each F_2; and S is the representative value of the highest grade.

2.4. Statistical Analysis of Phenotypic Data

The laboratory study was conducted at the experimental station of the Chongqing Academy of Agricultural Sciences. The phenotypic data collected for the disease parameters were considered and analyzed as individual traits. The resistance traits were recorded for the F_2 population and parents. The traits means were calculated using DPS 18.10 (DPS, China).

2.5. DNA Extraction, SLAF Library Construction, and High-Throughput Sequencing

An improved CTAB method was utilized to extract genomic DNA from the young leaves of two parental lines and 150 F_2 individuals that were at the five- to six-leaf stage [45]. We employed an ND-1000 spectrophotometer (NanoDrop, Wilmington, DE, USA) and performed 1.0% agarose gel electrophoresis to respectively measure DNA concentration and quality. The SLAF-seq library was constructed as detailed previously by Sun et al. [33], with only a few small changes. The restriction enzyme *Hae*III (New England Biolabs, NEB, USA) was utilized for digestion of the genomic DNA of the parental lines and individuals of the F_2 population. We added polyA tails to the $3'$ ends of the digested fragments, which were then connected to duplex-labelled sequencing adapters and PCR amplified. PCR was performed with the diluted restriction-ligation DNA sample, Q5® High-Fidelity DNA polymerase (NEB), dNTPs, and PCR primers (forward, $5'$-AATGATACGGCGACCACCGA-$3'$ and reverse, $5'$-CAAGCAGAAGACGGCATACG-$3'$). The PCR products were purified with Agencourt AMPure XP beads (Beckman Coulter, High Wycombe, UK) and then resolved on a 2% agarose gel. Fragments that were 314 to 364 bp in size were separated and purified with a QIAquick gel extraction kit (Qiagen, Hilden, Germany). SLAF-seq was then conducted on an Illumina High-Seq 2500 sequencing platform (Illumina, San Diego, CA, USA) at Beijing Biomarker Technologies Corp. (Beijing, China, http://www.biomarker.com.cn, accessed on: 8 January 2019). We employed the *Oryza sativa* L. genome as reference for quality control and conducted library construction and sequencing using similar settings as that for the pepper mapping population.

2.6. SLAF-seq Data Grouping and Genotyping

In this study, reads with a quality score below Q30 (quality score < 30e) were filtered out. After that, high-quality reads were mapped to the pepper reference genome utilizing

BWA software, with the paired-end mapped reads at the identical position and >95% identity divided into a single SLAF locus. In each SLAF, a polymorphism locus was observed between the parents, of which most were SNPs. All of the polymorphism SLAF loci were then genotyped with consistency at SNP loci of the offspring and parents. SLAFs that consisted of more than eight SNPs were screened out, and then the parental SLAFs with a sequencing depth of <10-fold were discarded. A high-density linkage map was then created using polymorphic SLAFs showing parental homozygosis (aa × bb).

2.7. High-Density Linkage Map Construction

We quantified the modified logarithm of odds (MLOD) value between two adjacent markers and markers with MLOD values < 5 were filtered out. Then, the SLAF markers were assigned to chromosomes (Chr), and 12 Chr were obtained. Simultaneously, we analyzed the linear array of markers in every Chr using HighMap software [46] and then estimated the genetic distances between a pair of adjacent markers.

2.8. QTL Mapping of P. capsici Resistance and Candidate Gene Prediction

QTL analysis was identified by r/QTL software using CIM methods [47,48]. The LOD score thresholds for evaluating the statistical significance of the QTL effects were established using 1000 permutations ($p < 0.05$). The predicted genes within the target QTL interval were determined by comparison with the annotated Zunla-1 and CM334 reference genomes (http://peppersequence.genomics.cn, accessed on: 20 January 2019). The function of genes identified in the candidate regions was manually determined by BLASTX (https://blast.ncbi.nlm.nih.gov/, accessed on: 20 January 2019). In addition, the predicted genes were further annotated based on KEGG (https://www.kegg.jp/kegg/, accessed on: 20 January 2019), COG (http://www.ncbi.nlm.nih.gov/COG/, accessed on: 20 January 2019), Swiss-Prot (http://www.ebi.ac.uk/uniprot/ accessed on: 20 January 2019), and NR (https://blast.ncbi.nlm.nih.gov/ accessed on: 20 January 2019) databases.

2.9. qRT-PCR Analysis

For expression analysis, we conducted qRT-PCR to investigate the expression pattern of five disease-resistant or defense-related genes for *P. capsici* resistance in pepper. Leaf samples were gathered from days 0, 1, 2, 3, 4, 5, 6, and 7 post inoculation with *P. capsici* in the resistant line "PI201234" and the susceptible line "Early calwonder." "Early calwonder" was defined as susceptible to three physiological races of *P. capsici*. Total RNAs were extracted utilizing the Plant RNA Kit (Tiangen DP441, China) as per the company's instructions. Subsequently, cDNAs were reverse-transcribed using TaKaRa Reverse Transcription Kit (Takara Biomedical Technology (Beijing) Co., Ltd., Beijing, China). Quantitative PCR was conducted on a 7500 Real-Time PCR System (Thermo Fisher Scientific, Waltham, MA, USA) using TB Green® Premix Ex Taq™ Kit (TaKaRa). The PCR program was as follows: Holding Stage Step 1: 95 °C 30 s, followed by 40 cycles of Step 1: 95 °C for 5 s, Step 2: 60 °C for 30 s, and 72 °C for 2 min. After the last cycle, the amplification was extended for 7 min at 72 °C. *AY572427* was used as internal control for qRT-PCR analysis. We employed the $2^{-\Delta\Delta T}$ method to determine relative expression levels of candidate genes, which were normalized to that of actin gene (*AY572427*). Each target sample was analyzed using three biological replicates. All values were reported as the mean ± standard deviation (n = 3), and the statistical significance of any differences was analyzed using a Student's *t*-test.

3. Results

3.1. Sequencing and Genotyping Based on SLAF-seq

In this study, genotyping of 150 F_2 individuals and their parents was performed using the SLAF-seq technology. The sequencing data generated in this work were sent to the NCBI SRA database (http://www.ncbi.nlm.nih.gov/sra/ accessed on: 20 October 2020) as accession no. PRJNA669602. Approximately 76.22 GB of raw bases and 381.15 Mb of paired-end reads were generated, of which 94.37% achieved or exceeded quality score of 30

(Q30), and GC (guanine-cytosine) content was 38.86% (Table 1). *Oryza sativa* L. was used as control for evaluating the effectiveness of library construction. In addition, 12,250,440 reads representing 139,046 SLAFs with average depths of 63.83 were obtained from the male parent (PI201234), and 13,232,257 reads representing 141,584 SLAFs with average depths of 72.32 were obtained from the female parent (1287) (Table 1). In the offspring (F_2 population), 2,371,153 reads that were representing 124,582 SLAFs with average depths of 14.66 were generated (Table 1).

Table 1. Specific-locus amplified fragment sequencing (SLAF)-seq data statistics of the *Capsicum* F_2 population.

Samples	Total Read	Total Bases	Q30 Percentage (%)	GC Content (%)	SLAF Number	Total Depth	Average Depth(X)
PI201234	12,250,440	2,449,757,552	93.98	38.42	139,046	8,875,578	63.83
1287	13,232,257	2,646,335,364	94.71	38.15	141,584	10,239,208	72.32
Offspring	2,371,153	474,202,396	94.37	38.36	124,582	1,825,928	14.66
Total	381,155,587	76,226,452,308	94.37	38.86	405,212	/	/

After filtration of low-depth SLAF tags, approximately 174,193 high-quality SLAF markers were obtained, of which 19.77% (34,432) were polymorphic SLAFs (Table 2). In addition, 25,839 of the 34,432 polymorphic SLAFs were cultured into eight segregation patterns (aa×bb, ab×cc, ab×cd, cc×ab, ef×eg, hk×hk, lm×ll, and nn×np) (Figure 1). As the parents were homozygous (i.e., with genotype aa or bb), 21,069 SLAFs exhibited the aa×bb segregation pattern and were successfully selected for map construction.

Table 2. Description on basic characteristics of the 12 linkage groups.

Linkage Group	SLAF Number	Polymorphic
Chr01	16,109	3221
Chr02	9259	1626
Chr03	15,231	3159
Chr04	12,696	1569
Chr05	13,024	2986
Chr06	12,887	2640
Chr07	11,667	1907
Chr08	9426	1263
Chr09	14,507	3250
Chr10	11,356	1687
Chr11	11,890	3937
Chr12	13,089	2343
Other	23,052	4844
Total	174,193	34,432

3.2. Genetic Map Construction

After four-step filtering, our final map contained 5565 markers on 12 Chrs, which were designated Chr01-Chr12 using HighMap software and presented in Table 2 and Figure 2. The linkage map encompassed a total of 1535.69 cM and exhibited an average marker distance of 0.28 cM (Figure 2). The largest Chr was Chr03, which consisted of 444 markers, showed a length of 169.18 cM, and an average marker-to-marker distance of 0.38 cM, while the smallest Chr was Chr05 that consisted of 460 markers, showed a length of 99.98 cM, and an average marker-to-marker distance of 0.22 cM (Table 3). The extent of linkage between markers was represented by the percentage of "Gaps ≤ 5 cM," which ranged from 99.08% to 100%, and an average of 99.70% (Table 3). The largest gap on this linkage map was situated on Chr10 at 9.99 cM, whereas the smallest gap was 3.94 cM at Chr08 (Table 3).

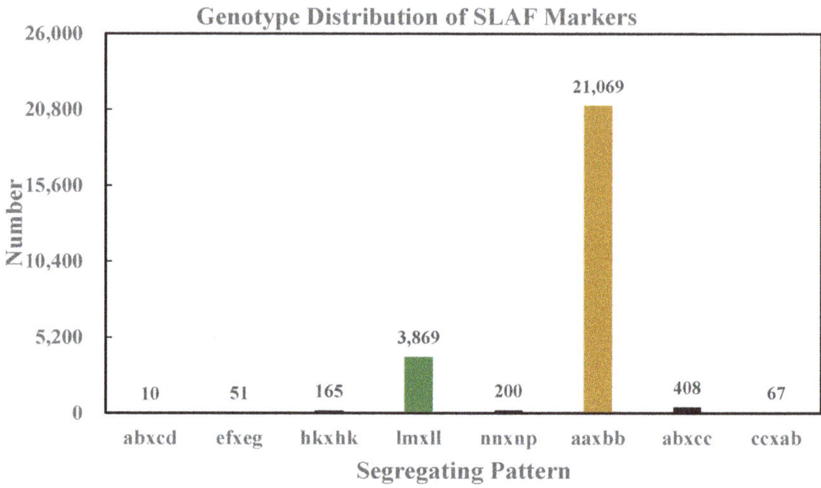

Figure 1. Specific-locus amplified fragment sequencing (SLAF) polymorphism analysis. Marker count in eight segregation patterns. The *x*-axis represents eight segregation patterns for the polymorphic SLAF markers, and the *y*-axis shows the number of markers.

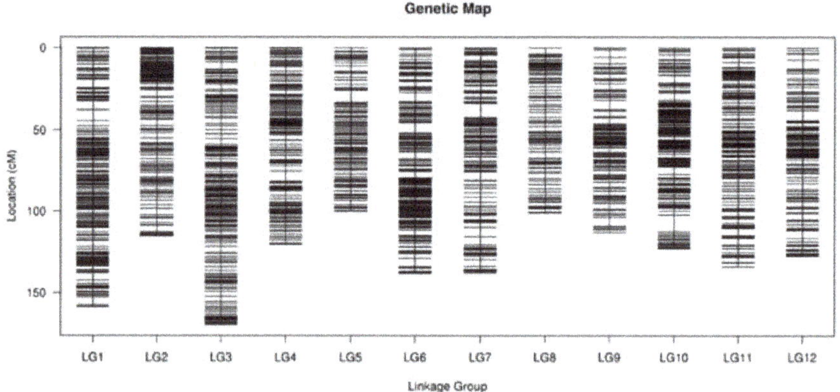

Figure 2. Distribution of SLAF markers across 12 pepper linkage groups. The black bar indicates a SLAF marker. The *x*-axis shows the linkage group number, whereas the *y*-axis represents genetic distance.

3.3. Quality and Accuracy of the Genetic Map

The quality and accuracy of the genetic map were assessed based on collinearity between the genetic and physical maps. The average integrity of each marker was 99.91% (Figure 3). Furthermore, among the 12 linkage groups, Chr03 showed the highest collinearity, with a correlation coefficient of 0.9979, and the average Spearman's rank correlation coefficient was 0.9758 (Table 3). On average, the coverage of these markers was 108.74-fold in PI201234 (male parent), 126.25-fold in 1287 (female parent), and 22.73-fold in every F_2 individual (Table 4), thereby indicating genotyping accuracy. Furthermore, collinearity with the physical map was utilized to examine the quality of the genetic map. Figure 4 shows that most of the genetically mapped loci were collinear with their physical positions on the reference genome sequence of *C. annuum* cv. Zunla-1 v2.0 [45]. Every correlation coefficient of 12 linkage groups was also assessed. The correlation coefficients of the 12 linkage groups were all close to 1, which indicated relatively high collinearity between linkage groups and the pepper reference genome (Figure 3).

Table 3. Basic information of the 12 linkage groups. The closer the Spearman's rank correlation coefficient is to 1, the better the collinearity.

Linkage Group	Marker Number	Average Distance between Markers (cM)	Size (cM)	Gaps ≤ 5	Max Gap (cM)	Correlation Coefficient
Chr01	437	0.36	158.32	99.08%	6.77	0.9968
Chr02	415	0.28	114.98	100.00%	4.57	0.9273
Chr03	444	0.38	169.18	100.00%	4.63	0.9979
Chr04	336	0.36	120.20	99.70%	6.38	0.9948
Chr05	460	0.22	99.98	99.78%	7.29	0.9054
Chr06	722	0.19	137.92	99.58%	6.31	0.9977
Chr07	517	0.27	137.66	99.22%	8.71	0.9712
Chr08	373	0.27	100.81	100.00%	3.94	0.9568
Chr09	414	0.27	112.88	99.52%	5.82	0.9875
Chr10	532	0.23	122.63	99.81%	9.99	0.9963
Chr11	458	0.29	133.80	100.00%	4.70	0.9974
Chr12	457	0.28	127.33	99.56%	6.38	0.9803
Maximum	722	0.38	169.18	100.00%	9.99	0.9979
Minimum	336	0.19	99.98	99.08%	3.94	0.9054
Total	5565	0.28	1535.69	/	/	/
Average	463.75	/	127.97	99.70%	/	0.9758

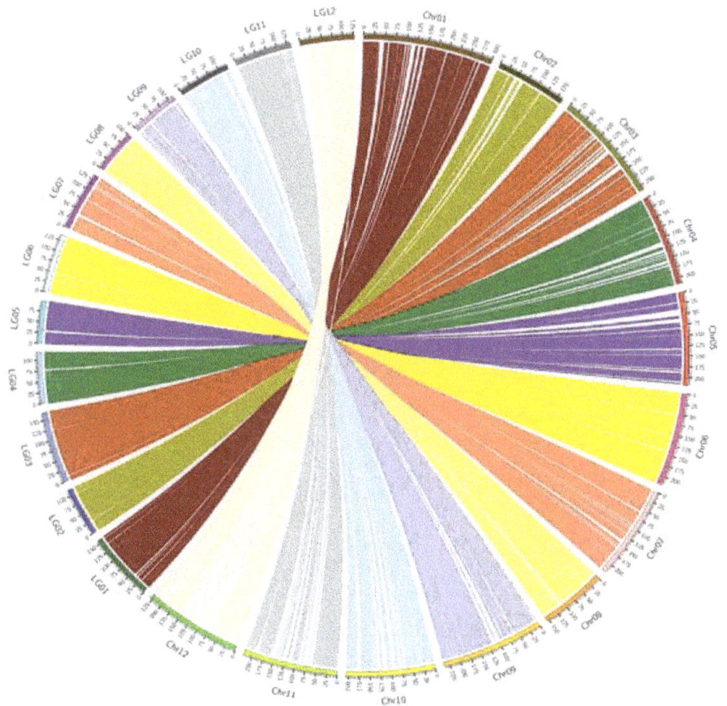

Figure 3. Collinearity between genetic and physical maps. The correlation between the pepper chromosomes (Chr) and the linkage group (LG) of the genetic map is illustrated.

Table 4. Details on the depth of mapped markers.

Samples	Marker Numbers	Total Depth(X)	Average Depth(X)
PI201234	5565	605,153	108.74
1287	5565	702,568	126.25
Offspring	5513	125,293	22.73

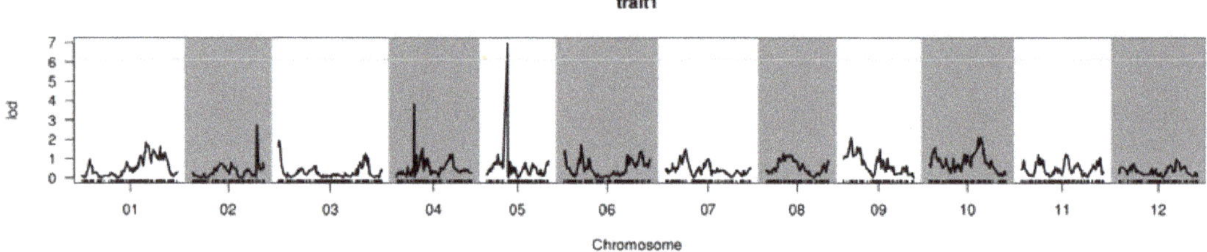
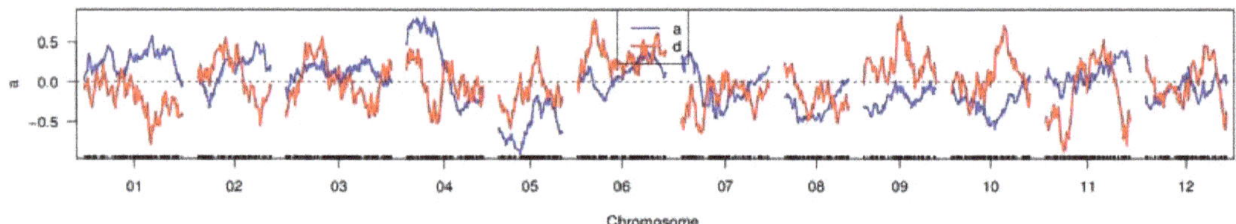

Figure 4. Quantitative trait locus (QTL) analysis of the *P. capsici* resistance trait of pepper. The *x*-axis indicates linkage group in pepper, and the *y*-axis presents LOD values. The blue line indicates the additive effect, and the red line represents the dominant effect.

3.4. Phenotypic Analysis of P. capsici Resistance

In 2008, the disease indices (DIs) of 150 F_2 populations were determined (Table 5). The highest DI value was recorded in the susceptible 1287 (84.3), while the lowest ID value was recorded in the resistance parent PI201234 (7.9). The DI values of the F_2 population varied between 0.00 and 100. The average DI value of the F_2 population was 47.2. The variation was 0.75, and skewness and kurtosis value of the DI in the F_2 population was small, indicating that the population was suitable for QTL identification.

Table 5. Descriptive statistics of disease index and the whole population of the parents.

Traits	F_2 Population								Parent		
	Min	Max	Range	Average	Standard Error	Var	Skew	Kurt	PI201234	1287	Midparent
Disease index (100%)	0	1	0–1	0.4772	0.029	0.75	−0.17	−1.43	7.9	84.3	46.1

3.5. QTL Mapping of P. capsici Resistance

In present study, the maximum LOD value of 6.972 was used as the threshold to determine the existence of QTL. Based on the high-density genetic map, a single major QTL for the *P. capsici* resistance trait was identified in the F_2 population and designated as *CQPc5.1* (Table 6, Figure 4), which explained 11.76% of the observed phenotypic variance. *CQPc5.1* was localized in 17.9–19.4 Mb on Chr05, which encompassed a genetic distance of about 0.35 cM, as well as a physical distance of about 1.47 Mb on Chr05 (Table 6).

Table 6. Quantitative trait locus (QTL) analysis of the *P. capsici* resistance trait in F$_2$ populations.

QTL	LOD Threshold	Chr ID	Physical Distance Interval (bp)	Genetic Distance Interval (cM)	Max LOD	ADD	DOM	PVE (%)
CQPc5.1	6.125	05	17,967,630–19,446,349	33.103–33.448	6.972	−0.897	0.079	11.758

Note: LOD, logarithm of odds. Maximum LOD score (QTL peak). ADD, additive effects. DOM, dominance effects. PVE, phenotypic variance explained.

3.6. Candidate Gene Prediction and qRT-PCR Analysis

According to the annotations of the *C. annuum* cv. Zunla-1 v2.0 genome, 23 predicted candidate genes were determined in the physical interval of *CQPc5.1* on Chr05 (Table 7). Among these, nine candidate genes were identified in the COG database, including 11 genes with KEGG annotations and 12 genes with Swiss-Prot annotations. Furthermore, 5 of the 23 genes were related to disease resistance or defense, and thus might be involved in *P. capsici* resistance in pepper; the *Capana05g000595* gene was annotated as disease resistance protein, RPP8-like; two genes (*Capana05g000596* and *Capana05g000597*) were annotated as disease resistance protein, RPP13-like; *Capana05g000598* was annotated as likely LRR receptor-like serine/threonine-protein kinase; and *Capana05g000604* was annotated as an F-box/LRR-repeat protein. These five genes were then analyzed by qRT-PCR. The primer sequences are listed in Table 8. The results showed that three genes (i.e., *Capana05g000595*, *Capana05g000596*, and *Capana05g000597*) were up-regulated in "PI201234," and expression levels peaked 2–3 days after pathogen inoculation (Figure 5). Five genes were up-regulated in "Early calwonder" after pathogen inoculation, and expression levels peaked at 5 days. In "Early calwonder," the expression of *Capana05g000604* gradually increased over time; however, it was expressed at a markedly lower level in "PI201234."

Table 7. Details of the annotated candidate genes.

Gene	Start	Stop	COG	KEGG	Swiss-Prot	Nr
Capana05g000592	18,024,840	18,035,001	–	K17550 (protein phosphatase 1 regulatory subunit 7)	Protein phosphatase 1 regulatory inhibitor subunit PPP1R7 homolog	PREDICTED: protein phosphatase 1 regulatory subunit pprA-like
Capana05g000594	18,317,198	18,327,621	–	K13099 (CD2 antigen cytoplasmic tail-binding protein 2)	–	PREDICTED: CD2 antigen cytoplasmic tail-binding protein 2
Capana05g000595	18,357,315	18,357,857	–	–	Disease resistance RPP8-like protein	Hypothetical protein T459_14155
Capana05g000596	18,358,568	18,359,167	General function prediction only	–	Putative disease resistance RPP13-like protein	Hypothetical protein T459_14156
Capana05g000597	18,359,457	18,359,987	–	–	Disease resistance protein RPP13	Hypothetical protein BC332_12877
Capana05g000598	18,387,075	18,390,244	Transcription	–	Probable LRR receptor-like serine/threonine-protein kinase At3g47570	PREDICTED: probable LRR receptor-like serine/threonine-protein kinase At3g47570
Capana05g000599	18,390,817	18,396,058	Carbohydrate transport and metabolism	K05298 (glyceraldehyde-3-phosphate dehydrogenase (NADP+))	Glyceraldehyde-3-phosphate dehydrogenase A, chloroplastic (Fragment)	PREDICTED: glyceraldehyde-3-phosphate dehydrogenase A, chloroplastic
Capana05g000600	18,401,537	18,405,948	General function prediction only	–	Protein high chlorophyll fluorescent 107	PREDICTED: protein high chlorophyll fluorescent 107
Capana05g000601	18,410,078	18,412,165	Post-translational modification, protein turnover, chaperones	K00587 (protein-S-isoprenylcysteine O-methyltransferase)	Protein-S-isoprenylcysteine O-methyltransferase B	PREDICTED: protein-S-isoprenylcysteine O-methyltransferase A-like isoform X1
Capana05g000602	18,413,946	18,415,745	–	–	Pentatricopeptide repeat-containing protein At2g13600 OS = Arabidopsis thaliana	PREDICTED: pentatricopeptide repeat-containing protein At2g13600
Capana05g000603	18,424,439	18,427,748	Coenzyme transport and metabolism	–	FAD synthetase 1, chloroplastic	PREDICTED: FAD synthetase 1, chloroplastic-like
Capana05g000604	18,449,805	18,454,056	Transcription	K10268 (F-box and leucine-rich repeat protein)	F-box/LRR-repeat protein 4	PREDICTED: F-box/LRR-repeat protein 20

Table 7. *Cont.*

Gene	Start	Stop	COG	KEGG	Swiss-Prot	Nr
Capana05g000605	18,709,353	18,710,879	–	–	UPF0481 protein At3g47200	PREDICTED: putative UPF0481 protein At3g02645 isoform X1
Capana05g000607	18,749,830	18,750,435	–	–	–	PREDICTED: uncharacterized protein LOC107870380 isoform X1
Capana05g000608	18,775,198	18,776,266	Cell wall/membrane/envelope biogenesis	K18819 (inositol 3-alpha-galactosyltransferase)	Galactinol synthase 2	Galactinol synthase 2
Capana05g000609	18,823,996	18,827,956	–	–	–	Hypothetical protein CQW23_12126
Capana05g000611	18,835,570	18,835,953	–	K13495 (cis-zeatin O-glucosyltransferase)	Zeatin O-xylosyltransferase	PREDICTED: zeatin O-glucosyltransferase
Capana05g000612	18,851,478	18,852,197	–	K13495 (cis-zeatin O-glucosyltransferase)	Zeatin O-glucosyltransferase	PREDICTED: zeatin O-xylosyltransferase-like
Capana05g000613	18,857,046	18,857,423	–	K13495 (cis-zeatin O-glucosyltransferase)	Zeatin O-glucosyltransferase	Hypothetical protein T459_14175
Capana05g000614	18,879,064	18,879,435	–	–	Putative cis-zeatin O-glucosyltransferase	PREDICTED: zeatin O-xylosyltransferase-like
Capana05g000615	18,879,694	18,880,182	–	K13495 (cis-zeatin O-glucosyltransferase)	Zeatin O-xylosyltransferase	Hypothetical protein T459_14174
Capana05g000617	19,193,439	19,194,854	–	K13495 (cis-zeatin O-glucosyltransferase)	Zeatin O-glucosyltransferase	PREDICTED: zeatin O-glucosyltransferase-like
Capana05g000618	19,354,157	19,355,500	Transcription	–	Receptor-like protein Cf-9	Hypothetical protein T459_14173
Total			9	11	20	23

Note: COG, Clusters of Orthologous Groups. KEGG, Kyoto Encyclopedia of Genes and Genom.

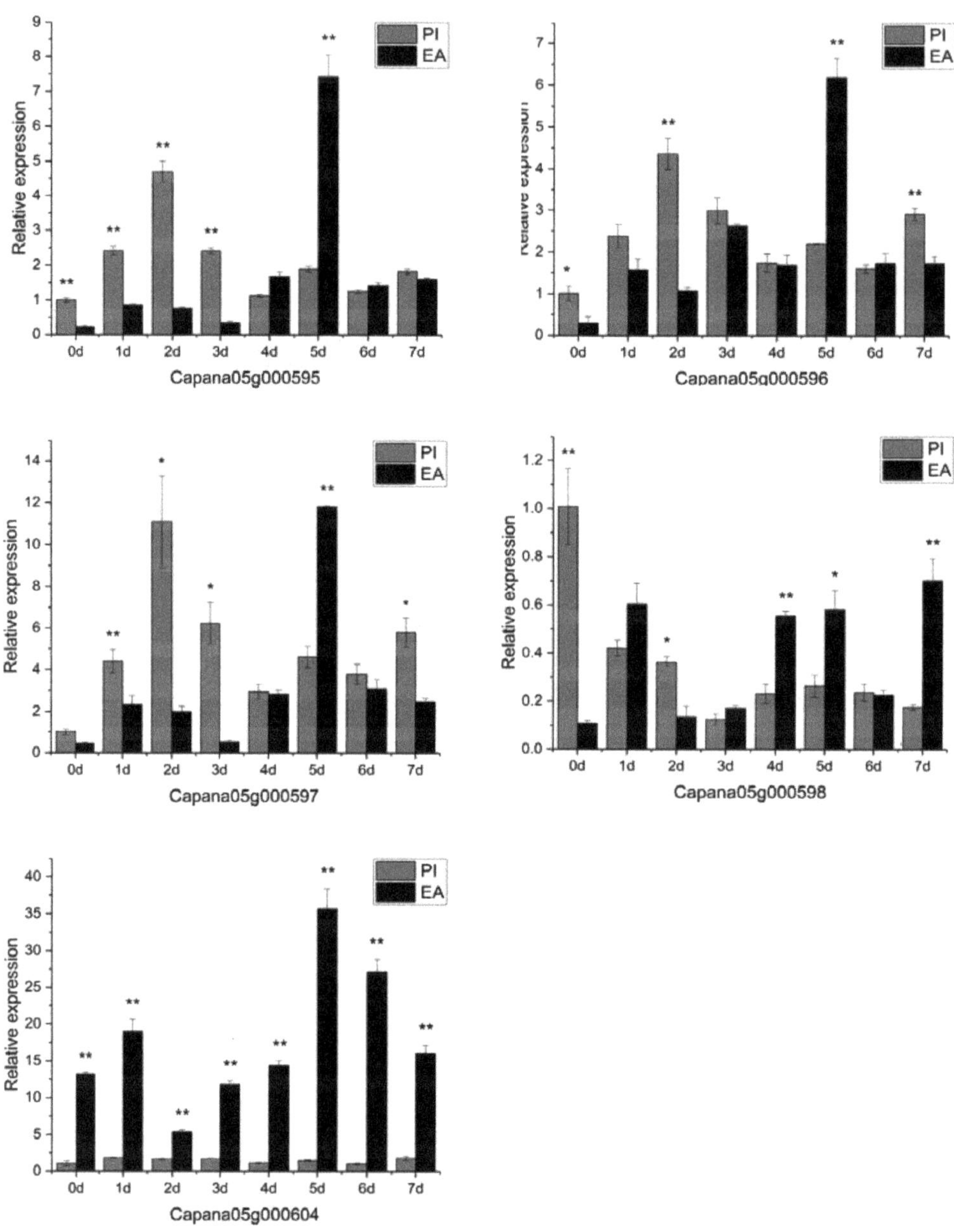

Figure 5. qPCR analysis of five genes in PI201234 and Early calwonder that are related to disease resistance or defense. PI: PI201234 inoculated using *P. capsici* zoospore suspension; EA: Early calwonder inoculated using *P. capsici* zoospore suspension. The *x*-axis shows the time points of sample collection, d: days post-inoculation. The *y*-axis shows the relative expression quantity of genes. Gene expression was normalized to that of actin, and the data were expressed as the mean ± standard deviation of two biological replicates and three technical replicates. A Student's *t*-test was used to analyze statistical significance of differences. * 0.05 level of significance; ** 0.01 level of significance.

Table 8. Information on genes employed in qRT-PCR analysis.

Gene Name	Primer Sequence (5′-3′)	PCR Product Size (bp)	TM (°C)
Capana05g000595	F:AAGGAGGCATTTAGCCGCAA R:TGTCTCAAGGCGAGCAACAT	115	59.0
Capana05g000596	F:CTGCAAGAAAGCGTGTCAGG R:AGCCTCCACATCTTTCCACC	98	59.0
Capana05g000597	F:CAATCCCTCAAGCGACGAGT R:CCAGGTCGGACCGATTGTTA	121	55.0
Capana05g000598	F:ACCTTCCGTGGTGAAATCCC R:CGATCCGCGTAACAGGTTTG	190	55.0
Capana05g000604	F:TTAGCTGTTGCTGAGGGGTG R:GCTTGCGTCCAGAGAGACAAA	163	59.0
Actin (AY572427)	F:AGCAACTGGGACGATATGGAGAAG R:AAGAGACAACACCGCCTGAATAGC	198	50.0

4. Discussion

4.1. Genetic Map Constructed of Pepper

Phytophthora blight caused by *P. capsici* is one of the most serious diseases in pepper, inducing a significant reduction in yield and quality [49]. Despite decades of genetic research on the resistance of pepper to *P. capsici*, no resistant cultivars have been established to date. At present, as a key tool, genetic linkage maps are not only used in plant genetics, but also to identify genomic regions that are related to agronomic and qualitative traits through QTL mapping. Recently, SLAF-seq has been utilized in the creation of genetic linkage maps of pepper, and a number of high-density genetic maps have been successfully created. For instance, Zhu et al. (2019) identified six QTLs using a molecular genetic linkage map via SLAF-seq in relation to flowering time and number of flowers per node in pepper, which consisted of a total of 9038 markers at an average spacing of 0.18 cm that were distributed across 12 linkage groups, and the total distance was 1586.78 cM [44]. In the same year, Zhang et al. (2019) identified two major pepper QTLs (*Ffn2.1* and *Ffn2.2*) that were strongly correlated with FFN using a high-density genetic map, which included 9328 SLAF markers from 12 linkage groups, showing a total genetic distance of 2009.69 cM, as well as an average distance of 0.22 cM [40]. Sun et al. (2020) reported two QTLs that were related to aphid survival (*Rmpas-1*) and reproduction (*Rmprp-1*) using a genetic linkage map that included 167 SNP markers [50]. In this work, we constructed a genetic map using the SLAF-seq technology and according to a F_2 population. The map consisted of 5565 markers that assigned 12 linkage groups, spanning a total length of 1535.69 cM, and showed a mean genetic distance of 0.28 cM. This genetic map exhibited adequate coverage of the polymorphic markers in regions of interest, and the mapped QTLs showed positional accuracy.

4.2. Identification QTL with the Resistance to P. capsici Traits

Previous studies have showed that the major QTLs related to resistance to *P. capsici* are situated on Chr05, despite the use of various resistant lines, pepper populations, or *P. capsici* isolates [23,26,27,32,51]. Mallard et al. (2013) utilized published pepper genome information and identified three major QTLs, namely, *Pc5.1*, *Pc5.2*, and *Pc5.3*, which were localized to the 22.4–24.6, 53.0–162.6, and 9.7–13.3 Mb regions on Chr05, respectively [31]. Siddique et al. (2019) reported three major QTLs on Chr05, namely, *QTL5.1* (18.7–19.5 Mb), *QTL5.2* (27.3–29.2 Mb), and *QTL5.3* (34.6–37 Mb) that were related to resistance to three *P. capsici* isolates on using combined traditional QTL mapping with GWAS [30]. Here, we performed *P. capsici* resistance QTL analysis of pepper. We detected a major QTL *CQPc5.1* based on a high-density linkage map of F_2 plants. *CQPc5.1* was localized to the

17.9–19.4 Mb region on Chr05, with a genetic distance of 33.103–33.448 cM. In an earlier study, Collard et al. (2005) documented that a QTL is only described as "major" when it accounts for >10% of the PVE [52]. In the present study, the phenotypic variance of *CQPc5.1* was 11.58%. In addition, the position of *CQPc5.1* on Chr05 differs from *Pc5.1*, *Pc5.2*, and *Pc5.3*, yet the location of *CQPc5.1* that was identified in this work coincides with that of the earlier determined locus *QTL5.1* [30]. However, the physical location of *CQPc5.1* on the chromosome is closer than that of *QTL5.1*, so we infer that *CQPc5.1* represents a more accurate mapping of resistance to *P. capsici* in pepper.

4.3. Candidate Gene Prediction

Here, we identified five genes that are related to disease resistance in the *CQPc5.1* QTL region. We identified three genes annotated as disease-resistance protein RPP13-like; *RPP13* was a singleton NBS-LRR gene located in *CQPc5.1* on Chr05. *Capana05g000595* gene was identified as disease resistance protein RPP-8. Two genes (*Capana05g000596* and *Capana05g000597*) were annotated to be disease resistance protein RPP13-like. *RPP13* is a CC (coiled-coil)-NBS-LRR domain-containing *R* gene that controls resistance to *Peronospora parasitica* oomycete pathogen in *Arabidopsis thaliana* [53,54]. These two candidate genes encode *RPP13*-like NBS-LRR proteins and serve as potential candidates for *P. capsici* resistance in pepper. *Capana05g000598* was annotated as a probable LRR receptor-like serine/threonine-protein kinase. *Capana05g000604* was annotated to be an F-box/LRR-repeat protein. Several LRR domain proteins have been determined to participate in defense responses to infiltrating pathogens [12,55–57].

4.4. Candidate Gene qRT-PCR Analysis

Our qRT-PCR outcomes indicate that the expression patterns of three genes (*Capana05g000595*, *Capana05g000596*, and *Capana05g000597*) are up-regulated in both the resistant "PI201234" and susceptible "Early calwonder" lines after pathogen inoculation. *Capana05g000598* was down-regulated in "PI201234" with pathogen inoculation and up-regulated in "Early calwonder." We infer that *Capana05g000598* may have the part of the negative regulator of resistance to *P. capsici* in PI201234. *Capana05g000604* was up-regulated in "Early calwonder" at post-infection, while its expression level was significantly lower throughout in "PI201234". Interestingly, the expression of five genes in "PI201234" peaked 2–3 days after pathogen infection, in contrast, expression in "Early calwonder peaked at 5 day after pathogen infection." Therefore, we deduced that *Capana05g000595*, *Capana05g000596*, and *Capana05g000597* might be related to resistance to *P. capsici*. These three genes are highly associated with *CQPc5.1*, but functional validation has not been reported. Therefore, it is essential to conduct functional analysis of these genes to verify their molecular functions in *P. capsici* resistance in pepper. The result of this study would provide information for the next stage of research such as gene functional analysis, pyramiding breeding, and marker-assisted selection (MAS) as well.

Author Contributions: Conceptualization, Y.-F.L.; methodology, Y.-F.L., S.-C.Z., and C.-P.W.; investigation, Y.-F.L., S.-C.Z., and X.-M.Y.; writing—original draft preparation, Y.-F.L.; writing—review and editing, S.-C.Z., Q.-Z.H., and R.-Z.H.; funding acquisition, R.-Z.H. All authors have read and agreed to the published version of the manuscript.

Funding: This research was financially supported by the National Key R&D Program of China (2017YFD0101903).

Institutional Review Board Statement: Not applicable for studies not involving humans or animals.

Informed Consent Statement: Not applicable.

Data Availability Statement: The sequencing data accession number (PRJNA669602).

Acknowledgments: We acknowledge Yu Pan for her efforts in revising the manuscript.

Conflicts of Interest: The authors declare no conflict of interest.

References

1. Chaim, A.B.; Grube, R.C.; Lapidot, M.; Jahn, M.; Paran, I. Identification of quantitative trait loci associated with resistance to cucumber mosaic virus in *Capsicum annuum*. *Theor. Appl. Genet.* **2001**, *102*, 1213–1220. [CrossRef]
2. Cai, W.-Q.; Fang, R.-X.; Shang, H.-S.; Wang, X.; Zhang, F.-L.; Li, Y.-R.; Zhang, J.-C.; Cheng, X.-Y.; Wang, G.-L.; Mang, K.-Q. Development of CMV-and TMV-resistant chili pepper: Field performance and biosafety assessment. *Mol. Breed.* **2003**, *11*, 25–35. [CrossRef]
3. Hong, J.K.; Yang, H.J.; Jung, H.; Yoon, D.J.; Sang, M.K.; Jeun, Y.-C. Application of Volatile Antifungal Plant Essential Oils for Controlling Pepper Fruit Anthracnose by *Colletotrichum gloeosporioides*. *Plant Pathol. J.* **2015**, *31*, 269–277. [CrossRef] [PubMed]
4. Bosland, P.W. A Seedling Screen for Phytophthora Root Rot of Pepper, *Capsicum annuum*. *Plant Dis.* **1991**, *75*, 1048. [CrossRef]
5. Parra, G.; Ristaino, J.B. Resistance to Mefenoxam and Metalaxyl Among Field Isolates of *Phytophthora capsici* Causing Phytophthora Blight of Bell Pepper. *Plant Dis.* **2001**, *85*, 1069–1075. [CrossRef]
6. Oelke, L.M.; Bosland, P.W.; Steiner, R. Differentiation of Race Specific Resistance to Phytophthora Root Rot and Foliar Blight in *Capsicum annuum*. *J. Am. Soc. Hortic. Sci.* **2003**, *128*, 213–218. [CrossRef]
7. Lefebvre, V.; Palloix, A. Both epistatic and additive effects of QTLs are involved in polygenic induced resistance to disease: A case study, the interaction pepper-*Phytophthora capsici* Leonian. *Theor. Appl. Genet.* **1996**, *93*, 503–511. [CrossRef]
8. Foster, J.M.; Hausbeck, M.K. Resistance of Pepper to Phytophthora Crown, Root, and Fruit Rot Is Affected by Isolate Virulence. *Plant Dis.* **2010**, *94*, 24–30. [CrossRef] [PubMed]
9. Barksdale, T.H.; Papavizas, G.C.; Johnston, S.A. Resistance to foliar blight and crown rot of pepper caused by *Phytophthora capsici*. *Plant Dis.* **1984**, *68*, 506–509. [CrossRef]
10. Flett, S.; Ashcroft, W.; Jerie, P.; Taylor, P. Control of Phytophthora root rot in processing tomatoes by metalaxyl and fosetyl-Al. *Aust. J. Exp. Agric.* **1991**, *31*, 279–283. [CrossRef]
11. Polizzi, G.; Agosteo, G.E.; Cartia, G. Soil solarization for the control of *Phytophthora capsici* on pepper. *Acta Hortic.* **1994**, *366*, 331–338. [CrossRef]
12. Lehmann, P. Structure and evolution of plant disease resistance genes. *J. Appl. Genet.* **2002**, *43*, 403–414.
13. Stieg, J.R.; Walters, S.A.; Bond, J.P.; Babadoost, M. Effects of fungicides and cultivar resistance for *Phytophthora capsici* control in bell pepper production. *HortScience* **2006**, *41*, 1076. [CrossRef]
14. Black, L. Studies on Phytophthora blight in pepper. In *Talekar NS (ed) AVRDC Report 1998*; Asian Vegetable Research and Development Center: Shanhua, Taiwan, 1999; pp. 25–27.
15. Ortega, R.G.; Espanol, C.P.; Zueco, J.C. Genetics of Resistance to *Phytophthora capsici* in the Pepper Line 'SCM-334'. *Plant Breed.* **1991**, *107*, 50–55. [CrossRef]
16. Kim, B.S. Characteristics of bacterial spot resistant lines and Phytophthora Blight resistant lines of *Capsicum pepper*. *Hortic. Environ. Biotechnol.* **1988**, *29*, 247–252.
17. Smith, P.G.; Kimble, K.A.; Grogan, R.G.; Millett, A.H. Inheritance of resistance in peppers to Phytophthora root rot. *Phytopathology* **1967**, *57*, 377–379.
18. Sugita, T.; Yamaguchi, K.; Kinoshita, T.; Yuji, K.; Sugimura, Y.; Nagata, R.; Kawasaki, S.; Todoroki, A. QTL analysis for resistance to Phytophthora Blight (*Phytophthora capsici* Leon.) using an intraspecific Doubled-Haploid population of *Capsicum annuum*. *Breed. Sci.* **2006**, *56*, 137–145. [CrossRef]
19. Kim, B.S.; Kwon, Y.S. Inheritance of resistance to Phytophthora Blight and to bacterial spot in pepper. *J. Korean Soc. Hortic. Sci.* **1990**, *7*, 17–24.
20. Saini, S.S.; Sharma, P.P. Inheritance of resistance to fruit rot (*Phytophthora capsici* Leon.) and induction of resistance in bell pepper (*Capsicum annuum* L.). *Euphytica* **1978**, *27*, 721–723. [CrossRef]
21. Wang, P.; Wang, L.; Guo, J.; Yang, W.; Shen, H. Molecular mapping of a gene conferring resistance to *Phytophthora capsici* Leonian race 2 in pepper line PI201234 (*Capsicum annuum* L.). *Mol. Breed.* **2016**, *36*, 1–11. [CrossRef]
22. Reifschneider, F.J.B.; Boiteux, L.S.; Vecchia, P.T.D.; Poulos, J.M.; Kuroda, N. Inheritance of adult-plant resistance to *Phytophthora capsici* in pepper. *Euphytica* **1992**, *62*, 45–49. [CrossRef]
23. Thabuis, A.; Palloix, A.; Pflieger, S.; Daubèze, A.-M.; Caranta, C.; Lefebvre, V. Comparative mapping of Phytophthora resistance loci in pepper germplasm: Evidence for conserved resistance loci across Solanaceae and for a large genetic diversity. *Theor. Appl. Genet.* **2003**, *106*, 1473–1485. [CrossRef] [PubMed]
24. Pflieger, S.; Palloix, A.; Caranta, C.; Blattes, A.; Lefebvre, V. Defense response genes co-localize with quantitative disease resistance loci in pepper. *Theor. Appl. Genet.* **2001**, *103*, 920–929. [CrossRef]
25. Minamiyama, Y.; Tsuro, M.; Kubo, T.; Hirai, M. QTL Analysis for Resistance to *Phytophthora capsici* in Pepper Using a High Density SSR-based Map. *Breed. Sci.* **2007**, *57*, 129–134. [CrossRef]
26. Kim, H.-J.; Nahm, S.-H.; Lee, H.-R.; Yoon, G.-B.; Kim, K.-T.; Kang, B.-C.; Choi, D.; Kweon, O.Y.; Cho, M.-C.; Kwon, J.-K.; et al. BAC-derived markers converted from RFLP linked to *Phytophthora capsici* resistance in pepper (*Capsicum annuum* L.). *Theor. Appl. Genet.* **2008**, *118*, 15–27. [CrossRef] [PubMed]
27. Truong, H.T.H.; Kim, K.T.; Kim, D.W.; Kim, S.; Chae, Y.; Park, J.H.; Oh, D.G.; Cho, M.C. Identification of isolate-specific resistance QTLs to phytophthora root rot using an intraspecific recombinant inbred line population of pepper (*Capsicum annuum*). *Plant Pathol.* **2011**, *61*, 48–56. [CrossRef]

28. Rehrig, W.Z.; Ashrafi, H.; Hill, T.; Prince, J.; Deynze, A.V. CaDMR1 Co-segregates with QTL Pc5.1 for resistance to *Phytophthora capsici* in pepper (*Capsicum annuum*). *Plant Genome* **2014**, *7*, 1–12. [CrossRef]
29. Kim, N.; Kang, W.H.; Lee, J.; Yeom, S.I. Development of clustered resistance gene analogs-based markers of resistance to *Phytophthora capsici* in chili pepper. *BioMed Res. Int.* **2019**, *2019*, 1–12.
30. Siddique, M.I.; Lee, H.Y.; Ro, N.Y.; Han, K.; Venkatesh, J.; Solomon, A.A.-O.; Patil, A.A.-O.; Changkwian, A.; Kwon, J.K.; Kang, B.C. Identifying candidate genes for *Phytophthora capsici* resistance in pepper (*Capsicum annuum*) via genotyping-by-sequencing-based QTL mapping and genome-wide association study. *Sci. Rep.* **2019**, *9*, 9962. [CrossRef]
31. Mallard, S.; Cantet, M.; Massire, A.; Bachellez, A.; Ewert, S.; Lefebvre, V. A key QTL cluster is conserved among accessions and exhibits broad-spectrum resistance to *Phytophthora capsici*: A valuable locus for pepper breeding. *Mol. Breed.* **2013**, *32*, 349–364. [CrossRef]
32. Bonnet, J.; Danan, S.; Boudet, C.; Barchi, L.; Sage-Palloix, A.-M.; Caromel, B.; Palloix, A.; Lefebvre, V. Are the polygenic architectures of resistance to *Phytophthora capsici* and *P. parasitica* independent in pepper? *Theor. Appl. Genet.* **2007**, *115*, 253–264. [CrossRef]
33. Sun, X.; Liu, D.; Zhang, X.; Li, W.; Liu, H.; Hong, W.; Jiang, C.; Guan, N.; Ma, C.; Zeng, H.; et al. SLAF-seq: An Efficient Method of Large-Scale De Novo SNP Discovery and Genotyping Using High-Throughput Sequencing. *PLoS ONE* **2013**, *8*, e58700. [CrossRef] [PubMed]
34. Li, B.; Tian, L.; Zhang, J.; Huang, L.; Han, F.; Yan, S.; Wang, L.; Zheng, H.; Sun, J. Construction of a high-density genetic map based on large-scale markers developed by specific length amplified fragment sequencing (SLAF-seq) and its application to QTL analysis for isoflavone content in Glycine max. *BMC Genom.* **2014**, *15*, 1086. [CrossRef]
35. Zhang, Z.; Shang, H.; Shi, Y.; Huang, L.; Li, J.; Ge, Q.; Gong, J.; Liu, A.; Chen, T.; Wang, D. Construction of a high-density genetic map by specific locus amplified fragment sequencing (SLAF-seq) and its application to Quantitative Trait Loci (QTL) analysis for boll weight in upland cotton (*Gossypium hirsutum*). *BMC Plant Biol.* **2016**, *16*, 79. [CrossRef] [PubMed]
36. Qi, Z.; Huang, L.; Zhu, R.; Xin, D.; Liu, C.; Han, X.; Jiang, H.; Hong, W.; Hu, G.; Zheng, H.; et al. A High-Density Genetic Map for Soybean Based on Specific Length Amplified Fragment Sequencing. *PLoS ONE* **2014**, *9*, e104871. [CrossRef] [PubMed]
37. Zhang, Y.; Wang, L.; Xin, H.; Li, D.; Ma, C.; Ding, X.; Hong, W.; Zhang, X. Construction of a high-density genetic map for sesame based on large scale marker development by specific length amplified fragment (SLAF) sequencing. *BMC Plant Biol.* **2013**, *13*, 141. [CrossRef] [PubMed]
38. Xu, X.; Lu, L.; Zhu, B.; Xu, Q.; Qi, X.; Chen, X. QTL mapping of cucumber fruit flesh thickness by SLAF-seq. *Sci. Rep.* **2015**, *5*, 15829. [CrossRef] [PubMed]
39. Zhu, Y.; Yin, Y.; Yang, K.; Li, J.; Sang, Y.; Huang, L.; Fan, S. Construction of a high-density genetic map using specific length amplified fragment markers and identification of a quantitative trait locus for anthracnose resistance in walnut (*Juglans regia* L.). *BMC Genom.* **2015**, *16*, 1–13. [CrossRef] [PubMed]
40. Zhang, X.-F.; Wang, G.-Y.; Dong, T.-T.; Chen, B.; Du, H.-S.; Li, C.-B.; Zhang, F.-L.; Zhang, H.-Y.; Xu, Y.; Wang, Q.; et al. High-density genetic map construction and QTL mapping of first flower node in pepper (*Capsicum annuum* L.). *BMC Plant Biol.* **2019**, *19*, 167. [CrossRef] [PubMed]
41. Hu, X.H.; Zhang, S.Z.; Miao, H.R.; Cui, F.G.; Shen, Y.; Yang, W.Q.; Xu, T.T.; Chen, N.; Chi, X.Y.; Zhang, Z.M.; et al. High-Density Genetic Map Construction and Identification of QTLs Controlling Oleic and Linoleic Acid in Peanut using SLAF-seq and SSRs. *Sci. Rep.* **2018**, *8*, 5479. [CrossRef]
42. Wang, L.; Yang, X.; Cui, S.; Zhao, N.; Li, L.; Hou, M.; Mu, G.; Liu, L.; Li, Z. High-density genetic map development and QTL mapping for concentration degree of floret flowering date in cultivated peanut (*Arachis hypogaea* L.). *Mol. Breed.* **2020**, *40*, 1–14. [CrossRef]
43. Guo, G.; Wang, S.; Liu, J.; Pan, B.; Diao, W.; Ge, W.; Gao, C.; Snyder, J.C. Rapid identification of QTLs underlying resistance to Cucumber mosaic virus in pepper (*Capsicum frutescens*). *Theor. Appl. Genet.* **2016**, *130*, 41–52. [CrossRef]
44. Zhu, Z.; Sun, B.; Wei, J.; Cai, W.; Huang, Z.; Chen, C.; Cao, B.; Chen, G.; Lei, J. Construction of a high density genetic map of an interspecific cross of *Capsicum chinense* and *Capsicum annuum* and QTL analysis of floral traits. *Sci. Rep.* **2019**, *9*, 1–14. [CrossRef]
45. Murray, M.G.; Thompson, C.L.; Wendel, J.F. Rapid isolation of high molecular weight plant DNA. *Nucleic Acids Res.* **1980**, *8*, 4321–4325. [CrossRef] [PubMed]
46. Li, R.; Li, Y.; Kristiansen, K.; Wang, J. SOAP: Short oligonucleotide alignment program. *Bioinformatics* **2008**, *24*, 713–714. [CrossRef] [PubMed]
47. Peichel, C.L.; Nereng, K.S.; Ohgi, K.A.; Cole, B.L.E.; Colosimo, P.F.; Buerkle, C.A.; Schluter, D.; Kingsley, D.M. The genetic architecture of divergence between threespine stickleback species. *Nat. Cell Biol.* **2001**, *414*, 901–905. [CrossRef] [PubMed]
48. Wang, W.; Huang, S.; Liu, Y.; Fang, Z.; Yang, L.; Hua, W.; Yuan, S.; Liu, S.; Sun, J.; Zhuang, M.; et al. Construction and analysis of a high-density genetic linkage map in cabbage (*Brassica oleracea* L. var. capitata). *BMC Genom.* **2012**, *13*, 523. [CrossRef]
49. Burdon, J.J.; Thrall, P.H. Coevolution of Plants and Their Pathogens in Natural Habitats. *Science* **2009**, *324*, 755–756. [CrossRef] [PubMed]
50. Sun, M.; Voorrips, R.E.; Westende, W.V.; van Kaauwen, M.; Visser, R.G.F.; Vosman, B. Aphid resistance in Capsicum maps to a locus containing LRR-RLK gene analogues. *Theor. Appl. Genet.* **2020**, *133*, 227–237. [CrossRef]

51. Ogundiwin, E.A.; Berke, T.F.; Massoudi, M.; Black, L.L.; Huestis, G.; Choi, D.; Lee, S.; Prince, J.P. Construction of 2 intra-specific linkage maps and identification of resistance QTLs for *Phytophthora capsici* root-rot and foliar-blight diseases of pepper (*Capsicum annuum* L.). *Genome* **2005**, *48*, 698–711. [CrossRef]
52. Collard, B.C.Y.; Jahufer, M.Z.Z.; Brouwer, J.B.; Pang, E.C.K. An introduction to markers, quantitative trait loci (QTL) mapping and marker-assisted selection for crop improvement: The basic concepts. *Euphytica* **2005**, *142*, 169–196. [CrossRef]
53. Rose, L.E.; Bittner-Eddy, P.D.; Langley, C.H.; Holub, E.B.; Michelmore, R.W.; Beynon, J.L. The Maintenance of Extreme Amino Acid Diversity at the Disease Resistance Gene, RPP13, in *Arabidopsis thaliana*. *Genetics* **2004**, *166*, 1517–1527. [CrossRef] [PubMed]
54. Serra, H.; Choi, K.; Zhao, X.; Blackwell, A.R.; Kim, J.; Henderson, I.R. Interhomolog polymorphism shapes meiotic crossover within the Arabidopsis RAC1 and RPP13 disease resistance genes. *PLoS Genet.* **2018**, *14*, e1007843. [CrossRef]
55. Alder, M.N.; Rogozin, I.B.; Iyer, L.M.; Glazko, G.V.; Cooper, M.D.; Pancer, Z. Diversity and Function of Adaptive Immune Receptors in a Jawless Vertebrate. *Science* **2005**, *310*, 1970–1973. [CrossRef] [PubMed]
56. Shanmugam, V. Role of extracytoplasmic leucine rich repeat proteins in plant defence mechanisms. *Microbiol. Res.* **2005**, *160*, 83–94. [CrossRef]
57. Zhang, X.S.; Choi, J.H.; Heinz, J.; Chetty, C.S. Domain-Specific Positive Selection Contributes to the Evolution of Arabidopsis Leucine-Rich Repeat Receptor-Like Kinase (LRR RLK) Genes. *J. Mol. Evol.* **2006**, *63*, 612–621. [CrossRef]

 *horticulturae*

Article

High-Throughput SSR Marker Development and the Analysis of Genetic Diversity in *Capsicum frutescens*

Yangmin Zhong [1], Yuan Cheng [1], Meiying Ruan [1], Qingjing Ye [1], Rongqing Wang [1], Zhuping Yao [1], Guozhi Zhou [1], Jia Liu [2], Jiahong Yu [1,2] and Hongjian Wan [1,3,*]

[1] State Key Laboratory for Managing Biotic and Chemical Threats to the Quality and Safety of Agro-Products, Institute of Vegetables, Zhejiang Academy of Agricultural Sciences, Hangzhou 310021, China; yangminzhong@hotmail.com (Y.Z.); chengyuan1005@126.com (Y.C.); ruanmy@163.com (M.R.); yeqingjing2003@126.com (Q.Y.); rongqingw2012@gmail.com (R.W.); zjyzp@163.com (Z.Y.); Chinazhougz@163.com (G.Z.); jiahong624887@163.com (J.Y.)

[2] Wulanchabu Academy of Agricultural and Husbandry Sciences, Wulanchabu 012000, China; liu-l-jia@163.com

[3] China-Australia Research Centre for Crop Improvement, Zhejiang Academy of Agricultural Sciences, Hangzhou 310021, China

* Correspondence: wanhongjian@sina.com; Tel.: +86-571-86407677; Fax: +86-571-86400997

Abstract: *Capsicum frutescens*, one of the domesticated species of pepper grown worldwide, is thought to be highly advantageous due to its strong resistance against plant pathogenesis, high productivity, and intense aroma. However, a shortage of molecular markers limits the efficiency and accuracy of genetic breeding for pepper. With the newly developed next-generation sequencing technology, genome sequences of *C. frutescens* can be generated, which are now available for identifying SSR markers via data mining. In this study, a total of 278,425 SSRs were detected from the pepper genome using MISA software. It was observed that trinucleotides were the dominant repeat motif. This was followed by dinucleotides, tetranucleotides, pentanucleotides, and the hexanucleotides repeat types. (AT)n (TTG)n (AAAT)n (AAATA)n (TATAGA)n is known to be the most common repeat motifs corresponding to dinucleotide to hexanucleotide repeats, respectively. In addition, a total of 240 SSR primers evenly distributed over all 12 chromosomes were designed and screened against 8 *C. frutescens* cultivars. Of these, 33 SSR markers that have high polymorphism, have been scrutinized for 147 accessions from 25 countries. The dendrogram constructed clustered these accessions into seven major groups. The groups were found to be consistent with their origins. The results obtained in this study provided resources of SSR molecular markers and insight into genetic diversity of the *C. frutescens*.

Keywords: *C. frutescens*; SSR; germplasm; genetic diversity

Citation: Zhong, Y.; Cheng, Y.; Ruan, M.; Ye, Q.; Wang, R.; Yao, Z.; Zhou, G.; Liu, J.; Yu, J.; Wan, H. High-Throughput SSR Marker Development and the Analysis of Genetic Diversity in *Capsicum frutescens*. Horticulturae **2021**, 7, 187. https://doi.org/10.3390/horticulturae7070187

Academic Editor: Yuyang Zhang

Received: 1 June 2021
Accepted: 3 July 2021
Published: 8 July 2021

Publisher's Note: MDPI stays neutral with regard to jurisdictional claims in published maps and institutional affiliations.

Copyright: © 2021 by the authors. Licensee MDPI, Basel, Switzerland. This article is an open access article distributed under the terms and conditions of the Creative Commons Attribution (CC BY) license (https://creativecommons.org/licenses/by/4.0/).

1. Introduction

Pepper is an indispensable spice, as well as an important vegetable crop which is cultivated around the globe. It originated in South America and belongs to the genus *Capsicum* (Solanaceae) [1–3]. This genus has many varieties of cultivated and wild species. However, only five of the species are commonly cited in the current literature as domesticated and culinary species. These include *C. annuum*, *C. chinense*, *C. frutescens*, *C. baccatum*, and *C. pubescens* [4]. Among those, *C. annuum* is considered to be the predominant species, which is comprised of many commercial varieties with major variations in the size, shape, color of the fruit, and, in particular, the pungency. However, after a long period of artificial selection, continuous cultivation, and domestication, the characteristic performance of pepper has tended to become diversified with narrowing genetic backgrounds. Consequently, the reduced genetic diversity index has engendered a straggle in the production of the piquant/hot pepper varieties. This has entailed searching and restoring potent traits from related wild species/wild pepper varieties in order to improve the quality traits [5]. Furthermore, the enriched genetic diversity in the other four species, along with the rest of the wild species, has significantly rendered the possibility of improving *C. annuum* through various new technologies [6]. The genetic diversity analyses of the aforementioned accessions have provided excellent research resources, thereby utilizing their advantages and also the potential of reasonably allocating parent traits, which will definitely guarantee future breeding efficiency [7].

C. frutescens is a domesticated annual/perennial shrub crop bearing small erect, wheat-shaped fruit, which are often strongly spicy in flavor. The unripe fruit are green to pale yellow, which turn red as ripening occurs. This species has many wild types, particularly one of the most commonly reported wild pepper species in China, which are mainly distributed in tropical areas such as Yunnan and Hainan provinces in China. A member of *C. frutescens* is locally known as 'Xiaomijiao' (Kunming Institute of Botany, Chinese Academy of Sciences, 1979) [8]. In addition, this plant species is known to be highly resistant to various biotic and abiotic stresses [8–11]. It also has the ability to withstand pathogenic attacks [9]. *C. frutescens* is characterized by high yield capacity, delicate flesh fragrance, and is also of ethnomedicinal importance. Therefore, investigating the genetic diversity of the germplasm resources of *C. frutescens* can potentially provide a basis for the improvement of existing cultivated pepper in China.

Simple sequence repeats (SSRs), or microsatellites, are a group of tandemly repeated DNA sequences comprised of one to six nucleotide units. These are ubiquitous in the genomes of prokaryotic and eukaryotic organisms [12,13]. SSRs are the most highly recognized genetic markers that are actively employed in plant breeding due to their distinguished traits of co-dominant inheritance, multi-allelic nature, extensive genome coverage, high abundance, and especially high reproducibility [12,14]. The polymorphism resulting from SSRs can be detected from amplifying genes by employing primers flanking the repeated motifs [15]. SSR markers are widely used when examining plants and animals from different aspects, such as analyzing population characteristics [16], functional diversity [17–19], constructing linkage maps [20,21], DNA fingerprinting [22–25], and assisted breeding techniques [26–28]. There are many SSR markers which have been developed and successfully used. However, it is imperative to develop a few with strong stability and highly polymorphic SSR markers in order to comply with such intricate studies as high-density genetic mapping, genome comparative mapping, and genome-wide association analysis.

It has been found that examining the diversity of the pepper germplasm via assessing the morphological characteristics is not only very stringent, but often results in misguidance. Furthermore, such attributes may be equally impacted by environmental conditions. However, molecular markers have been found to have several advantages in representing colossal amounts of information, as well as better stability and high analysis efficiency when used to study the genetic diversity of plant germplasm resources. Therefore, molecular markers are widely used in genetic diversity research [29]. For example, in recent years, molecular markers have been extensively employed for analyzing the genetic diversity of pepper germplasm resources. In addition, among the many types of molecular markers, SSRs are highly preferred for examining diversity due to their high polymorphism, ample repeatability, and co-dominance traits [30]. These characteristics have been supported by many earlier reports. For example, in the study conducted by Gu [31] regarding pepper diversity, 1904 pepper materials were accommodated into two categories by employing 29 pairs of polytropic SSR markers. Similarly, Zhang [32] used SSR labeling technology to analyze 372 pepper materials and then cluster them into three groups. The results were found to be consistent with the botanical characteristics. Chen [5], when comparing the efficiency of SRAP and SSR for the genetic diversity analyses of eight pepper germplasm, found that the SSRs were more efficient, with higher polymorphism and detection ability. In another related study, 14 pairs of SSR polymorphic primers were employed by Li [33], who also analyzed the genetic diversity of 169 pepper materials. It was observed that when classifying them into seven groups and validating the findings, the results were in exact congruence with the basic classification of the pepper species.

Pepper germplasm resources are the prime material basis for the breeding and production of pepper crops. However, the genetic background of the pepper germplasm resources in China is relatively narrow. Therefore, it is indicated that an emphasis should be placed on gaining significant data resources with efficient utilization values from the collection and proper exploitation of wild germplasm resources. Among the available resources, 'Xiaomijiao' is the only wild *C. frutescens* plant found in China. As a result, this research focused on the objectives of studying the development of SSR molecular markers for *C. frutescens* on the basis of the 'Xiaomijiao' sequence data (unpublished), and then analyzed the genetic diversity of the germplasm resources collected from the genome levels through SSR markers. These new polymorphic microsatellite markers provide the basis for further population research. In this paper, the genetic diversity of 147 *C. frutescens* germplasm from 25 countries was analyzed to understand the genetic relationships and genetic composition of various accessions, so as to provide basis for more effective utilization of these germplasm resources in the future.

2. Materials and Methods

2.1. Plant Materials and DNA Extraction

A total number of 147 pepper (*C. frutescens*) accessions, which had been collected from 25 countries, were used in this study. The details of the samples encompassing their English names, cultivation regions, DNA concentrations (ng/uL), and the type of selected species (wild/cultivated) are summarized in Supplementary File S1. Following the sampling processes, young leaves were immediately frozen under liquid nitrogen and transferred to −80 °C conditions for future DNA extraction. Then, following the process described by Murray and Thompson [34], CTAB (Cetyl Trimethyl Ammonium Bromide) methods were used for genomic DNA extraction. In addition, the quality of isolated DNA was ascertained using an NanoDropOneC Microvolume UV-Vis Spectrophotometer. Finally, the DNA concentrations were adjusted to 10 to 35 ng/μL for use in the subsequent polymerase chain reactions (PCR).

2.2. Source of Genic Sequences, SSR Identification and Primer Design

In this study, the wild pepper 'Xiaomijiao' (*C. frutescens*) was sequenced at Beijing Nuohe Zhiyuan Technology Co., Ltd. The genome sequencing of 'Xiaomijiao' (*C. frutescens*) was performed with Illumina HiSeq4000 (300x coverage) and PacBio Sequel (30x coverage). The assembled sequences, which totaled 2.95 G bases, were used in this study to characterize the distribution of microsatellites in the pepper genome. The completeness of the 2.95 Gbp assembly is supported by the mapping of over 99% of ~3 million EST reads (generated using HiSeq4000 technology) from 'Xiaomijiao' (*C. frutescens*) leaf, stem and root tissues. A MISA (MIcroSAtellite) SSR identification tool program was employed for the sequence identification [35]. It was confirmed that 2 to 6 nucleotide motifs could be considered for identifying the presence of microsatellites. The minimum repeating units for the dinucleotides, trinucleotides, tetranucleotides, pentanucleotides, and hexanucleotides were defined as 6, 5, 4, 4, and 4, respectively. We allowed up to 5 nucleotide mismatches at the 5′ end of the primer, but no mismatches at the 3′ end, and a minimum of 80% overall match homology. For a given primer pair, we considered that a specific amplicon was generated if both forward and reverse primers were mapped to the same chromosomes/scaffold. Then, based on the MISA results, Primer 5 software was employed for designing the SSR primers. They generated amplicon sizes of 100 to 300 bp with the following criteria: 22 to 25 bp lengths with 40 to 70% GC content levels; 45 to 65 °C melting temperature (Tm); and the remaining parameters used the program's default values. Eight pepper cultivars, referred to as GRIF 9194, GRIF 9316, PI 439309, PI 439489, PI 631142, VI029462, VI062180, and LJ091, respectively, were selected for validating the primers via PCR and electrophoretic techniques. A total number of 240 SSR markers, which were distributed among 12 linkage groups, were screened. Finally, 33 markers were identified as being evenly distributed along the linkage groups, which produced clear bands with high polymorphism. These markers were further used to analyze all of the accessions, as detailed in Table 1.

Table 1. Details of 33 SSR primers used in the study.

No.	Primers	Forward Primer Sequence (5′-3′)	Reverse Primer Sequence (5′-3′)	Repeat Unit	Repeat Number
1	Chr1SSR8	GACTATAGCAAACATGTCCCCAG	CCAGATTTTGTGGATCCTATTGA	TCA	4
2	Chr1SSR12	ATTCAAAGAGGGCATGATGTAGA	TGTCAACTTAGTGGGTGGGTTAG	GGTAGG	6
3	Chr1SSR18	TTAGTGTTGTCAAAATACCCGTG	CAATAAACATATCACACGTGCAAC	ATAG	6
4	Chr2SSR12	GAAAGATCCGTCTAAGCAAACAA	AAATCTCATGATTTTCGGTGATG	AAT	4
5	Chr2SSR14	TGAATTTCGGAGTGTTACGTAGAG	GCGAAGGTGAGTCTGTTCAATTA	TAA	4
6	Chr2SSR15	AATTTGAAGAGCGTGCATAAAAA	ACCATACCATACCATGGAAACAA	ATGGT	4
7	Chr3SSR5	ATAAAATAGACCCCACCCACTTC	GGGCTGCTGTACTAAAGAAGAAGA	TTC	4
8	Chr3SSR6	CTCTCTAGAATGAAAAGTGCCGA	TTGTCGATTTGTTCTTCTTCCAT	CAC	4
9	Chr3SSR13	ACCCTAAAAGCTATGGAGTTGCT	CCCACCATTCTTCTAAACTTTCC	AC	6
10	Chr4SSR11	CCTACAAGTGAGGTCTGAGGAGA	CAGCAAGTTGGAGAAACTAATGC	ACCCT	4
11	Chr4SSR14	AGTGGAAAGTGCTGTTACGATGT	GCGTGATAATTTTTCCACAAGAA	TTTA	5
12	Chr4SSR20	AAAACAACACGACACACAGCTTA	TATATTTTTCTTGGGAACGAGCA	TATAT	6
13	Chr5SSR13	TAAAACCTTGTCACATGTACCCC	AAATGAATAAATCCTCTGCATGG	TGCA	4
14	Chr5SSR14	TAAGTTGTTCGAGAAATCAGACG	CTGTCTCGTATTGAAGGGTGTTT	CGGTGA	4
15	Chr5SSR16	CACTGTGAGAGCAACTTTCTGTG	CACTATTTTCTCATGCACTTTACCA	ACGGGC	4
16	Chr6SSR12	TTCGACCTCCGTATCACTATCAG	CGTACTCTATCGCTTGTTGCTTT	ACTC	7
17	Chr6SSR17	TCTGAAAAATCCTCGGCTAAAGT	ACTGTCCCACCTTACATCCCTAT	AAAAG	5
18	Chr6SSR19	CAGAGGCAGTTAGGTAGTAGCGA	TGCTAAACCCACCTTCAGTCTTA	GGATTC	4
19	Chr6SSR20	TCTGCTTTCCCACAGTATCTCTC	TCAACAGATAAGCGTCAAGTGAA	TCACCA	4
20	Chr7SSR2	CTGCTTAAAAAGTTGAAGATGAGAA	TTGATACGCTAATAAAATTGTTGAAA	AAT	4
21	Chr7SSR15	GCCTGGCATGTTTTGTATTTCTA	TTTGGTGACCGACAATATAAAGG	ATT	4
22	Chr8SSR6	TGAGTCAAGAAAACTTGCAGAAA	TTTGAAAATAATTGAAGTTCCGC	AAAGCA	4
23	Chr8SSR13	TAGATGTTGAACCCCTATTGGAA	GGTAGAGGGTAGAGTGTACGCAG	TACCAC	4
24	Chr8SSR15	GGTGTCATGCGTAAGCTCATAGT	GGTGTCATGCGTAAGCTCATAGT	GAGTTG	4
25	Chr9SSR16	CCCCACCGATGAATTTAGTAGA	TGATGATGTGTCATGGTGTATGA	CCCCCA	4
26	Chr9SSR17	ATTCCCCATATCGAAACTTCTTC	AAATCAAGGAGACGATTGTTGAA	CTGGTG	5
27	Chr10SSR12	ACGAGAGTTTGCTTTCTTTTCCT	TCAGAGGTAGAGGTATGGACTGC	AAC	4
28	Chr10SSR14	TTTTTCAGGCTTTTGTGTATGTAAA	CGGAAACAACCTCTCTACTTCAG	CAA	5
29	Chr11SSR4	AATTTTCAACAACAAACTCCACG	CAGCAGTGAGGATGAAAAAGTTTA	GAG	4
30	Chr11SSR14	TTTTAACTTTTTGTAATTTCGTGTCA	ATCCAATTTTGTTAGGCCTATTG	ATCAAA	4
31	Chr11SSR16	ACGAGTGAACACGTACATGAAAA	GGTAGGAGCTAGGAGTAGTGACG	TTA	6
32	Chr12SSR10	AAAAGGGACTTGTTTTCCTATCA	AAATGGGAATGCGATTATCTAAA	AAAAAT	4
33	Chr12SSR19	TCTTTCCTTCGGATTAAGTTTCC	GAGGCAAGAAATAAGAGATGCCT	CAAATC	4

2.3. PCR Amplification and Polyacrylamide Gel Electrophoresis

PCR were carried out in the final volume of the 10 μL reaction mixture (1 μL DNA, 1 μL forward primer, 1 μL reverse primer (100 ng/μL), 5 μL of 2 × T5 Super PCR Mix, and 2 μL nuclease-free water) on a thermocycler (Applied Biosystems, StepOnePlus ABI7500) with the following reaction conditions: initial denaturation at 94 °C for 3 min, then 30 cycles at 94 °C for 30 s; annealing at 55 °C for 30 s, extension at 72 °C for 1 min; and final extension at 72 °C for 5 min. Electrophoretic analysis was completed in order to assess the PCR amplicons, employing 8% polyacrylamide gel and 0.5 × TBE buffer at a constant voltage of 180 V, 150 mA, and 50 W for three hours, along with a 100 bp DNA ladder. When the electrophoresis was completed, the gel was carefully retrieved, rinsed with sterile water, and kept incubated with 1% silver nitrate ($AgNO_3$) (1 L) for 20 min under shaking conditions. Following the incubation, the gel was washed 3 times with sterile water and immersed in 1 L of developing solution (1.5% sodium hydroxide (NaOH) and 4 mL formaldehydes (CHHO)) until the bands were clearly visible (approximately 5 min). Then, based on the number of clearly visible bands, the alleles in each pepper variety were visually determined.

2.4. Data Statistics and Analysis Results

The scoring was given as 1 (presence) and 0 (absence) for the amplified fragments in each microsatellite loci, and data matrixes were constructed accordingly. Then, employing Popgen (version 1.32) software, several indices were calculated, such as the observed number of alleles (Na), effective number of alleles (Ne), observed heterozygosity (Ho), expected heterozygosity (He), and the Shannon information index (I) [36]. The major allele frequency (MAF), polymorphism information content (PIC), and gene diversity index were calculated using PowerMarker (version 3.0) software [37]. In addition, the cluster analysis of the germplasms was based on the Nei genetic distance [38] and a neighbor joining (NJ) method was used to construct a dendrogram via PowerMarker (version 3.0) software. The dendrogram tree was visualized and edited using MEGA7 (version 7.0) [39].

3. Results

3.1. Distribution of the SSRs in the Capsicum Frutescens Genome

The search results of the genome sequences of *C. frutescens* resulted in a total number of 278,425 SSR loci being identified. (AT)n(TTG)n(AAAT)n(AAATA)n(TATAGA)n is the most common repeat motifs corresponding to the dinucleotide to hexanucleotide repeat, respectively. The SSR repeat types were found to be different. For example, the dominant amongst the 1638 SSR repeats were the trinucleotides and dinucleotides, which accounted for 57.1% (158,967) and 34.1% (94,916), respectively. The remaining was occupied by tetranucleotide, pentanucleotide, and hexanucleotide repeat motifs, accounting for 5.9% (16,351), 1.6% (4531), and 1.3% (3660), respectively. Taken together, it was found that the majority of the SSR repeat motifs along the entirety of genome sequences were trinucleotide repeats, and the hexanucleotide repeats were the fewest, as illustrated in Figure 1A. The frequencies of each of SSR motif types along the entire *C. frutescens* genome were also detected. Among the dinucleotide motifs, AT/TA was observed to be the most common (69.91%; 66,356). This was followed by AC/GT (9.14%; 8671) and TC/GA (7.88%; 7476). Meanwhile, the CG/GC motif repeats were rarely observed (0.05%). The trinucleotide repeat motif consisted of 30 different types. The predominant motifs were TTG/CAA and AAT/ATT, which accounted for 12.41 and 10.02%, respectively (Figure 1B). In addition, AAAT/ATTT (13.85%) were the predominantly found tetranucleotide repeats (Figure 1C).

The statistical data showed that the SSR loci were widely distributed on all 12 chromosomes of the *C. frutescens* genome. These were mainly found in Chr3 (25,014), followed by Chr1 (23,644), Chr12 (23,240), Chr9 (22,761), Chr11 (22,617), Chr7 (22,066), Chr5 (21,976), Chr6 (21,448), Chr8 (21,286), Chr4 (20,939), Chr10 (20,333), and Chr2 (16,579), respectively. In addition, 16,522 SSR loci were unable to allocate in the chromosomes. It was found that, while analyzing the distribution frequency of the SSR loci/Mb, the results revealed that the number of SSR loci/Mb on each chromosome ranged from 91.27 to 104.13. However, the number of SSR loci on Chr3 was the highest. The density of the SSR loci on Chr3 was found to be the third highest, with an average of 97.45 SSR loci/Mb. Moreover, although the number of SSR loci on Chr2 was found to the fewest, the density of the SSR loci was the highest overall (104.13 SSR loci/Mb), as shown in Figure 1D.

3.2. Analysis of the SSR Repeat Motif Types and Frequencies

As shown in Figure 1E, the distributions of the SSRs were also examined from the aspect of the number of repeat units. It was observed that for all the SSR types, the SSR frequencies decreased as the number of repeat units increased. Meanwhile, the change rates became more gradual for the dinucleotides when compared with the longer repeat motif types. The dominant numbers of repeats in the pepper SSR loci ranged between 4 and 10, with the exception of a few (more than 10). At the same time, the majority of the observed repeat times were 6, accounting for 26.20% (45,076). The dinucleotides were found to be the most abundant number of repeats, accounting for 55.16% (94,916), whereas the number of repeats for remaining trinucleotide, tetranucleotide, pentanucleotide, and hexanucleotide were determined to be 30.58% (52,614), 9.50% (16,351), 2.63% (4531), and 2.13% (3660), respectively, as shown in Figure 1E.

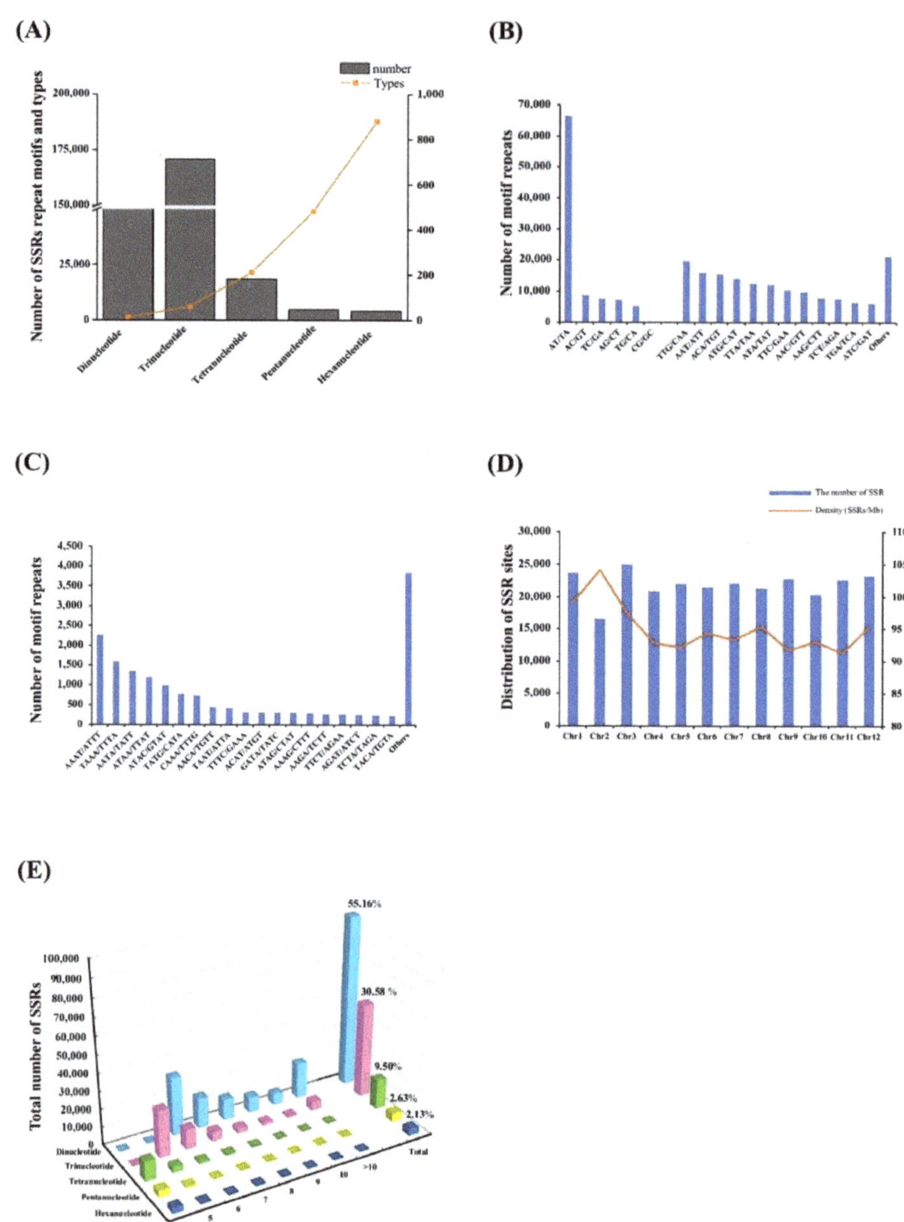

Figure 1. Total number, distribution, frequency, and distinct motif types of SSRs in the entire genome of *C. frutescens*: (**A**) Distribution of the SSR in the 'Xiaomijiao' genome; (**B**) Number of dinucleotide and trinucleotide repeat motifs in the genome; (**C**) Number of tetranucleotide repeat motifs in the genome; (**D**) Distribution and frequency of the SSR loci in the genome; (**E**) Relative frequency (%) of the SSR types with different repeat numbers in the 'Xiaomijiao' genome.

3.3. Primer Design of the Pepper Plant Genomic SSR Markers

For the designing of the primers in the current study, a total of 240 SSRs, which were distributed on different chromosomes, were selected. Then, the reliability was evaluated on eight pepper cultivars. It was determined that out of the total primer sets tested, 41 were successfully amplified showing full length polymorphisms. The remaining 199 primers were found to be either non-polymorphic, non-specific amplification with ambiguous bands, or not amplified, as evidenced from the gel results. Of the 41 amplified primer sets, only 33 (13.75%) were found to have generated both polymorphic and unambiguous bands on the gel. Therefore, those primer sets were selected for further analysis, as detailed in Table 1. Among the 33 SSR loci, 1 was observed to be dinucleotides, 11 were trinucleotides, 4 were tetranucleotides, 4 were pentanucleotides, and 13 were hexanucleotides, respectively.

3.4. Polymorphism Analysis with SSR

The results of polyacrylamide gel electrophoresis of several highly polymorphic SSR markers are shown in Figure 2C. In total, 91 alleles were obtained with the aforementioned 33 amplified SSR primers. Among those markers, the Number of Alleles (Na) per locus ranged from 2 (Chr1SSR8, Chr2SSR12, Chr2SSR14, Chr3SSR5, Chr3SSR6, Chr4SSR11, Chr5SSR13, Chr5SSR14, Chr6SSR12, Chr6SSR20, Chr7SSR2, Chr8SSR6, Chr9SSR16, Chr10SSR14, Chr11SSR14, and Chr11SSR16) to 6 (Chr4SSR20), with an observed average of 2.8 alleles. The Effective Number of Allele (Ne) per locus ranged from 1.0288 (Chr7SSR2) to 3.6226 (Chr7SSR15), with an observed average of 1.7055 alleles. The major allele frequency (MAF) ranged from a low of 0.3231 (Chr7SSR15) to a high reaching 0.9858 (Chr7SSR2), with an average of 0.7547. In addition, the Observed Heterozygosity (Ho) ranged from 0.000 (Chr1SSR8, Chr4SSR11, Chr6SSR17, Chr7SSR2, Chr10SSR12, Chr11SSR4, and Chr11SSR14) to 0.9863 (Chr5SSR16), with an average of 0.0989 observed. It was also determined that the Expected Heterozygosity (He) ranged from 0.0281 (Chr7SSR2) to 0.7264 (Chr7SSR15), with an average of 0.3313. Also, the Shannon information index (I) ranged between 0.0744 and 1.3273, with an average of 0.5758, and the polymorphic information content (PIC) ranged from 0.0276 to 0.6718, with a mean value of 0.2893. In the present study, the calculation of the mean gene diversity confirmed it to be 0.3300 for all of the 147 types of material, as shown in Table 2. It was observed that the different markers had displayed different polymorphism. For example, primer Chr7SSR15 was found to be the most informative (PIC value: 0.6718), whereas primer Chr7SSR2 was the least informative (PIC value: 0.0276). Therefore, this study concluded that when considered altogether, the performances of the selected SSR markers were very effective in detecting genetic diversity.

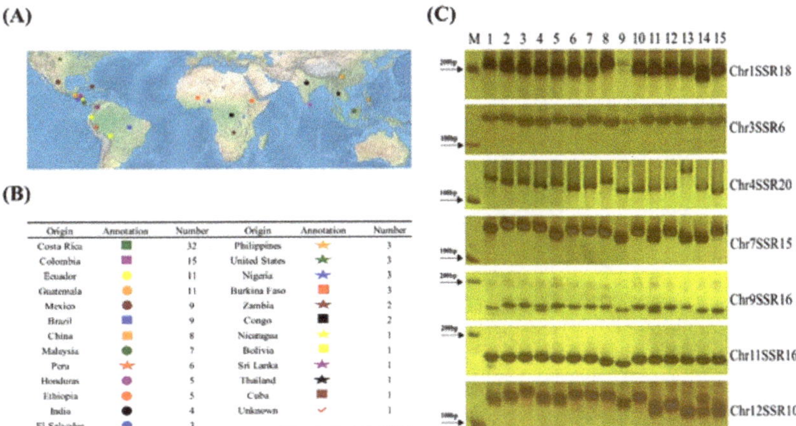

(A)

(B)

Origin	Annotation	Number	Origin	Annotation	Number
Costa Rica		32	Philippines		3
Colombia		15	United States		3
Ecuador		11	Nigeria		3
Guatemala		11	Burkina Faso		3
Mexico		9	Zambia		2
Brazil		9	Congo		2
China		8	Nicaragua		1
Malaysia		7	Bolivia		1
Peru		6	Sri Lanka		1
Honduras		5	Thailand		1
Ethiopia		5	Cuba		1
India		4	Unknown		1
El Salvador		3			

(C)

Figure 2. Regional distribution of the 147 pepper (*C. frutescens*) cultivars, and the detailed information material collected from these countries. (**A**) Regional distribution of the 147 pepper cultivars in this study, mainly collected from Central and South America, Africa, and Asia; (**B**) Labeling of 25 countries on the map, and information material collected from these countries (including an unidentified source); (**C**) Polyacrylamide gel electrophoresis analysis of 7 SSR markers with high polymorphism on samples 1 to 15.

Table 2. Polymorphism analysis of the 147 *C. frutescens* accessions with SSR primers.

Primer	Allele Size	Na	Ne	MAF	Ho	He	I	PIC	Gene Diversity
Chr1SSR8	134	2.0000	1.0615	0.9701	0.0000	0.0584	0.1342	0.0562	0.0579
Chr1SSR12	294	3.0000	1.2132	0.9048	0.1837	0.1763	0.3680	0.1665	0.1757
Chr1SSR18	278	4.0000	2.7999	0.4532	0.0360	0.6452	1.1545	0.5758	0.6428
Chr2SSR12	286	2.0000	1.1104	0.9476	0.0350	0.0997	0.2057	0.0945	0.0994
Chr2SSR14	294	2.0000	1.1764	0.9184	0.0272	0.1504	0.2827	0.1387	0.1499
Chr2SSR15	278	3.0000	1.3743	0.8417	0.2014	0.2733	0.4998	0.2452	0.2723
Chr3SSR5	286	2.0000	1.1259	0.9406	0.1189	0.1122	0.2254	0.1056	0.1118
Chr3SSR6	290	2.0000	1.3729	0.8379	0.0345	0.2725	0.4431	0.2347	0.2716
Chr3SSR13	272	3.0000	1.4543	0.8162	0.2059	0.3135	0.5784	0.2837	0.3124
Chr4SSR11	290	2.0000	1.1012	0.9517	0.0000	0.0922	0.1934	0.0877	0.0919
Chr4SSR14	204	3.0000	2.1276	0.5392	0.0588	0.5326	0.8242	0.4252	0.5300
Chr4SSR20	292	6.0000	2.3599	0.5822	0.0137	0.5782	1.0946	0.5175	0.5763
Chr5SSR13	282	2.0000	1.4302	0.8156	0.3404	0.3019	0.4780	0.2556	0.3008
Chr5SSR14	278	2.0000	1.0516	0.9748	0.0216	0.0493	0.1176	0.0479	0.0491
Chr5SSR16	292	3.0000	2.1992	0.5068	0.9863	0.5472	0.8578	0.4426	0.5453
Chr6SSR12	248	2.0000	1.9541	0.5766	0.0081	0.4902	0.6814	0.3691	0.4883
Chr6SSR17	288	3.0000	1.1656	0.9236	0.0000	0.1426	0.2931	0.1338	0.1421
Chr6SSR19	294	3.0000	1.3276	0.8571	0.0136	0.2476	0.4375	0.2195	0.2467
Chr6SSR20	280	2.0000	1.0894	0.9571	0.0857	0.0823	0.1769	0.0787	0.0820
Chr7SSR2	282	2.0000	1.0288	0.9858	0.0000	0.0281	0.0744	0.0276	0.0280
Chr7SSR15	294	4.0000	3.6226	0.3231	0.0476	0.7264	1.3273	0.6718	0.7240
Chr8SSR6	288	2.0000	1.5732	0.7604	0.0486	0.3656	0.5506	0.2980	0.3644
Chr8SSR13	294	3.0000	1.4308	0.8265	0.1361	0.3021	0.5775	0.2795	0.3011
Chr8SSR15	286	4.0000	3.4282	0.4021	0.0699	0.7108	1.3023	0.6566	0.7083
Chr9SSR16	250	2.0000	1.9678	0.5640	0.0560	0.4938	0.6849	0.3709	0.4918
Chr9SSR17	294	3.0000	1.6031	0.7721	0.0748	0.3775	0.6873	0.3430	0.3762
Chr10SSR12	270	4.0000	3.5014	0.3556	0.0000	0.7171	1.3058	0.6606	0.7144
Chr10SSR14	292	2.0000	1.5026	0.7877	0.0411	0.3356	0.5170	0.2785	0.3345
Chr11SSR4	292	3.0000	1.3060	0.8699	0.0000	0.2351	0.4723	0.2206	0.2343
Chr11SSR14	76	2.0000	1.2321	0.8947	0.0000	0.1909	0.3365	0.1706	0.1884
Chr11SSR16	294	2.0000	1.7306	0.6973	0.0884	0.4236	0.6132	0.3331	0.4222
Chr12SSR10	278	3.0000	2.5179	0.4928	0.1007	0.6050	0.9921	0.5226	0.6028
Chr12SSR19	294	4.0000	1.3399	0.8571	0.2313	0.2546	0.5140	0.2366	0.2537
Mean	271	2.7576	1.7055	0.7547	0.0989	0.3313	0.5758	0.2893	0.3300

3.5. Genetic Diversity Analysis

The genetic diversity and phylogenetic relationships were determined using 147 collected pepper cultivars from 26 different countries around the world (Figure 2 and Supplementary File S1). This study adopted Nei genetic distance and neighbor-joining methods, and a dendrogram was constructed based on the genotypes detected by the newly developed SSR markers (Figure 3). These were clustered into seven main groups (designated in this study as Groups I, II, III, IV, V, VI, and VII), which were comprised of 18, 37, 32, 20, 21, 5, and 14 members, respectively.

The dendrogram not only reflected the phylogenetic relationships of the cultivars, but was also consistent with their places of origin. Remarkably, the dendrogram revealed that Group I had become a unique branch. Furthermore, Group II consisted of 37 accessions, majority of which were from Latin America, with the exception of 3 accessions (PI 281,419 and PI 281,420 from the Philippines, and PI 281,347 from India). Of plant material collected from 15 countries, 32 types comprised Group III. All of the materials collected in Africa were found to be clustered in that group. In addition, 4 Chinese cultivars were also assigned to Group III. Group IV consisted of 20 accessions, among which 7 were from Guatemala; 7 were from South America; 3 were from United States; 2 were from Mexico; and 1 was from an unknown country, which this study speculated may have been of North American origin. Group V was composed of 21 accessions which were from various geographical origins, such as North America, South America, and other Asian areas. The five members which were clustered in Group VI included four cultivars derived from Costa Rica, and one derived from El Salvador. In addition, 20 accessions derived from Latin America were clustered together in Group VII (Figure 3). These results suggested that the newly developed SSR markers were both stable and suitable for assessing the genetic relationships among *C. frutescens* cultivars.

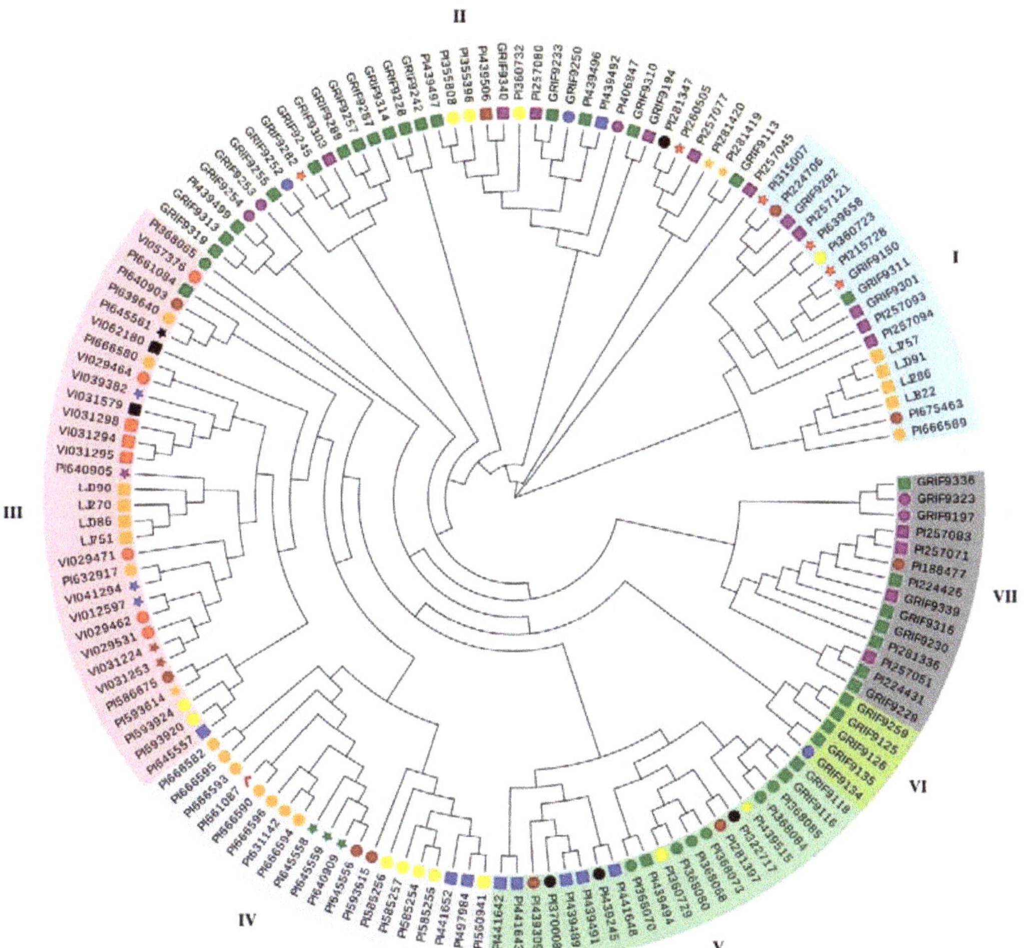

Figure 3. Dendrogram was constructed based on the genotypes from 33 SSR markers using neighbor-joining methods, and the icons indicated the information of material source, which was shown in Figure 2B.

4. Discussion

The genome wide analysis of SSRs could provide the opportunity to decipher the optimal functions of these repeats in the regulation and organization of a genome. Also, the potential uses of these markers, such as diversity and population analyses, evolutionary history, and genome and comparative mapping, are currently being explored [19–21,40–42]. In previous studies, based on the results of sequencing analyses, assessments of the high-quality genome sequences of *C. frutescens* were made possible, thereby accrediting the opportunity to develop suitable SSR primers. In the present study, it was observed that the dinucleotide and trinucleotide motif repeats were the most abundant, accounting for 85.74%. The remainder of the repeats (14.26%) were contributed by the tetranucleotides, pentanucleotides, and hexanucleotides, as illustrated in Figure 1E. This phenomenon has also been reported in other plants, such as *Radix codonopsis*, *Anthuriumand raeanum*, and *Camellia sinensis*, respectively [43–45]. Nevertheless, a few reports have found high abundance of tetranucleotide repeats in some plants, such as *Cucumis sativus*, *Medicago truncatula*, and *Vitis vinifera* [46–48]. The differences in the previous finding may have been due to the dissimilarities in paradigms adopted for the SSR identifications. Moreover, the examination of the SSR motif frequency manifested that the scatterings of the dinucleotides, trinucleotides, tetranucleotides, pentanucleotides, and hexanucleotides repeats were generally skewed toward fewer numbers of repeats. These findings indicated that there were predominantly fewer repeats along the pepper genome. The results obtained in this study indicated that higher repeats were found in dinucleotide and trinucleotide SSRs. However, repeats were fewer in number or absent among the tetranucleotide, pentanucleotide, and hexanucleotide SSRs (Figure 1A). Similarly, in other plants, such as citrus, watermelon, and tea, the same trends were observed [18,19,45]. This may be due to the obvious differences in the frequencies and types of SSR motifs. In the pepper genome, this study found AT/TA was the most common, while the CG/GC motif was very rare in the dinucleotide repeats (Figure 1B). These findings were consistent with the motif frequencies found among cucumber, strawberry, maize, *Radix codonopsis*, potato, plum, watermelon, and horseradish [19,43,48–53]. However, our results greatly differed from the motif frequencies observed in rice, citrus, onion, and *Atremisia frigida* [18,54–56], where AG/CT has been found to be the most dominant type. Similarly, TTG, AAT, and ATT were the prevailing motifs of the trinucleotide in this study, with CCG and AGG being the predominant motifs in the monocotyledons, such as barley, rice, and corn. It was found that the number of SSRs, along with their structure and repeat motifs, will greatly differ compared with those in plant species.

It has been found that SSR markers are very much beneficial in population genetics and molecular breeding. However, their effectiveness mainly relies on the marker quality and the accuracy of the experimentation [25]. In the present study, 240 selected SSR loci markers were scrutinized, resulting in 33 unique markers. It was found that when evaluating 147 pepper cultivars, these markers demonstrated remarkable and unambiguous amplification bands (Table 1). The screened SSR polymorphism primers accounted for 13.8% of the total. Previously, Li et al. obtained 17 pairs of SSR polymorphism primers with clear bands and high polymorphism from 152 pairs of SSR primers covering 12 chromosomes [33]. These accounted for 11.2% of the total number of SSR primers, which was slightly lower than that obtained in this result. Liu et al. evaluated 85 pairs of SSR polymorphic primers and 12 pairs were scrutinized [57]. These accounted for 14.1% of the total, which was similar to this study's research results. Wu et al. used three different peppers as templates to select 65 pairs of SSR polymorphism primers from 153 pairs of SSR primers [42]. These accounted for 42.5% of the total number, which was substantially higher than that obtained in this study. Therefore, it was determined that the proportion of SSR polymorphism primers screened in this study was relatively low, which may have been attributed to the small differences existing in these peppers However, although the proportion of polymorphic primers was low, the results could still be used to analyze the genetic diversity of the pepper population. In regard to pepper, many SSR markers have been developed and mapped to linkage groups [58–60], which provide a key basis for analyzing pepper genetic diversity. However, such factors as the number, size, and types of SSR markers, frequencies of the SSR motifs, as well as the various sampling schemes, apparently result in differences in genetic diversity [25,61].

It is of major significance for the collection and efficient utilization of germplasm resources to continue to carry out genetic diversity evaluations. Highly polymorphic, as well as stable markers, are the prerequisites for studying genetic relationships and diversity. Nevertheless, it was found that the SSR loci showed less diversity, as evidenced by low polymorphism information index and gene diversity when compared to earlier reports. For example, in the study conducted by Nicolaï [62], a PIC of 0.67 and a gene diversity of 0.7 were reported, which were 0.38% and 0.37% higher than those of the current study (PIC: 0.29; Gene Diversity: 0.33). However, this may have been due to the volume and types of test sampling. In the aforementioned study, 1352 accessions from 89 countries were utilized, including 11 species of *Capsicum*. However, this study only examined a single species, *C. frutescens*. Therefore, compared with Li's report [33], the amount of accessions used was approximate. The PIC was approximate (slightly higher than this study), but the number of markers was different. Therefore, it was considered that the hypothesis that the difference in genetic diversity was influenced by the number of SSR markers had been confirmed.

Previously, researchers reported that genetic diversity of some peppers (*Capsicum* spp.) accessions, including *Capsicum chinense*, *Capsicum annuum*. In 2016, 71 *C. chinense* accessions from different Brazilian geographic regions, using fruit morphological descriptors and AFLP molecular markers, were analyzed [63]. The results found no association between the morphological descriptors and AFLP markers [63]. In the same year, the researchers investigated patterns of molecular diversity using a transcriptome-based 48 single nucleotide polymorphisms (SNPs) in a large germplasm collection comprising 3821 accessions. Among the 11 species examined, *Capsicum annuum* showed the highest genetic diversity (HE = 0.44, I = 0.69), whereas the wild species *C. galapagoense* showed the lowest genetic diversity (HE = 0.06, I = 0.07). The *Capsicum* germplasm collection was divided into 10 clusters (cluster 1 to 10) based on population structure analysis, and five groups (group A to E) based on phylogenetic analysis [64]. The dendrogram constructed in this study from 147 pepper accessions using NJ methods, indicated that the genetic relatedness of the pepper cultivars clustered in the majority of the groups were in good agreement with their geographic origins. This study's analysis results were also consistent with the previous findings reported by Luo et al. and Jia et al. [7,65]. Moreover, the geographical sources of the Group III and Group V materials were found to be diverse, and not only attributed to Asian and African countries, but also to Latin American countries. It was observed that the pepper plants of the same geographical origin were not strictly divided into the same groups. For example, the eight pepper materials from China were grouped into Group I and Group III. These findings suggested that numerous complex migrations had occurred in the pepper genotype as the result of human migration, which had led to their adoption, acclimatization, and local selection.

5. Conclusions

In the present study, 278,425 SSRs were identified by searching the *C. frutescens* genome sequences. The AT/TA, TTG/CAA, and AAAT/ATTT were observed to be the most common repeat motifs in the dinucleotide, trinucleotide, and tetranucleotide repeats, respectively. Among them, dinucleotides were the most abundant number of repeats, accounting for 55.16%. In this research investigation, the genetic diversity of *C. frutescens* germplasms was investigated using 33 SSR markers, which were evenly distributed on all of the chromosomes. The 147 experimental materials used in this study were wild peppers with rich genetic diversity. Their eminent properties (withstanding pathogenic attacks, high yielding capacity, and so on) will potentially provide excellent genes for the acquisition of pepper breeding accessions. They may also provide a basis for improving the existing cultivated pepper species, as well as having important significance in expanding the narrow genetic basis of pepper breeding in China. In addition, it was considered that the genome-wide identification and development of SSR markers could be very useful and may possibly provide insights into various research areas regarding *C. frutescens* in the future, such as high-density genetic mapping, genome comparative mapping, and genome-wide association analyses.

Supplementary Materials: The following are available online at https://www.mdpi.com/article/10.3390/{h}orticulturae7070187/s1, Supplementary File S1, Detailed information for the 147 pepper (*C. frutescens*) cultivars used in this study.

Author Contributions: Y.Z., Y.C. and H.W. conceived and designed the research; M.R., Q.Y. and R.W. performed the experiments; Z.Y. and G.Z. analyzed the data and wrote the paper; Y.Z., J.L. and J.Y. revised the manuscript. All authors have read and agreed to the published version of the manuscript.

Funding: This research was partially supported by China Agriculture Research System of MOF and MARA (CARS-23-G-44), National Key Research and Development Program of China (2017YFE0114500, 2018YFD1000800) and National Natural Science Foundation of China (31772294, 31960601, 32060446).

Institutional Review Board Statement: Not applicable.

Informed Consent Statement: Not applicable.

Data Availability Statement: The data used for the analysis in this study are available within the article and supplementary materials.

Conflicts of Interest: The authors declare no conflict of interest.

References

1. Moscone, E.A.; Scaldaferro, M.; Grabiele, M.; Cecchini, N.; García, Y.S.; Jarret, R.; Daviña, J.; Ducasse, D.; Barboza, G.; Ehrendorfer, F. The evolution of chili peppers (*Capsicum*-solanaceae): A cytogenetic perspective. *Acta Hortic.* **2007**, *745*, 137–169. [CrossRef]
2. Olmstead, R.G.; Bohs, L.; Migid, H.A.; Santiago-Valentin, E.; Garcia, V.F.; Collier, S.M. A molecular phylogeny of the solanaceae. *TAXON* **2008**, *57*, 1159–1181. [CrossRef]
3. Qin, C.; Yu, C.; Shen, Y.; Fang, X.; Chen, L.; Min, J.; Cheng, J.; Zhao, S.; Xu, M.; Luo, Y.; et al. Whole-genome sequencing of cultivated and wild peppers provides insights into *Capsicum* domestication and specialization. *Proc. Natl. Acad. Sci. USA* **2014**, *111*, 5135–5140. [CrossRef]
4. Nee, M. Peppers, the domesticated *Capsicums*. *Brittonia* **1985**, *37*, 218. [CrossRef]
5. Chen, X.; Zhou, K.; Zong, H.; Fang, R. Genetic diversity of *Capsicum frutescens* in China as revealed by SRAP and SSR markers. *Acta Bot. Boreali-Occident. Sin.* **2012**, *32*, 2201–2205.
6. Zhao, H.; Cao, Y.; Zhang, Z.; Zhang, B.; Bai, R.; Zhao, Y.; Wang, L. Analysis and evaluation of genetic diversity of pepper (*Capsicum* spp.) Core Germplasm Resources in China. *China Veg.* **2018**, *1*, 25–34.
7. Jia, H.; Wei, X.; Yao, Q.; Yuan, Y.; Wang, Z.; Jiang, J.; Yang, S.; Zhao, Y.; Wang, B.; Zhang, X. Genetic diversity analysis of *Capsicum* genus germplasm resources using SSR markers. *Mol. Plant Breed.* **2017**, *15*, 353–363.
8. Deng, M.; Wen, J.; Zhu, H.; Zou, X. The hottest pepper variety in China. *Genet. Resour. Crop Evol.* **2009**, *56*, 605–608. [CrossRef]
9. Liu, S.; Li, W.; Wu, Y.; Chen, C.; Lei, J. De Novo Transcriptome Assembly in Chili Pepper (*Capsicum frutescens*) to Identify Genes Involved in the Biosynthesis of Capsaicinoids. *PLoS ONE* **2013**, *8*, e48156. [CrossRef]
10. Wei, J.; Zheng, J.; Yu, J.; Zhao, D.; Cheng, Y.; Ruan, M.; Ye, Q.; Yao, Z.; Wang, R.; Zhou, G.; et al. Production and identification of interspecific hybrids between pepper (*Capsicum annuum* L.) and the wild relative (*Capsicum frutescens* L.). *J. Agric. Sci. Technol.* **2019**, *21*, 761–769.
11. Wei, J.; Li, J.; Yu, J.; Cheng, Y.; Ruan, M.; Ye, Q.; Yao, Z.; Wang, R.; Zhou, G.; Deng, M.; et al. Construction of high-density bin map and QTL mapping of horticultural traits from an interspecific cross between *Capsicum annuum* and Chinese wild *Capsicum frutescens*. *Biotechnol. Biotechnol. Equip.* **2020**, *34*, 549–561. [CrossRef]
12. Tautz, D.; Renz, M. Simple sequences are ubiquitous repetitive components of eukaryotic genomes. *Nucleic Acids Res.* **1984**, *12*, 4127–4138. [CrossRef]
13. Tóth, G.; Gáspári, Z.; Jurka, J. Microsatellites in different eukaryotic genomes: Survey and analysis. *Genome Res.* **2000**, *10*, 967–981. [CrossRef]
14. Powell, W.; Machray, G.C.; Provan, J. Polymorphism revealed by simple sequence repeats. *Trends Plant Sci.* **1996**, *1*, 215–222. [CrossRef]
15. Chen, C.; Zhou, P.; Choi, Y.A.; Huang, S.; Gmitter, F.G. Mining and characterizing microsatellites from citrus ESTs. *Theor. Appl. Genet.* **2006**, *112*, 1248–1257. [CrossRef]
16. Cheng, Z.; Huang, H. SSR fingerprinting Chinese peach cultivars and landraces (*Prunus persica*) and analysis of their genetic relationships. *Sci. Hortic.* **2008**, *120*, 188–193. [CrossRef]
17. Mccouch, S.; Teytelman, L.; Xu, Y.; Lobos, K.; Clare, K.; Walton, M.; Fu, B.; Maghirang, R.; Li, Z.; Xing, Y.; et al. Development and mapping of 2240 new SSR markers for rice (*Oryza sativa* L.). *DNA Res.* **2002**, *9*, 199–207. [CrossRef]
18. Liu, S.; Li, W.; Long, D.; Hu, C.; Zhang, J. Development and characterization of genomic and expressed SSRs in citrus by genome-wide analysis. *PLoS ONE* **2013**, *8*, e75149. [CrossRef]

19. Zhu, H.; Song, P.; Dal-Hoe, K.; Guo, L.; Li, Y.; Sun, S.; Weng, Y.; Yang, L. Genome wide characterization of simple sequence repeats in watermelon genome and their application in comparative mapping and genetic diversity analysis. *BMC Genom.* **2016**, *17*, 557–583. [CrossRef]

20. Tan, L.; Wang, L.; Wei, K.; Zhang, C.; Wu, L.; Qi, G.; Cheng, H.; Zhang, Q.; Cui, Q.; Liang, J. Floral transcriptome sequencing for SSR marker development and linkage map construction in the tea plant (*Camellia sinensis*). *PLoS ONE* **2013**, *8*, e81611. [CrossRef]

21. Bali, S.; Mamgain, A.; Raina, S.N.; Yadava, S.K.; Bhat, V.; Das, S.; Pradhan, A.K.; Goel, S. Construction of a genetic linkage map and mapping of drought tolerance trait in Indian beveragial tea. *Mol. Breed.* **2015**, *35*, 1–20. [CrossRef]

22. Li, L.; He, W.; Ma, L.; Liu, P.; Xu, H.; Xu, J.; Zheng, X. Construction Chinese cabbage (*Brassica rapa* L.) core collection and its EST-SSR finger print database by EST-SSR molecular markers. *Genom. Appl. Biol.* **2009**, *28*, 76–88.

23. Sarao, N.K.; Vikal, Y.; Singh, K.; Joshi, M.A.; Sharma, R.C. SSR marker-based DNA fingerprinting and cultivar identification of rice (*Oryza sativa* L.) in Punjab state of India. *Plant Genet. Resour.* **2009**, *8*, 42–44. [CrossRef]

24. Tan, L.; Peng, M.; Xu, L.; Wang, L.; Chen, S.; Zou, Y.; Qi, G.; Cheng, H. Fingerprinting 128 Chinese clonal tea cultivars using SSR markers provides new insights into their pedigree relationships. *Tree Genet. Genomes* **2015**, *11*, 1–12. [CrossRef]

25. Liu, S.; Liu, H.; Wu, A.; Hou, Y.; An, Y.; Wei, C. Construction of fingerprinting for tea plant (*Camellia sinensis*) accessions using new genomic SSR markers. *Mol. Breed.* **2017**, *37*, 1–14. [CrossRef]

26. Ercisli, S.; Ipek, A.; Barut, E. SSR marker-based DNA fingerprinting and cultivar identification of olives (*Olea europaea*). *Biochem. Genet.* **2011**, *49*, 555–561. [CrossRef]

27. Hameed, U.; Pan, Y.B.; Muhammad, K.; Afghan, S.; Iqbal, J. Use of simple sequence repeat markers for DNA fingerprinting and diversity analysis of sugarcane (*Saccharum* spp.) cultivars resistant and susceptible to red rot. *Genet. Mol. Res.* **2012**, *11*, 1195–1204. [CrossRef] [PubMed]

28. Lanteri, S.; Portis, E.; Acquadro, A.; Mauro, R.; Mauromicale, G. Morphology and SSR fingerprinting of newly developed *Cynara cardunculus* genotypes exploitable as ornamentals. *Euphytica* **2012**, *184*, 311–321. [CrossRef]

29. Zhebentyayeva, T.; Reighard, G.; Gorina, V.; Abbott, A. Simple sequence repeat (SSR) analysis for assessment of genetic variability in apricot germplasm. *Theor. Appl. Genet.* **2003**, *106*, 435–444. [CrossRef]

30. Grover, A.; Sharma, P.C. Development and use of molecular markers: Past and present. *Crit. Rev. Biotechnol.* **2016**, *36*, 290–302. [CrossRef]

31. Gu, X.; Cao, Y.; Zhang, Z.; Zhang, B.; Zhao, H.; Zhang, X.; Wang, H.; Li, X.; Wang, L. Genetic diversity and population structure analysis of *Capsicum* germplasm accessions. *J. Integr. Agric.* **2019**, *18*, 1312–1320. [CrossRef]

32. Zhang, X.; Zhang, Z.; Gu, X.; Mao, S.; Li, X.; Chadoeuf, J.; Palloix, A.; Wang, L.; Zhang, B. Genetic diversity of pepper (*Capsicum* spp.) germplasm resources in China reflects selection for cultivar types and spatial distribution. *J. Integr. Agric.* **2016**, *15*, 1991–2001. [CrossRef]

33. Li, Y.; Zhao, H.; Wang, Y.; Jiang, J.; Meng, X.; Wei, X.; Li, J. Genetic diversity analysis of 169 accessions of *Capsicum*. *J. Henan Agric. Sci.* **2018**, *47*, 91–97.

34. Murray, M.G.; Thompson, W.F. Rapid isolation of high molecular weight plant DNA. *Nucleic Acids Res.* **1980**, *8*, 4321–4325. [CrossRef]

35. Thiel, T.; Michalek, W.; Varshney, R.; Graner, A. Exploiting EST databases for the development and characterization of gene-derived SSR-markers in barley (*Hordeum vulgare* L.). *Theor. Appl. Genet.* **2003**, *106*, 411–422. [CrossRef]

36. Yeh, F.; Boyle, T.B. Population genetic analysis of co-dominant and dominant markers and quantitative traits. *Belg. J. Bot.* **1997**, *129*, 157.

37. Liu, K.; Muse, S.V. PowerMarker: An integrated analysis environment for genetic marker analysis. *Bioinformatics* **2005**, *21*, 2128–2129. [CrossRef]

38. Nei, M. Estimation of average heterozygosity and genetic distance from a small number of individuals. *Genetics* **1978**, *89*, 583–590. [CrossRef]

39. Kumar, S.; Stecher, G.; Tamura, K. MEGA7: Molecular evolutionary genetics analysis version 7.0 for bigger datasets. *Mol. Biol. Evol.* **2016**, *33*, 1870–1874. [CrossRef] [PubMed]

40. Zhao, D.; Yang, J.; Yang, S.; Kenji, K.; Luo, J. Genetic diversity and domestication origin of tea plant *Camellia taliensis* (Theaceae) as revealed by microsatellite markers. *BMC Plant Biol.* **2014**, *14*, 14. [CrossRef]

41. Göl, Ş.; Göktay, M.; Allmer, J.; Doğanlar, S.; Frary, A. Newly developed SSR markers reveal genetic diversity and geographical clustering in spinach (*Spinacia oleracea*). *Mol. Genet. Genom.* **2017**, *292*, 847–855. [CrossRef] [PubMed]

42. Wu, J.; Cheng, F.; Cai, C.; Zhong, Y.; Jie, X. Association mapping for floral traits in cultivated *Paeonia rockii* based on SSR markers. *Mol. Genet. Genom.* **2017**, *292*, 187–200. [CrossRef] [PubMed]

43. Wang, D.; Cao, L.; Gao, J. Data mining of simple sequence repeats in *Codonopsis pilosula* transcriptome. *Chin. Tradit. Herb. Drugs* **2014**, *45*, 2390–2394.

44. Yu, Y.; Tian, D.; Pan, X.; Jin, L.; Ge, Y. Mining and developing SSR molecular markers based on transcriptome sequences of *Anthurium*. *Mol. Plant Breed.* **2015**, *13*, 1349–1354.

45. Liu, S.; An, Y.; Li, F.; Li, S.; Liu, L.; Zhou, Q.; Zhao, S.; Wei, C. Genome-wide identification of simple sequence repeats and development of polymorphic SSR markers for genetic studies in tea plant (*Camellia sinensis*). *Mol. Breed.* **2018**, *38*, 1–13. [CrossRef]

46. Mun, J.; Kim, D.; Choi, H.; Gish, J.; Debéllé, F.; Mudge, J.; Denny, R.; Endré, G.; Saurat, O.; Dudez, A.; et al. Distribution of microsatellites in the genome of *Medicago truncatula*: A resource of genetic markers that integrate genetic and physical maps. *Genetics* **2006**, *172*, 2541–2555. [CrossRef]

47. Jaillon, O.; Aury, J.; Noel, B.; Policriti, A.; Clépet, C.; Casagrande, A.; Choisne, N.; Aubourg, S.; Vitulo, N.; Jubin, C.; et al. The grapevine genome sequence suggests ancestral hexaploidization in major angiosperm phyla. *Nature* **2007**, *449*, 463–467.

48. Cavagnaro, P.F.; Senalik, D.A.; Yang, L.M.; Simon, P.W.; Harkins, T.T.; Kodira, C.D.; Huang, S.W.; Weng, Y.Q. Genome-wide characterization of simple sequence repeats in cucumber (*Cucumis sativus* L.). *BMC Genom.* **2010**, *11*, 569–587. [CrossRef]

49. Dong, Q.; Wang, X.; Zhao, M.; Song, C.; Ge, A.; Wang, J. Development of EST-derived SSR markers and their application in strawberry genetic diversity analysis. *Sci. Agric. Sin.* **2011**, *44*, 3603–3612.

50. Xu, J.; Liu, L.; Xu, Y.; Chen, C.; Rong, T.; Ali, F.; Zhou, S.; Wu, F.; Liu, Y.; Wang, J.; et al. Development and characterization of simple sequence repeat markers providing genome-wide coverage and high resolution in maize. *DNA Res.* **2013**, *20*, 497–509. [CrossRef]

51. Gong, L.; Cheng, Y.; Gan, X.; Chen, Y.; Nie, F.; Zhang, L.; Shi, L.; Song, Y.; Guo, Z.; Wang, F. Mining EST-SSRs based on potato transcriptome and analyzing their polymorphism. *Mol. Plant Breed.* **2015**, *13*, 1535–1544.

52. Fang, Z.; Ye, X.; Zhou, D.; Jiang, C.; Pan, S. Analysis on SSR information in 'Furongli' plum transcriptome and development of molecular markers in *Prunus salicina* Lindl. *J. Fruit Sci.* **2016**, *33*, 416–424.

53. Zhang, Y.; Wei, X.; Wang, X.; Pan, S.; Ye, X. SSR loci in transcriptome of *Moringa oleifera* lam. *Fujian J. Agric. Sci.* **2017**, *32*, 955–958.

54. Zhang, Z.; Deng, Y.; Tan, J.; Hu, S.; Yu, J.; Xue, Q. A Genome-wide microsatellite polymorphism database for the indica and japonica Rice. *DNA Res.* **2007**, *14*, 37–45. [CrossRef] [PubMed]

55. Li, M.; Zhang, S.; Deng, P.; Hou, X.; Wang, J. Analysis on SSR information in transcriptome of onion and the polymorphism. *Acta Hortic. Sin.* **2015**, *42*, 1103–1111.

56. Yue, C.; Chen, C.; Guo, F.; Li, H.; Sun, H.; Pei, D.; Ma, X.; Chen, F.; Yang, H.; Li, Q. Data mining of simple sequence repeats in transcriptome sequences of mongolia medicinal plant artemisia frigida willd. *J. Agric. Sci. Technol.* **2016**, *18*, 31–43.

57. Liu, Z.; Yang, Y.; Sun, J.; Liu, Z.; Cao, Z. Purity identification of Rela No.2 pepper and genetic diversity analysis of excellent pepper inbred lines. *Chin. J. Trop. Crop.* **2014**, *35*, 847–853.

58. Rivera, A.; Monteagudo, A.B.; Igartua, E.; Taboada, A.; García-Ulloa, A.; Pomar, F.; Riveiro-Leira, M.; Silvar, C. Assessing genetic and phenotypic diversity in pepper (*Capsicum annuum* L.) landraces from North-West Spain. *Sci. Hortic.* **2016**, *203*, 1–11. [CrossRef]

59. Yuan, X.; Zhou, K.; Wu, Y.; Fang, R.; Chen, X. Genetic Diversity and population structure analysis of pepper core collections. *Mol. Plant Breed.* **2019**, *17*, 3090–3104.

60. Uncu, A.T. Genome-wide identification of simple sequence repeat (SSR) markers in *Capsicum chinense* Jacq. with high potential for use in pepper introgression breeding. *Biologia* **2019**, *74*, 119–126. [CrossRef]

61. Yao, M.; Ma, C.; Qiao, T.; Jin, J.; Chen, L. Diversity distribution and population structure of tea germplasms in China revealed by EST-SSR markers. *Tree Genet. Genomes* **2012**, *8*, 205–220. [CrossRef]

62. Nicolaï, M.; Cantet, M.; Lefebvre, V.; Sage-Palloix, A.M.; Palloix, A. Genotyping a large collection of pepper (*Capsicum* spp.) with SSR loci brings new evidence for the wild origin of cultivated *C. annuum* and the structuring of genetic diversity by human selection of cultivar types. *Genet. Resour. Crop Evol.* **2013**, *60*, 2375–2390. [CrossRef]

63. Baba, V.Y.; Rocha, K.; Gomes, G.P.; Ruas, C.F.; Ruas, P.; Rodrigues, R.; Gonçalves, L. Genetic diversity of *Capsicum chinense* accessions based on fruit morphological characterization and AFLP markers. *Genet. Resour. Crop Evol.* **2015**, *63*, 1371–1381. [CrossRef]

64. Lee, H.; Ro, N.; Jeong, H.; Kwon, J.; Jo, J.; Ha, Y.; Jung, A.; Han, J.; Venkatesh, J.; Kang, B. Genetic diversity and population structure analysis to construct a core collection from a large *Capsicum* germplasm. *BMC Genet.* **2016**, *17*, 142–154. [CrossRef]

65. Luo, Y.; Li, J.; Li, M. Analysis of genetic diversity of *Capsicum* germplasm resources by using SSR markers. *Biotechnol. Bull.* **2006**, *S1*, 337–341.

Article

Evaluation of 130 Eggplant (*Solanum melongena* L.) Genotypes for Future Breeding Program Based on Qualitative and Quantitative Traits, and Various Genetic Parameters

Md. Shalim Uddin [1], Masum Billah [2], Rozina Afroz [1], Sajia Rahman [1], Nasrin Jahan [1], Md. Golam Hossain [1], Shamim Ara Bagum [1], Md. Sorof Uddin [1], Abul Bashar Mohammad Khaldun [1], Md. Golam Azam [1], Neelima Hossain [1], Mohammad Abdul Latif Akanda [1], Majid Alhomrani [3], Ahmed Gaber [4,*] and Akbar Hossain [5,*]

[1] Bangladesh Agricultural Research Institute, Gazipur 1701, Bangladesh; shalimuddin40@gmail.com (M.S.U.); rafrozbd@yahoo.com (R.A.); nazbau@yahoo.com (S.R.); nasrin.jahan83@gmail.com (N.J.); hossainbari07@gmail.com (M.G.H.); happyshalim@yahoo.com (S.A.B.); sorof79@gmail.com (M.S.U.); abkhaldun@gmail.com (A.B.M.K.); kbdrashedbari@gmail.com (M.G.A.); neelima.hossain@yahoo.com (N.H.); alatifakanda@gmail.com (M.A.L.A.)
[2] Institute of Cotton Research, Chinese Academy of Agricultural Sciences, Anyang 455000, China; kazimasum.agpstu20@gmail.com
[3] Department of Clinical Laboratories Sciences, The Faculty of Applied Medical Sciences, Taif University, P.O. Box 11099, Taif 21944, Saudi Arabia; m.alhomrani@tu.edu.sa
[4] Department of Biology, College of Science, Taif University, P.O. Box 11099, Taif 21944, Saudi Arabia
[5] Department of Agronomy, Bangladesh Wheat and Maize Research Institute, Dinajpur 5200, Bangladesh
* Correspondence: a.gaber@tu.edu.sa (A.G.); akbarhossainwrc@gmail.com (A.H.)

Citation: Uddin, M.S.; Billah, M.; Afroz, R.; Rahman, S.; Jahan, N.; Hossain, M.G.; Bagum, S.A.; Uddin, M.S.; Khaldun, A.B.M.; Azam, M.G.; et al. Evaluation of 130 Eggplant (*Solanum melongena* L.) Genotypes for Future Breeding Program Based on Qualitative and Quantitative Traits, and Various Genetic Parameters. *Horticulturae* **2021**, *7*, 376. https://doi.org/10.3390/horticulturae7100376

Academic Editor: Yuyang Zhang

Received: 2 September 2021
Accepted: 3 October 2021
Published: 8 October 2021

Publisher's Note: MDPI stays neutral with regard to jurisdictional claims in published maps and institutional affiliations.

Copyright: © 2021 by the authors. Licensee MDPI, Basel, Switzerland. This article is an open access article distributed under the terms and conditions of the Creative Commons Attribution (CC BY) license (https://creativecommons.org/licenses/by/4.0/).

Abstract: Eggplant is an essential widespread year-round fruit vegetable. This study was conducted using 130 local germplasm of brinjal to select diverse parents based on the multiple traits selection index for the future breeding program. This selection was performed focusing on 14 qualitative and 10 quantitative traits variation and genetic parameters namely, phenotypic and genotypic variance (PV and GV) and genotypic and phenotypic coefficients of variation (GCV and PCV), broad-sense heritability (hBS), genetic advance, traits association, genotype by trait biplot (G × T), heatmap analysis and multi-trait index based on factor analysis and genotype-ideotype distance (MGIDI). Descriptive statistics and analysis of variance revealed a wide range of variability for morpho-physiological traits. Estimated hBS for all the measured traits ranged from 10.6% to 93%, indicating that all the traits were highly inheritable. Genetic variances were low to high for most morpho-physiological traits, indicating complex genetic architecture. Yield per plant was significantly correlated with fruit diameter, fruits per plant, percent fruits infestation by brinjal shoot and fruit borer, and fruit weight traits indicating that direct selection based on fruit number and fruit weight might be sufficient for improvement of other traits. The first two principal components (PCs) explained about 81.27% of the total variation among lines for 38 brinjal morpho-physiological traits. Genotype by trait (G × T) biplot revealed superior genotypes with combinations of favorable traits. The average genetic distance was 3.53, ranging from 0.25 to 20.01, indicating high levels of variability among the germplasm. The heat map was also used to know the relationship matrix among all the brinjal genotypes. MGIDI is an appropriate method of selection based on multiple trait information. Based on the fourteen qualitative and ten quantitative traits and evaluation of various genetic parameters, the germplasm G80, G54, G66, and G120 might be considered as best parents for the future breeding program for eggplant improvement.

Keywords: eggplant; heritability; genetic advance; multi-trait selection; principal component analysis

1. Introduction

Eggplant or brinjal (*Solanum melongena* L.; 2n = 2x = 24) is considered a rich member of the species Solanaceae, which contains approximately 1300 species. It can be grown in diversified climatic conditions of various ecological regions. It possesses high species richness with considerable flexibility of phenotypic adaptability that made the species the most important vegetable economically. Eggplant is a general term for various Solanum species cultivated for their fruits, including the East Asian aubergine (*S. melongena* L.) and the two African native eggplants, Scarlet (*S. aethiopicum* L.) and Gboma (*S. macrocarpon* L.) [1]. Eggplant has become prominent due to its health-promoting properties. Therefore, eggplant and its relatives have numerous medicinal applications, with 77 distinct medicinal properties [2]. Eggplant is a widespread vegetable that grown from the subtropics to the Mediterranean region, popularly in Asia, Africa, and the southern part of the USA, with significant production in 2019 globally (55.15 million tons). Asia produces more than 90% of global eggplant production with 87% of the growing area coverage [3,4]. It ranks second most-produced vegetable after potato in Bangladesh.

It occupies roughly 15% of total vegetable farmland and produces about 8% of total vegetable production [5]. However, wild forms grow in sympatry with landraces and cultivars throughout their distinct areas of origin and domestication. Natural gene flow between wild and cultivated materials, followed by natural and human selection, has resulted in intermediate phenotypes that correlate with many wild features. The contribution of features of breeding importance to diversity is unequal. The variables that contributed the most to the divergence between accessions in Indian landraces of *S. melongena* were yield per plant, fruit width, number of long-styled flowers per plant, flowering earliness, total phenolic content, and ascorbic acid content [6]. Nevertheless, the results depend on the sample size used. To date, there has not been any large-scale study of a representative sample containing the complete phenotypic diversity of each cultivated eggplant. In Bangladesh, eggplant is grown throughout the country, however, the yield is not sufficient due to the lack of improved and desired variety and remarkable infestation of insect pests. However, morphological characterization has been useful in studying the relationship and diversity of various eggplant varieties. The European Eggplant Genetic Resources Network (EGGNET) defined the morphological characterization for eggplant [7], which has been validated and used in the characterization of eggplant breeding materials in numerous studies [8–10]. Therefore, creating variation through mutation, hybridization, and biotechnology approaches is an expensive and time-consuming method [8]. Consequently, characterizing collected germplasm (populations) is required to identify lines suitable for new variety development [11]. Plant breeders are interested in genetic diversity studies based on qualitative and quantitative traits because such traits can be scored quickly and easily using low-cost methods.

The phenotypic variation of fruits, plants, and other interesting traits has been demonstrated in many articles on Solanum or *S. aethiopicum* or two or more eggplant species [12–19]. Moreover, summarizing the phenotypic diversity of eggplants following the Mendelian or quantitative heredity patterns of traits of interest have been widely studied in many reports [20–22]. Different scientists [23–25] studied in-depth genetic diversity, heritability and genetic advance in eggplant genotypes. Consequently, breeders face a challenge in selecting genotypes that combine high yields in multiple attributes, which requires a reliable decision support tool. In plant breeding studies, a strong selection approach can save a lot of time and resources.

The Smith–Hazel Index (SH index) is widely utilized in plant breeding as a multi-trait selection index [26]. Reversing a phenotypic covariance matrix and a vector of economic weights is required to calculate the SH index. As a result of the presence of multicollinearity, poorly conditioned matrices and biased index coefficients would occur, affecting genetic gain estimations [26]. To account for the multicollinearity issue in multi-trait indexes, a combination of multivariate approaches is effective in overcoming their limits. FAI-BLUP is a factor analysis-based model in which each ideotype's factorial scores are created based on desirable and undesirable elements [26]. Then, depending on the genotype-ideotype distance, a geographic probability is calculated allowing genotype ranking. Olivoto and Nardino [27] offered a new multi-trait genotype-132 ideotype distance index (MGIDI) and the entire current index is combined with the exercise calculations in the R-Metan package, which contains all the functions required for genotype selection in plant breeding programs.

In Bangladesh, however, numerous genotypes of eggplant are available. The Bangladesh Agricultural Research Institute (BARI Plant)'s Genetic Resource Center (PGRC) collects and preserves several types of eggplant germplasm from all around Bangladesh. Studying the level of accessible diversity in a crop development program is a crucial stage in crop improvement, which can be accomplished through the collection and evaluation of germplasm. Therefore, the present study was carried out to determine the inherent variation of local eggplant germplasm to identify the promising germplasm that exhibits genetic diversity for crop improvement programs through advanced multi-disciplinary analysis.

2. Material and Methods

2.1. Experimental Site

The experiment was executed at PGRC of BARI at Gazipur, Bangladesh during winter (Rabi season) 2019–2020 at 23.988929 N latitude, 90.412393 E longitudes and 8.40 m above sea level. The soil in the test field was silty clay with a pH of 6.

2.2. Experimental Materials, Treatments, Desing and Procedures

A total of 130 genotypes (126 locally collected germplasm and Bangladesh has developed and released four Bt brinjal varieties expressing Cry1Ac gene (Bt brinjal), viz., BARI Bt brinjal-1,2,3 and 4 variety as a check) were used in the experiment (Table S1).

All 130 genotypes were arranged in an augmented randomized complete block design (augmented RCBD) with four check varieties and seven blocks was followed in this study. All check varieties received seven replications, giving a total of 154 experimental plots. The plot size was 3 m × 2.1 m. Each genotype was implanted in three rows per plot. The spacing was 70 × 60 cm.

Direct seeding was completed with inside the well-organized seedbeds on 12 November 2019. Thirty-day-old seedlings were transplanted in the organized pits of the predominant experimental plot on 22 December 2019.

Fertilizer doses were 10 tons ha^{-1} Cowdung, 210 kg ha^{-1} Urea, 33 kg ha^{-1} triple superphosphate (TSP), 200 kg ha^{-1}, MP and 5 kg ha^{-1} Borax ([28]. The total requirement of Cow-dung, TSP and Borax was applied during final land preparation about one week before transplanting. Urea and MP were supplied in the three equal splits [29]. Four times weeding and mulching were carried out in the first 25 days of mid-December. Sumithion 60 EC at 2.5 ml L^{-1}, Sevin 75 WP at 0.1 g pit^{-1} and Vertimac 18 EC at 1.2 ml L^{-1} were sprayed for controlling insect and mite, respectively. The data was noted as per the descriptor developed by IBPGR, 1990.

2.3. Data Recorded

Fourteen qualitative traits and ten quantitative traits (Table 1) were considered during the morphological characterization based on EGGNET [7] and IBPGR descriptors [30].

Quantitative Traits Measuring

Data on the number of days required from planting to the first opening of the flower (for example early genotypes took <91 days), optimum genotypes took 91–105 days, and late genotypes took >105 days).

Data on plant height was recorded from five randomly selected plants at the edible fruiting stage (for example, short (~30 cm), intermediate (~60 cm), and tall (61–100 cm).

Table 1. Fourteen qualitative traits and ten quantitative traits.

Sl. No.	Trait	Code	Sl. No.	Trait	Code
	Qualitative Traits			**Quantitative Traits**	
1.	Plant Growth Habit	PGH	15.	Days to First Flowering	DFF (Day)
2.	Leaf Blade Lobing	LBL	16.	Plant Height	PH (cm)
3.	Leaf Blade Tip Angle	LBTA	17.	Fruit Diameter	FD (cm)
4.	Leaf Prickles	LP	18.	Fruit Length	FL (cm)
5.	Leaf Hairs	LH	19.	Fruit Weight	FW (g)
6.	Corolla Color	CC	20.	Normalized Difference Vegetation Index	NDVI
7.	Fruit Calyx Prickles	FCP	21.	Single Leaf Area	SLA (cm2)
8.	Fruit Color Distribution	FCD	22.	Soil Plant Analyses Development	SPAD
9.	Fruit Curvature	FC	23.	Total Number of Fruits	TF
10.	Fruit Apex Shape	FAS	24.	Yield Per Plant	YPP (kg)
11.	Fruit Cross Section	FCS			
12.	Fruit Color at Ripening	FCR			
13.	Fruit Flesh Density	FFD			
14.	Fruit Position	FP			

The normalized difference vegetation index (NDVI) was measured by hand green seeker (Trimble) and green seeker RT100 (Agri Optics). Quantification of two natural light sources (near infrared-NIR and red light) is measured by NDVI. These two natural lights have individual mechanisms on vegetation. For example, NIR is reflected by the vegetation, whereas red light is absorbed by the vegetation.

The NDVI formula is:

$$NDVI = \frac{NIR - Red}{NIR + Red}$$

The chlorophyll content in plant leaves was determined with a SPAD meter (Model: SPAD-502). The SPAD value was carried from the middle portion of the leaf of the tagged main shoot at the first flowering stage through using a self-calibrating Minolta chlorophyll meter. Measurements at each experimental plot consisted of an average of five readings [31].

Leaf area was calculated with an Automatic Leaf Area Meter (Leaf area meter-LICOR-3300, USA) at the first flowering stage of a single leaf. The leaf area index (LAI) of the crop at different growth stages was calculated using the equation as described by [32].

Data on fruit length, fruit diameter, mean number of edible fruits, and mean weight of edible fruits were measures from 10 randomly selected fruits.

Length of fruit was measured from base of calyx to tip of fruit (for example, very short (<1 cm), short (3–5 cm), Intermediate (6–10 cm), long (11–20 cm) and very long (>20 cm))

Fruit diameter was estimated as small (2–3 cm), intermediate (4–5 cm), large (6–10 cm), and very large (>10 cm)

The mean number of edible fruits was very low (<7), low (7–12), intermediate (13–18), high (19–24), very high (>25)).

The mean weight of edible fruit was low (<30 g), intermediate (30–60 g) and high (>60 g)).

The multi-trait index based on factor analysis and genotype-ideotype distance (MGIDI) proposed by [27] was used to select the novel donors with high performing under optimum and low nitrogen conditions. We also compared the result of the MGIDI index with the result of the Smith–Hazel (SH) index proposed by Smith (29) and Hazel (30) multiple trait index based on factor analysis and ideotype-design (FAI-BLUP) index proposed by [33]. The MGIDI [33] was computed as follows:

$$MTSI_i = \left[\sum_{j=1}^{f} \left(\gamma_{ij} - \gamma_j \right)^2 \right]^{0.5}$$

where MGIDI*i* is the distance index of multi-trait genotype-ideotype for the *i*th genotype, γij is the score of the ith genotype in the jth factor I = 1, 2, ... , g; j = 1, 2, ... , f), g and f are the number of genotypes and factors, respectively, and Fj is the ideotype's jth score. This means that the genotype with the lowest MGIDI is closest to the ideotype, and as a result, it has the ideal values for all the analyzed features.

2.4. Statistical Analysis

Statistical analyses were performed under the R-statistics platform (software version 4.0.2) [34]. Analyses of variance (ANOVA) for each trait were assessed by using the R package 'augmented RCBD' [35]. Phenotypic and genetic variance (PV and GV), along with genotypic and phenotypic coefficients of variation (GCV and PCV), were calculated using the formula provided by [36]. Broad sense heritability (hBS), genetic advance (GA) calculation as formula elucidated in [37]. Components of the phenotypic variance of each trait were estimated using restricted maximum likelihood methods. For the estimation of variance components of linear mixed-effect "lmer", lme4 package was employed. R package ggplot2, scales and GGally were used for heatmap analysis.

The hierarchical clustering was performed using Spearman's rank correlation algorithm. Principal component analysis (PCA) was performed using R package ggplot2, ggfortify, usethis, devtools, plyr, scales and grid. Using a two-way matrix of 10 characteristics and 130 genotypes, a G × T biplot was constructed. The first two PCs were plotted. Genotypes were schemed according to scores on each PC, and traits were plotted based on the eigenvectors on each PC. The genotypic, phenotypic variance and broad-sense heritability were estimated using Agricola R-package [38] with "metan" package. Mathematic figures were plotted using the ggplot2 package [39].

3. Results

3.1. Qualitative Traits

To determine the variability of the examined germplasm, different qualitative traits were evaluated (Tables 2 and 3).

Table 2. Variability in growth and foliage traits in eggplant germplasm.

Trait Names	Descriptor State	No. of Germplasm	Germplasm (%)
Plant Growth Habit	Upright	23	17.69
	Intermediate	64	49.23
	Prostrate	43	33.08
Leaf Blade Lobing	Weak	59	45.38
	Intermediate	43	33.08
	Strong	23	17.69
	Very strong	5	3.85
Leaf Blade Tip Angle	Very acute	6	4.62
	Acute	28	21.54
	Intermediate	64	49.23
	Obtuse	32	24.62
Leaf Prickles	Very few	13	10.00
	Few	38	29.23
	Intermediate	72	55.38
	Many	5	3.85
	Very many	2	1.54

Table 2. *Cont.*

Trait Names	Descriptor State	No. of Germplasm	Germplasm (%)
Leaf Hairs	Very many	128	98.46
	Few	2	1.54
CorollaColor	Light Violet	41	31.54
	Pale Violet	84	64.62
	White	5	3.85

Table 3. Variability in fruit traits in eggplant germplasm.

Descriptor Name	Descriptor State	No. of Germplasm	Germplasm (%)
Fruit Calyx Prickles	Very few	3	2.31
	Few	31	23.85
	Intermediate	49	37.69
	Many	32	24.62
	Very many	15	11.54
Fruit Color Distribution	Uniform	18	13.85
	Mottled	21	16.15
	Netted	18	13.85
	Striped	73	56.15
Fruit Curvature	None (fruit straight)	68	52.31
	Slightly curved	37	28.46
	Curved	25	19.23
Fruit Apex Shape	Rounded	39	30.00
	Depressed	91	70.00
Fruit Cross Section	Circular (no grooves)	130	100.00
Fruit Color at Ripening	Milky white	2	1.54
	Lilac grey	7	5.38
	Purple	32	24.62
	Green with mottled at the distal end	51	39.23
	Green with a yellowish stripe	37	28.46
	Purple with light green at the distal end	3	2.31
Fruit Flesh Density	Very loose (spongy)	74	56.92
	Loose (crumbly)	24	18.46
	Average density	21	16.15
	Dense	7	5.38
	Very dense	4	3.08
Fruit Position	Pendant	130	100.00

3.2. Variability in Growth and Foliage

All the traits studied related to plant growth habits and foliage exhibited noticeable variation among the germplasm except leaf hairs and corolla Color (Table 2). Plant growth habit was observed in upright (17.69%), intermediate (49.23%) and prostrate (33.08%). Leaf-blade lobing were exhibited four categories as Weak (45.38%), Intermediate (33.08%), Strong (17.69%), and very strong (3.85%). The leaf blade tip angles were also exhibited as four categories such as very acute (4.62%), Acute (21.54%), Intermediate (49.23%) and Obtuse (24.62%). Leaf prickles were found in maximum variation such as very few (10.0%), few (29.23%), intermediate (55.38%), many (3.58%) and very many (1.54%) in the studied germplasm. Leaf hairs and showed minimum variation. Corolla Colors were found light violet (31.54%), pale violet (64.62%) and white (3.85%). All the fruits traits displayed distinctive variation among the genotypes except fruit cross-section and fruit position (Table 3). The maximum variation was identified in overall leaf prickles, fruit calyx prickles, fruit color at ripening and fruit flesh density.

3.3. The Analysis of Variance and Frequency Distribution of Quantitative Traits

The germplasm panel, consisting of 126 accessions with four check varieties, was tested for characterized through different morphological traits including 14 qualitative and 10 quantitative traits. Analysis of variance (ANOVA) showed high significant variation among the accessions with a check for all the investigated traits. ANOVA for all the traits also revealed highly significant differences among the check, accession by check interaction as well as among the accession (Table 4) excluding normalized difference vegetation index (NDVI), plant height (PH), single leaf area (SLA), and yield per plant (YPP) for check varieties. Adjusted blocks are insignificant for all the traits. All the traits matched with normal distribution except for some traits skewed left and some were right (Figures 1 and 2).

Table 4. Analysis of variance of the tested quantitative traits.

Traits	Source of Variation					
	Accession (G) with C (df = 129)	Check (C) (df = 3)	Accession (G) vs. C (df = 1)	Accession (G) (df = 125)	Adjusted Block (B) (df = 6)	Residuals (df = 18)
DFF	59.61 **	161.62 **	0.7 ns	57.63 **	20.31 ns	16.67
FD	390.72 **	586.56 **	766.74 **	383.01 **	31.44 ns	27.71
FL	3.91 **	5.87 **	7.67 **	3.83 **	0.31 ns	0.28
FW	8.63 **	49.04 **	0.36 ns	7.72 **	1.67 ns	1.27
NDVI	0.01 **	0.01 ns	0.07 **	0.01 **	0.0023 ns	0.0028
PH	169.94 **	18.57 ns	4014.75 **	142.81 **	3.14 ns	7.57
SLA	1572.19 **	71.58 ns	40431.19 **	1297.33 **	8.38 ns	22.74
SPAD	59.63 **	235.78 **	637.78 **	50.78 **	17.9 ns	9.96
TF	104.55 **	128.42 **	68.89 **	104.26 **	2.54 ns	4.25
YPP	9.59 **	0.9 ns	3.89 ns	9.84 **	0.97 ns	1.12

ns, non-significant at $p > 0.05$; **, significant at $p <= 0.01$, DFF = Days to first flowering (day), FD = fruit diameter (cm), FL = Fruit length (cm), FW = Fruit weight (g), YPP = Yield plant^{-1} (kg), NDVI = Normalized difference vegetation index, PH = Plant height (cm), SLA = Single leaf area (cm^2), SPAD = Soil plant analyses development, TF = Total number of fruits.

3.4. Descriptive Statistic of the Traits

For morpho-physiological traits evaluated in this study, their descriptive statistics including means and standard error (std.Error), standard deviations (std.Dev), minimum (Min), maximum (Max), skewness and kurtosis are summarized in Table 5. All the phenotypic values are shown a wide range of variability. Plant height (PH) ranged from 87.82 to 139.57. The NDVI value ranged from 0.33 to 0.78, SPAD ranged from 25.37 to 66.97, single leaf area (SLA) ranged from 86.12 to 242.85 cm^2, DFF ranged from 70.21 to 80.5 days, FD ranged from 0.77 to 9.47 cm, FL ranged from 3.71 to 24.93 cm, FW ranged from 10.00 to 650.40 g, TF ranged from 10.21 to 49.21, and YPP ranged from 0.24 to 10.57 kg.

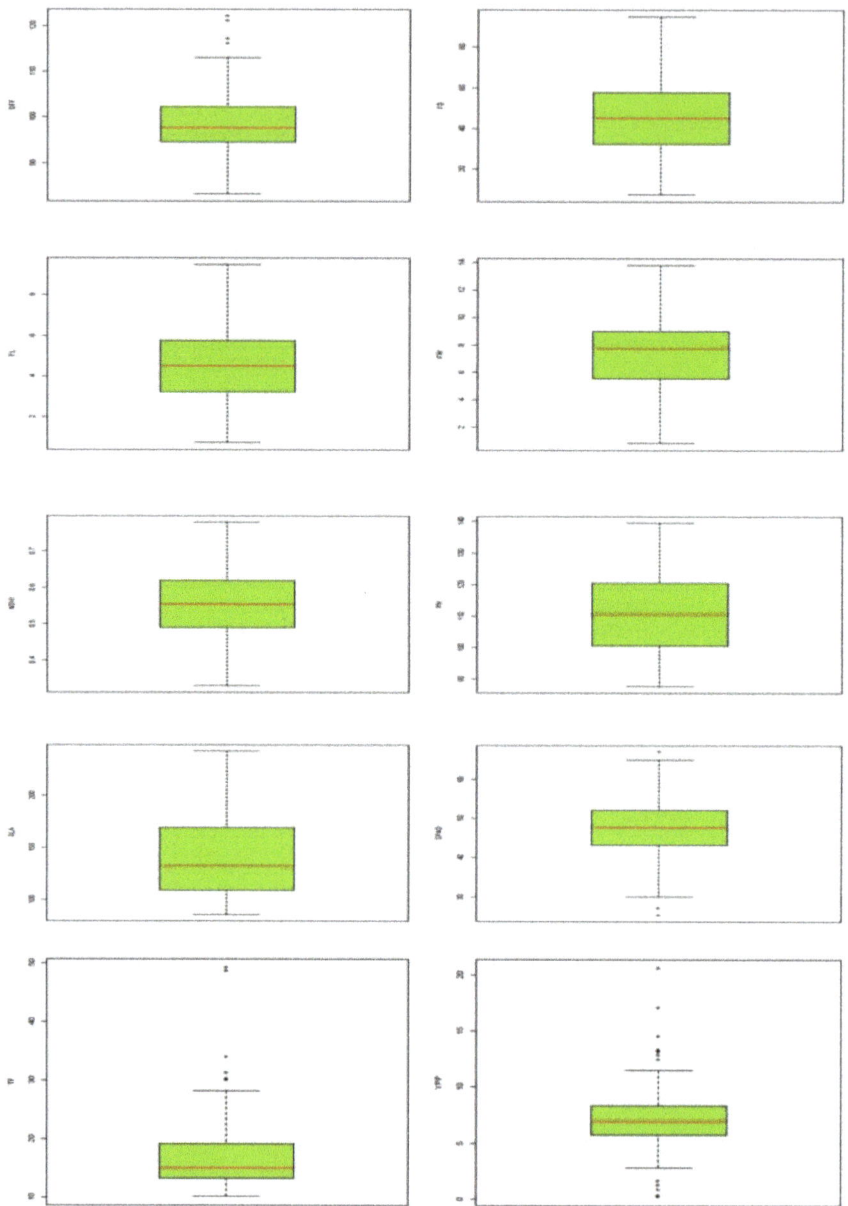

Figure 1. Box plots showing the pattern of the measured traits of germplasm. DFF = Days to first flowering (day), FD = Fruit diameter (cm), FL = Fruit length (cm), FW = Fruit weight (g), YPP =Yield plant^{-1} (kg), NDVI = Normalized difference vegetation index, PH = Plant height (cm), SLA = Single leaf area (cm^2), SPAD = Soil plant analyses development, TF = Total number of fruits.

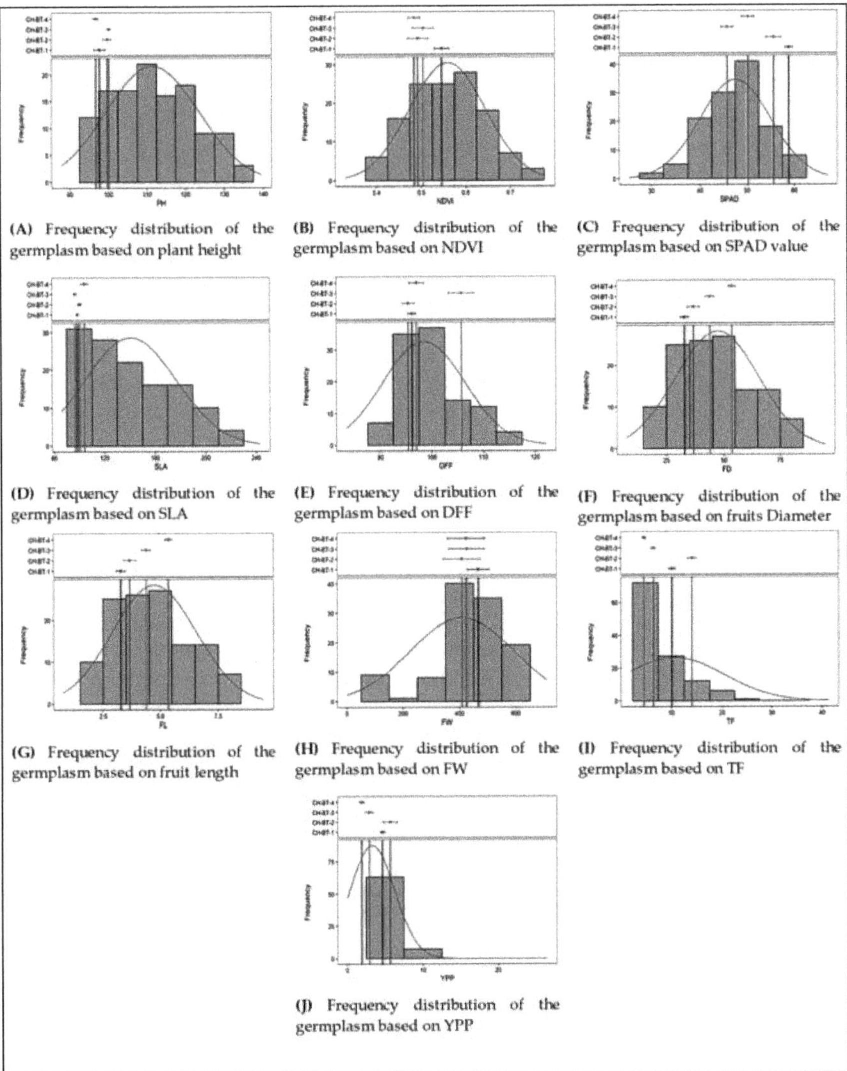

Figure 2. Frequency distribution of the germplasm based on quantitative traits: (**A**) Plant height, (**B**) NDVI, (**C**) SPAD value, (**D**) SLA, (**E**) DFF, (**F**) fruits diameter, (**G**) fruit length, (**H**) FW, (**I**) TF and (**J**) YPP of eggplant.

Table 5. Descriptive statistics of measured traits.

Trait	Mean	Std.Error	CV	Min	Max	Skewness	Kurtosis
PH	111.4	1.05	2.52	87.82	139.57	0.21 ns	2.25 *
NDVI	0.56	0.01	9.55	0.33	0.78	0.11 ns	2.7 ns
SPAD	47.5	0.66	6.54	25.37	66.97	−0.07 ns	3.37 ns
SLA	140.53	3.18	3.55	86.12	242.85	0.56 **	2.37 ns
DFF	98.4	0.7	4.15	83.32	122.07	0.46 *	3.27 ns
FL	47.25	1.6	11.35	3.71	24.73	0.47 *	2.71 ns
FD	4.72	0.16	11.35	0.77	9.47	0.47 *	2.71 ns
FW	7.55	0.23	39.52	10.00	650.40	0.14 ns	2.69 ns
TF	18.65	0.86	16.27	10.21	49.21	2.25 **	7.16 **
YPP	7.12	0.28	46.14	0.24	10.57	0.54 *	5.41 **

ns, non-significant; *, ** indicate the significance at 5% and 1% level of probability.

Skewness is a measure of the asymmetry and kurtosis is a measure of 'peakedness' of a distribution. The skewness and kurtosis were non-significant for all the traits except PH, TF and YPP indicating all the traits fitted with a normal distribution (Figure 2). The traits SLA, DFF, FD, TF and YPP were significant, and the distribution is positively skewed, which means that more accessions are below the mean than expected in a normal distribution. Only the trait SPAD was non-significant, and the distribution is negatively skewed, which means that more accessions above the mean than expected in a normal distribution (Figure 2). The traits PH, TF and YPP were significant and positive for kurtosis which means heavily leptokurtic distributions (Figure 2).

The results pertaining to genetic parameters viz., phenotypic coefficient of variation (PCV), genotypic coefficient of variation (GCV), broad-sense heritability (h^2 BS), and genetic advance as percent of the mean (GAM) for all the 10 traits are summarized in Table 6. TF had the highest PCV (54.76%) and GCV (53.64%), followed by FD, FL, FW, and YPP. High PCV and high GCV suggesting that these traits were under the influence of genetic control. The traits PH, NDVI, and SPAD were noted for moderate magnitudes of both PCV and GCV respectively. The traits DFF recorded for low magnitudes of both PCV and GCV, respectively.

Table 6. Estimation of statistical and genetic parameters of yield and its contributing traits of different eggplant germplasm.

Traits	PV	GV	GCV	GCV	PCV	PCV	hBS	hBS	GA	GAM	GAM
PH	142.81	135.24	10.44	Medium	10.73	Medium	94.7	High	23.35	20.96	High
NDVI	0.01	0.0047	12.26	Medium	15.46	Medium	62.92	High	0.11	20.07	High
SPAD	50.78	40.82	13.45	Medium	15	Medium	80.38	High	11.82	24.88	High
SLA	1297.3	1274.6	25.4	High	25.63	High	98.25	High	73	51.95	High
DFF	57.63	40.96	6.5	Low	7.71	Low	71.07	High	11.13	11.31	Medium
FD	383.01	355.3	39.9	High	41.42	High	92.77	High	37.45	79.27	High
FL	3.83	3.55	39.9	High	41.42	High	92.77	High	3.75	79.27	High
FW	7.72	6.46	33.65	High	36.8	High	83.61	High	4.79	63.47	High
TF	104.26	100.01	53.64	High	54.76	High	95.92	High	20.21	108.37	High
YPP	9.84	8.72	41.49	High	44.07	High	88.63	High	5.74	80.58	High

GV = Genotypic variance, PV = Phenotypic variance, GCV = Genotypic coefficients of variation, PCV = Phenotypic coefficients of variation, hBS = broad-sense heritability, GA = Genetic advance at 5% selection intensity, GAM = Genetic advance as the percentage of the mean at 5% selection intensity, DFF = Days to first flowering (day), FD = Fruit diameter (cm), FL = Fruit length (cm), FW = Fruit weight (g), YPP =Yield Plant^{-1} (kg), NDVI = Normalized difference vegetation index, PH = Plant height (cm), SLA = Single leaf area (cm^2), SPAD = Soil plant analyses development, TF = Total number of fruits.

3.5. Variability in Fruit Traits

All the traits studied related to the fruit of eggplant showed distinct variation among the germplasm except fruit cross-section and fruit position (Table 6). The maximum variation was found in 'fruit Color at ripening stage'. Six categories of fruit Color at ripening such as milky white (1.54%), lilac grey (5.38%), Purple (24.62%), green with mottled at the distal end (39.23%), and Green with yellowish stripe (28.46%) and purple with light green at the distal end (2.31%) were observed. The next higher variation was found in 'fruit flesh density'. Very loose (spongy), loose (crumbly), average density, dense, and very dense type of fruit flesh density was found where the majority of the germplasm exhibited average density type. On the other hand, fruit calyx prickles were found as very few (1.6%), few (13.9%), intermediate (36.5%), many (23.0%), and very many (25.0%). Fruit Color distribution was exhibited as uniform (50.8%), mottled (13.1%), netted (16.8%), and striped (27.9%). No fruits were found curved in 80.56% of the germplasm. Only 8.33 % germplasm showed curved and 11.11% showed slightly curved. Fruit apexes were exhibited as two categories as rounded (40.28%) and depressed (59.72%).

3.6. Analysis of Correlation Matrix

The phenotypic correlation analysis is being used to explore a linear relationship between various traits, which was visualized in the correlation matrix (Figure 3). In this analysis, DFF displayed a significant positive correlation with FL, while NDVI was correlated negatively. FW exhibited a strong negative correlation with TF and TF showed a negative correlation with YPP. FW and PH showed a moderate positive significant correlation with YPP (Figure 3).

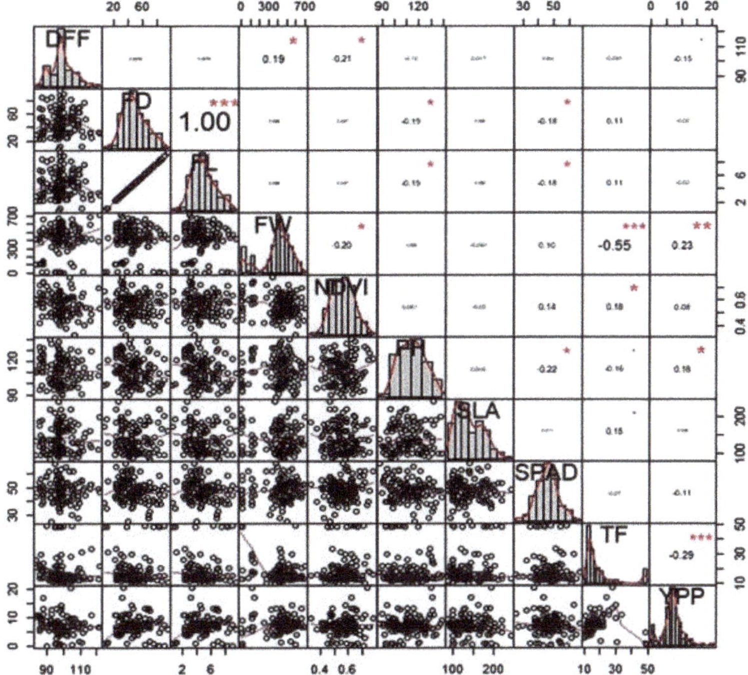

Figure 3. Correlation matrix, scatter plot and phenotypic frequency distribution of traits; * $p <= 0.05$; ** $p <= 0.01$; *** $p > 0.001$; DFF = Days to first flowering (day), FD = Fruit diameter (cm), FL = Fruit length (cm), FW = Fruit weight (g), YPP =Yield Plant^{-1} (kg), NDVI = Normalized difference vegetation index, PH = Plant height (cm), SLA = Single leaf area (cm^2), SPAD = Soil plant analyses development, TF = Total number of fruits.

3.7. Multivariate Analysis

Multivariate analysis is a tool to find patterns and relationships between several variables simultaneously. To understand the relationship among 130 eggplant genotypes with various morpho-physiological traits, principal component, biplot, and heatmap analysis were done which revealed different clusters of genotypes that performed better in different aspects. The genotypes by traits biplot were constructed from a two-way matrix of 10 morpho-physiological traits and 130 eggplant genotypes using the relative value of the trait (Figure 4). Again, biplot analysis showed the trait profiles of the genotypes, especially, those genotypes positioned far away from the origin and the results indicated a correlation between traits with genotypes. Again, traits on opposite sides of the origin are negatively correlated and traits near each other are positively correlated. Moreover, traits at 90° to each other are not correlated, concerning the origin. The principal component (PC) analysis identified a total of 10 principal components (PCs) for the morpho-physiological traits. Among them, the first two PC explained 82.26% of the entire morpho-physiological variations (Figure 4). This biplot revealed superior genotypes with higher levels of expression of favorable trait combinations. The total outcome proposed that TF, FD, SLA, FW, PH, and YPP could help to detect superior genotypes in elite germplasm.

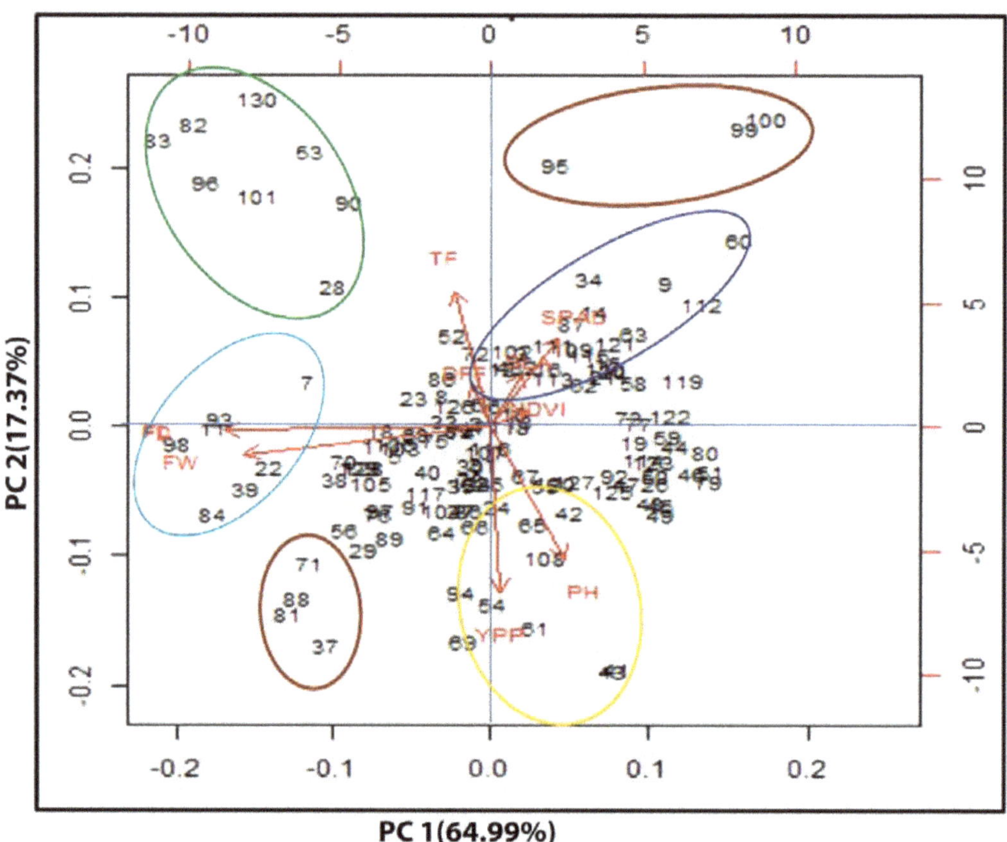

Figure 4. Genotypes by traits (G × T) biplot based on 130 germplasm and 10 quantitative traits of eggplant. DFF =Days to first flowering (day), FD = Fruit Diameter (cm), FL = Fruit Length (cm), FW = Fruit Weight (g), YPP =Yield Per Plant (kg), NDVI = Normalized Difference Vegetation Index, PH = Plant height (cm), SLA = Single Leaf Area (cm^2), SPAD = Soil Plant Analyses Development, TF = Total Number of Fruits.

3.8. Heatmap Analysis

The heatmap represented the overall performance of 10 observable traits among the 130 germplasm. A heatmap is a two-dimensional data visualization technique that uses color to show the scope of a phenomenon. Color variation by hue or intensity provides the reader with a visual representation of how the phenomenon is grouped or varies over space. It depicts the relative patterns of highly abundant features against a background of mostly low-abundance features (Figure 5).

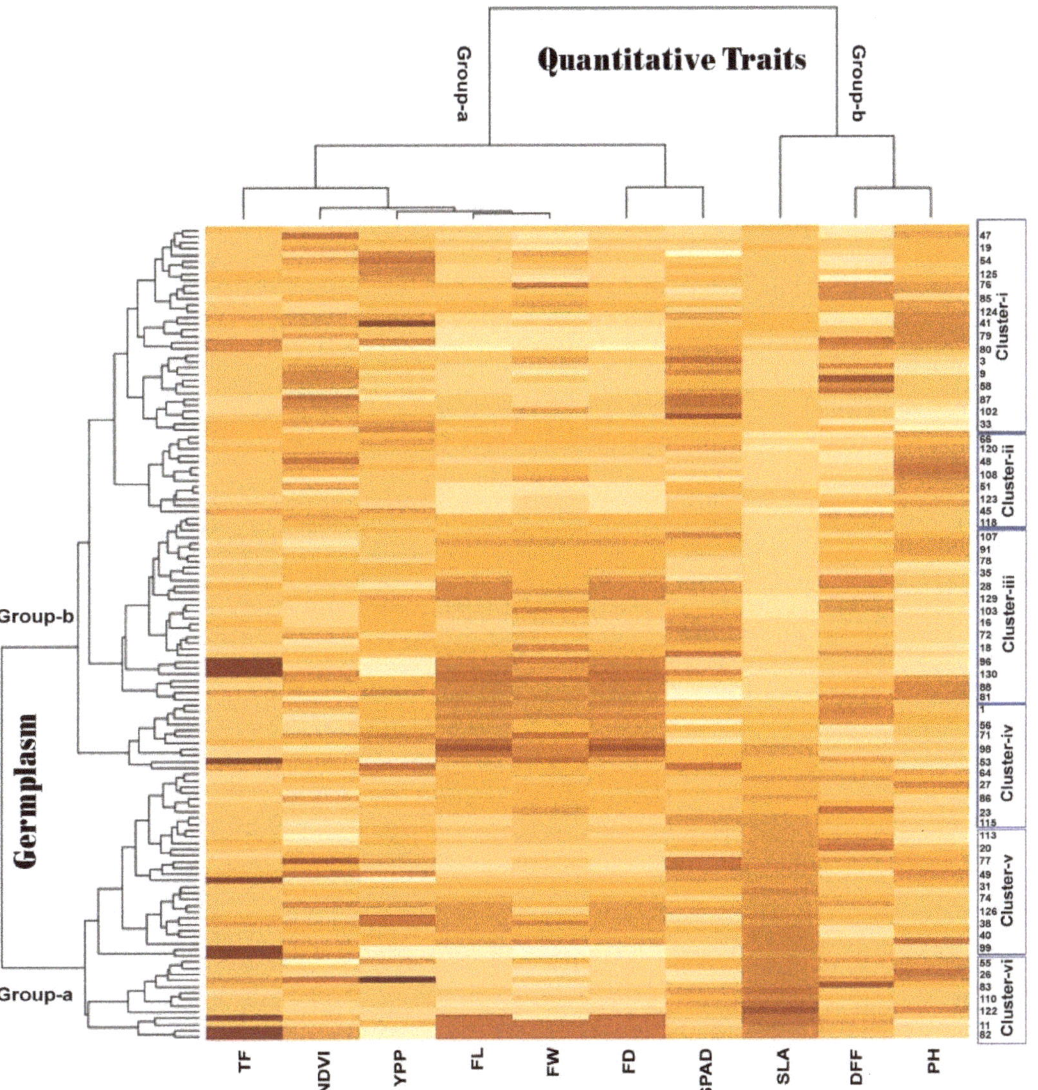

Figure 5. Heatmap showing the clustering pattern of 130 eggplant genotypes with 10 morpho-physiological traits. Heat map displaying the relationship matrix among Eggplant genotypes. The red diagonal represents a perfect relationship of each accession with itself. The symmetric off-diagonal elements represent the relationship measures for pairs of genotypes. The white of warmer colors on the diagonal show clusters of closely related genotypes. DFF = Days to first flowering (day), FD = Fruit diameter (cm), FL = Fruit length (cm), FW = Fruit weight (g), YPP = Yield plant^{-1} (kg), NDVI = Normalized difference vegetation index, PH = Plant height (cm), SLA = Single leaf area (cm^2), SPAD = Soil plant analyses development, TF = Total number of fruits.

A heatmap analysis of characteristics was performed to demonstrate a chromatic examination of the genotypes. The heatmap analysis produced two dendrograms: one in the vertical direction, representing the germplasm, and one in the horizontal direction, representing the traits that caused the diffusion. The red diagonal depicts each accession's perfect connection to itself. The relationship measurements for pairs of germplasm are represented by the symmetric off-diagonal elements. Based on the morpho-physiological properties of the germplasm studied, six clusters emerged through hierarchical clustering (Figure 5). Group (a) included 43 germplasm from clusters v (30 germplasm) and vi (13 germplasm), while group (b) included the remaining 87 germplasm from four clusters. Out of a total of 87 germplasm, cluster-i received 33, cluster-ii received 13, cluster-iii received 30, and cluster-iv received 11 (Figure 5). Dendrogram two also revealed two significant groups: group (a) is associated with seven traits (FD, FL, FW, YPP, NDVI, SPAD, and TF), while group (b) is associated with three traits (DFF, PH, and SLA) (b). Surprisingly, the dendrogram two groups and sub-groups revealed the disparity effects of different eggplants.

3.9. Multi-Trait Index Based on Factor Analysis and Genotype-Ideotype Distance (MGIDI)

The MGIDI index was intended to select the genotypes with respect to considering all measured traits. Based on the analysis, a highly significant genotypic effect was noted for 10 measured traits involving DFF (Day), PH (cm), FD, FL, FW, NDVI, SLA (cm^2), SPAD, TF, and YPP (kg) (Table 1). The broad-sense heritability (h^2) ranged from 71.07% (for DFF) to 98.25% (for SLA). All filtered traits are evaluated with high heritability values, which indicates that the selection gain of these traits is promising. Among selected traits, traits FD, FL, FW, TF, and YPP showed the highest genetic advanced mean. However, the genotypes selected using the MGIDI index were G80, G54, G66, G120, G46, G61, G65, G108, G4, G79, G42, G77, G47, G50, G51, G43, G44, G48, and G49 (Figure 6). The strengths and weaknesses of all the genotypes are shown in Figure 7.

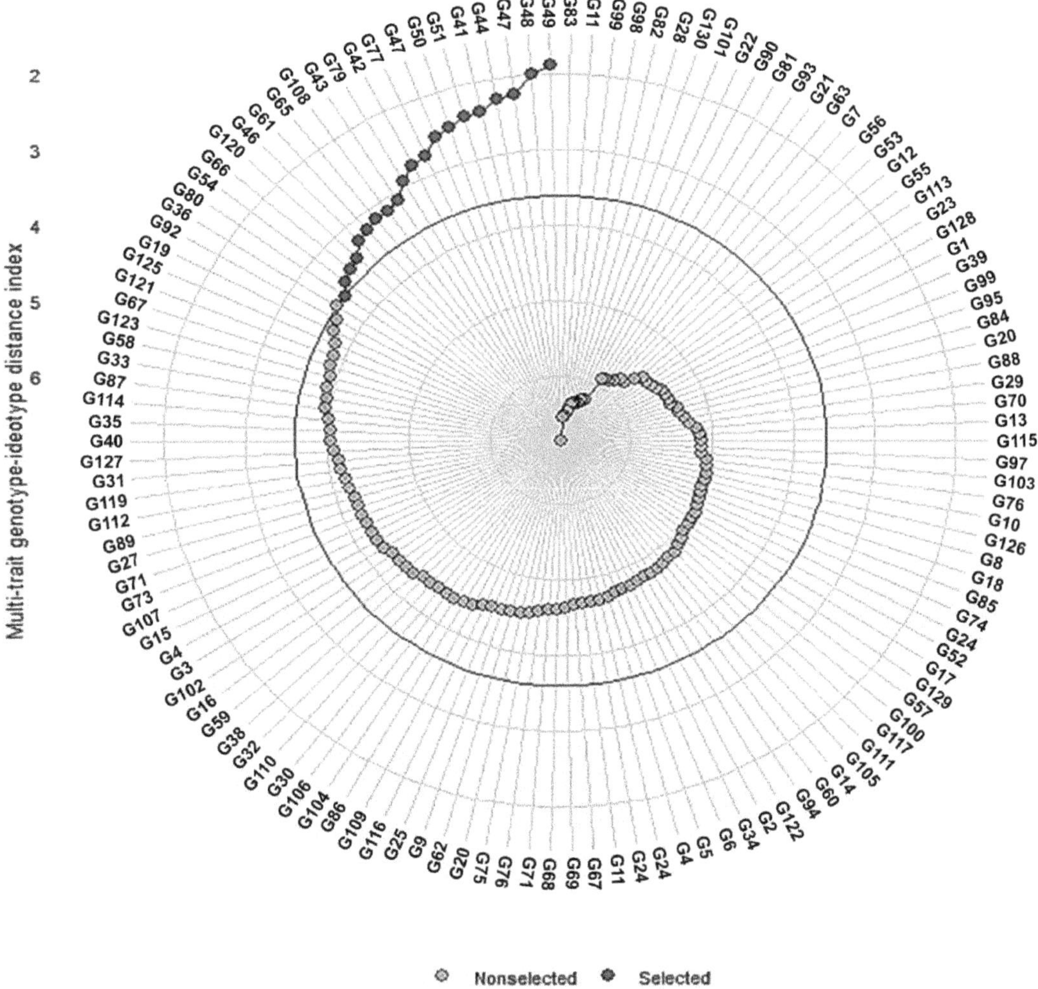

Figure 6. Germplasm ranking and selected germplasm from 130 local germplasm through multi-trait genotype-ideotypes distance index (MGIDI) considering 15% selection intensity.

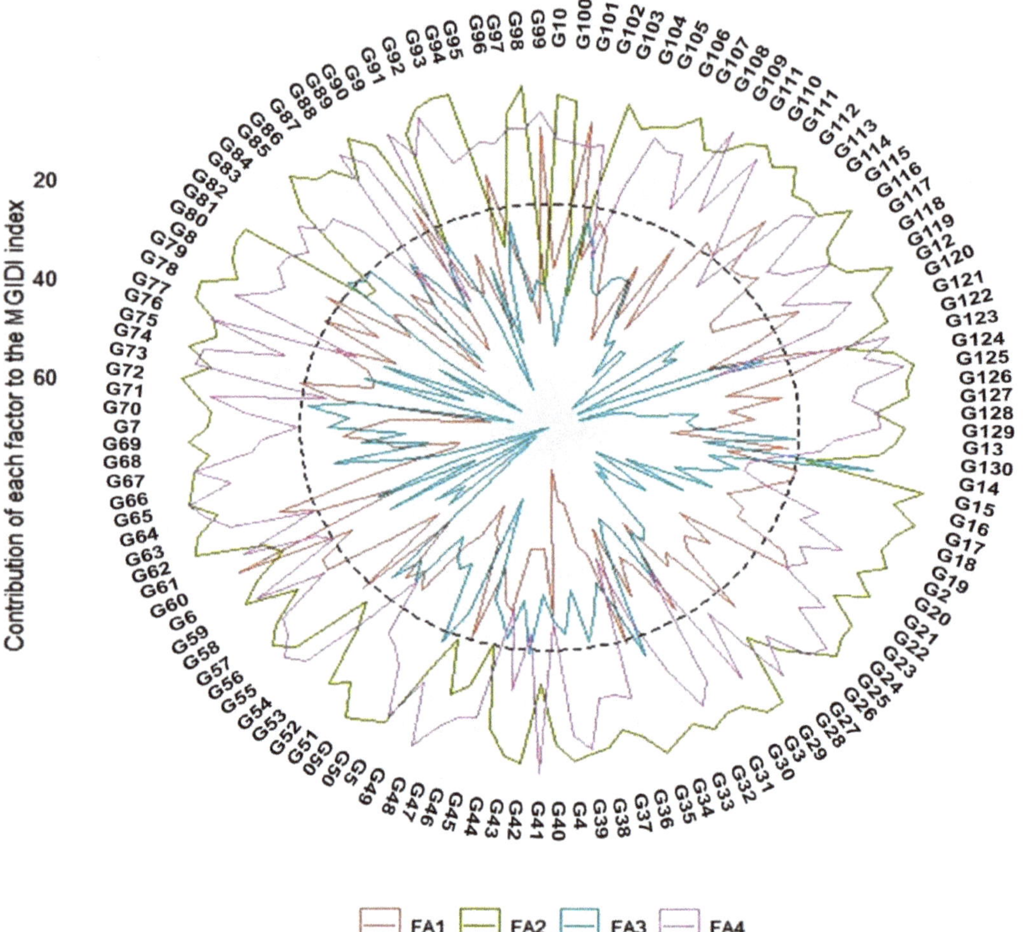

Figure 7. The strengths and weaknesses of all the genotypes.

4. Discussion

In the present study, numerous qualitative and quantitative features were assessed in 130 brinjal genotypes in order to discover superior genotypes. Genetic diversity research is critical for the successful evaluation, preservation, and use of germplasm resources [40]. The breeding strategy is mostly determined by the degree of genetic variation, and morphological characteristics are viewed as a critical initial step in characterizing and identifying plant genetic resources [41]. Screening for qualitative and quantitative features is crucial for determining a plant's socioeconomic preferences.

4.1. Qualitative Traits

All qualitative traits are found to have significant variations except leaf hairs, corolla color, fruit cross-section, and fruit position (Tables 2 and 3). Some other previous reports also published a similar type of fruit curvature [42–44]. Eggplant is a herbaceous plant, mostly upright in nature [45]. Our results demonstrated that 17.69% upright, 49.23% intermediate, and 33.08% prostrate growth habit among all the studied genotypes

(Table 2). Shekar et al. [46] clustered the eggplant plants into upright and intermediate. Islam et al. [42] saw 48% intermediate, 45% upright, and 7% prostrate growth habit at vegetative stage among the studied genotypes. They suggested that the plant growth habit is used for the identification of eggplant varieties.

As per leaf blade length, eggplant genotypes were divided into two groups, viz., intermediate and short [47]. They reported that 30 genotypes had intermediate leaf blade length while five genotypes showed short leaf blade length. Similarly, based on leaf blade width, the genotypes were divided into three types, viz., wide, intermediate, and narrow [45]. Osei et al. [13] found significant variation in leaf blade length and width in eggplant. In the current study, we distributed eggplant genotypes into three types, viz., very strong (5 genotypes), strong, intermediate (43 genotypes), and weak (59 genotypes), based on leaf blade lobing (Table 2). Sunseri et al. [48] reported 60% weak, 37% intermediate, and 3% strong leaf blade lobbing in eggplant. In our investigation, 6 genotypes had very acute, 28 had acute leaf blade tip angle followed by intermediate leaf blade tip angle, which includes 64 genotypes, while 32 genotypes each had obtuse. Sunseri et al. [48] observed 62% acute, 33% very acute, and 5% intermediate leaf blade tip angle in eggplant. Dash et al. [49] noticed acute leaf blade tip angle in many eggplant genotypes. A much higher variation was observed in leaf prickle (Table 2), where most of the genotype (72 genotypes) had intermediate prickle on the leaf. Sunseri et al. [48] noted that leaf prickles were absent in 25% of genotypes, whereas the remaining genotypes produced 40% very few, 25% few, and 5% both intermediate and many prickles. Tiwari et al. [50] noted prickles in stem, petiole, calyx including peduncle, and leaf including veins in eggplant. Many leaf hairs (98.46%) are found in most of the genotypes. Light violet (31.54%), pale violet (64.62%), and white (3.85%) colors were noted among the studied genotypes for corolla Color (Table 2).

In the present study, we described eight distinctive traits of eggplant fruits (Table 3). For traits related to the fruit calyx prickles, fruit color distribution, fruit color at ripening, and fruit flesh density much higher variation were observed in the cultivated eggplant (Table 3). Similar kinds of fruit calyx prickle distribution were also reported by [45]. It was noticed that 68 genotypes had no curvature on fruit; 37 genotypes showed slightly curved fruit and 25 genotypes had curved fruit. Sunseri et al. [48,51] reported sickle-shaped, snaked-shaped, curved, and U-shaped eggplant fruits along with fruits with no curvature. Eggplant genotypes were distributed into two types: rounded (39 genotypes) and depressed (91 genotypes), on the basis of fruit apex shape. Sunseri et al. [48] reported that 38, 34, and 28% of eggplant genotypes had depressed, rounded, and protruded types of fruit apex shape, respectively. Eggplant fruits traits variations are important in protection from UV irradiation, insect attack in plants as well as socio-economic value. More fruit color variation was noted by [52] and [53], which support our present study. Solaimana et al. [51] found uniform and stripped fruit color distribution. The variations in fruits color were also described by [51,54,55]. Tiwari et al. [50] divided the eggplant fruits into six color groups i.e., green (37.27%), purple (25.45%), milky white (13.62%), purple-black (12.72%), light purple or lilac grey (9.09%), and scarlet red (1.08%). Fruit flesh density is an important characteristic for the determination of fruit volume and weight. A wide range of variation was observed in fruit flesh density, the highest number of genotypes had very loose flesh density (74 genotypes).

4.2. Quantitative Traits

4.2.1. Genetic Components

The proposed index of 0–10% for low, 10–20% for moderate, and 20% for high variation was used to characterize the projected GCV and PCV values. In our experiment, closer PCV and GCV values were estimated in most of the traits which possibly were less influenced by the environment suggesting the reliability of selection based on these traits. The assessed 10 quantitative traits exhibited a wide range of variation and more or less similar results were observed by [48,53,56]. The selection procedure considers the differences between traits based on the degree of heredity. In order to understand the predicted selection benefits, assessing genetic progress may be an important method for improving crops. Several studies have shown that selection can effectively use available genetic variations with a certain degree of heritability to improve specific traits [26,57]. The concern of both heritability and genetic advance is more effective over the distinctively use of heritability. We found that the phenotype variance values of all traits were higher than the genotype variance, indicating that the environment regulates the expression of traits. The same kind of outcome also was gained from several studies for various traits in eggplant [51,58–61].

The evidence of the traits having essential potential in the assortment process due to low environmental impacts was calculated using strong GA with hBS for all yield-related traits except DFF. Heterosis breeding has the potential to improve traits with poor heritability and genetic advance [28]. Regarding the hBS and GA index [62], which were greater than 60% for high, 30–60% for moderate, and 0–30% for low, we discovered that all traits were highly heritable together with a high genetic advance mean value except DFF, implying that direct selection can be effective for eggplant crop improvement based on these traits with the effect of additive genes; a more or less similar result was obtained by many researchers [63,64]. Due to the strong influence of the environment on genetic effects, low to moderate heritability and genetic advanced values will inhibit the improvement of traits. Therefore, effective selection can only be achieved by selecting higher values of GCV, PCV, hBS, and GA, which means that the influence of additive genes is more stable than the influence of the environment. High GAM was also observed for all traits except DFF. This revealed that if the selection was carried out for the next generation for these features, a greater improvement in the population mean may be seen.

4.2.2. Correlation Matrix

The plant breeding correlation matrix is an outstanding method to assess the relationship between two or more variables. For higher genotype selection procedures, considering the correlation matrix can be a scaling measure [65]. In the correlation matrix, DFF displayed a strong positive correlation with FL, while NDVI was correlated negatively, FW and PH showed a moderate positive correlation with YPP (Figure 6). A moderate to positive significant association can be proposed; selection based on these traits can help increase the yield of this crop. More or less similar results were consistent with the study of [28,64,66,67].

4.3. Multivariate Statistical

Compared with univariate and bivariate statistical methods, multivariate statistical methods can analyze more than one relationship at the same time. There are many multivariate data analysis methods, each of which has a different purpose, such as Regression analysis, factor analysis, cluster analysis, analysis of variance, discriminant analysis, etc. [68]. Biplot analysis is usually employed to assess the component effects creating the genotypic variations.

The highest values indicate the highest influence of the trait on the total variation. Biplot analysis determines varietal stability in the multi-environmental trial [69]. It describes the relationship between different genotype traits. The association between morpho-physiological traits among the 130 genotypes was observed by the biplot analysis [70]. Again, the biplot analysis showed the trait profiles of the genotypes, especially, those genotypes positioned far away from the origin and the results indicated a correlation between traits with genotypes (Figure 4). An acute angle between two elements indicates a positive correlation, and an obtuse angle between two elements indicates a negative correlation. As a result, principal components (PC) analysis provides a good screening of available genotypes and aids in the selection of possible parents for crop breeding initiatives. In our data, the first two PC accounted for 82.36% of the overall variation (Figure 4). The yield potential of accessions was represented in PC 1; thus, the accessions contributing to this component are likely to undergo direct selection, or selected parents can be used in hybridization operations. These PC1 results are consistent with the results of the correlation analysis. The figure summarizes the information of the matrix in principal components, where the cosine of the angle between the vectors connecting the objects to the origin is proportional to the correlation coefficient between these objects. The heatmap shows the highest and lowest values of each genotype in different colors against all the traits comparing. The intensity of the color indicates the degree of high or low of the traits. Hierarchical clustering based on the morpho-physiological traits of the studied germplasm revealed six clusters (Figure 5). The heatmap analysis depicted the degree of correspondence among the morphological traits assessed in brinjal genotypes, and this result was consistently supported by [71,72].

4.4. Multi-Trait Index Based on Factor Analysis and Genotype-Ideotype Distance (MGIDI)

Experienced breeders often try to combine several desired traits into a new genotype to produce high performance. When measuring multiple traits, it is often difficult to select a genotype from the ideotypes. In this regard, various multivariate methods are widely used, such as principal component analysis, factor analysis, cluster analysis, and different samples to group measured traits or select test genotypes [73]. We used a two-way heat map clustering pattern and PCA to connect test genotypes and measured attributes in this study (Figures 6 and 7), however, we could not pick specific genotypes. To make the selection of genotypes with several features easier, [27] the recently introduced MGIDI (multi-trait genotype-ideotypes distance index) is a new method for genotype selection based on multiple trait information. The eggplant genotypes were ranked based on information on measured multiple traits (Figure 7). The MGIDI index selected genotypes G80, G54, G66, G120, G46, G61, G65, G108, G4, G79, G42, G77, G47, G50, G51, G43, G44, G48, and G49 as promising eggplant genotypes. Apart from these genotypes, G80 was very close to the cut point, which recommends that this genotype can exist desirable features. Hence, the researcher should pay particular attention to assessing genotypes that are very close to the cut point [27]. The application of the MGIDI index to plant crop research is predicted to grow rapidly. Similarly, This index was used to find the best strawberry genotype [74].

5. Conclusions

The current study clearly established that the improvement of brinjal yield and related traits can be obtained through selection with the valuation of different genetic parameters analysis. Most of the qualitative traits showed distinct variations among the germplasm. Qualitatively, the maximum variation was observed in leaf prickles, fruit calyx prickles, fruit color at ripening, and fruit flesh density. Quantitatively, the highest variation was observed in fruit yield per plant which was followed by fruit weight. Nevertheless, the present study has shown that selection with the evaluation of various analyzes of genetic parameters such as GCV, PCV, hBS, and GA can achieve an improved eggplant yield and related traits. It can be observed that practically all of the agronomic traits tested in this study exhibit significant variability based on the recorded data and additional analyses (heatmap analysis, correlation matrix, PCA, MGIDI analysis). Therefore, the germplasm G80, G54, G66, and G120 might be considered as best parents based on the qualitative and quantitative characters for the future breeding program. The present findings have a great genetic potential for the studied germplasm. Concurrently, the promising germplasm identified in the present study might be used in future breeding programs for eggplant improvement.

Supplementary Materials: The following are available online at https://www.mdpi.com/article/10.3390/horticulturae7100376/s1, Table S1: Experimental materials (130 genotypes) with estimated parameters.

Author Contributions: Conceptualization, M.S.U. (Md. Shalim Uddin), M.B., R.A., S.R., N.J., M.G.H., S.A.B., M.S.U. (Md. Sorof Uddin), A.B.M.K., M.G.A., N.H. and M.A.L.A.; methodology, M.S.U. (Md. Shalim Uddin), M.B., R.A. and S.R.; software, M.S.U. (Md. Shalim Uddin) and A.H.; validation, M.S.U. (Md. Shalim Uddin) and A.B.M.K.; formal analysis, M.S.U. (Md. Shalim Uddin) and A.H.; investigation, M.S.U. (Md. Shalim Uddin), M.B., R.A., S.R., N.J., M.G.H., S.A.B., M.S.U. (Md. Sorof

Uddin), A.B.M.K., M.G.A., N.H. and M.A.L.A.; resources, M.S.U. (Md. Shalim Uddin); data curation, M.S.U. (Md. Shalim Uddin) and A.H.; writing—original draft preparation, M.S.U. (Md. Shalim Uddin), M.B., R.A., S.R., N.J., M.G.H., S.A.B., Md. Sorof Uddin, A.B.M.K., M.G.A., N.H. and M.A.L.A.; writing—review and editing M.A., A.G. and A.H.; visualization, M.S.U. (Md. Shalim Uddin), M.B., R.A., S.R., N.J. and M.G.H.; supervision, M.S.U. (Md. Shalim Uddin); project administration, M.S.U. (Md. Shalim Uddin), M.A. and A.G.; funding acquisition, M.S.U. (Md. Shalim Uddin), M.A., A.G. and A.H. All authors have read and agreed to publish the current version of the manuscript.

Funding: The current work was funded by Project Implementation Unit, National Agricultural Technology Program-Phase II Project (NATP-2) using the research fund of WB, IFAD and GoB through Ministry of Agriculture and Taif University Researchers Supporting Project number (TURSP 2020/257), Taif University, Taif, Saudi Arabia.

Institutional Review Board Statement: Not applicable.

Informed Consent Statement: Not applicable.

Data Availability Statement: Data recorded in the current study are available in all Tables and Figures of the manuscript.

Acknowledgments: The research has been conducted in Bangladesh Agricultural Research Institute (BARI) under the Program Based Research Grant (PBRG), Sub-Project entitled "Collection and Characterization of Important Plant Genetic Resources in Bangladesh." The project was approved by Project Implementation Unit National Agricultural Technology Program-Phase II Project (NATP-2), Bangladesh Agricultural Research Council (BARC), Ministry of Agriculture, Bangladesh. The authors also extend their appreciation to Taif University for funding current work by Taif University Researchers Supporting Project number (TURSP 2020/257), Taif University, Taif, Saudi Arabia.

Conflicts of Interest: The authors declare no conflict of interest.

References

1. Aubriot, X.; Daunay, M.-C. Eggplants and relatives: From exploring their diversity and phylogenetic relationships to conservation challenges. In *The Eggplant Genome*; Springer: Berlin/Heidelberg, Germany, 2019; pp. 91–134.
2. Minardi, A. Food Security in Developing Countries: The Case of "Production of Appropriate Food: Sufficient, Safe and Sustainable" Project. Ph.D Thesis, (Ph.D. in Agro-Food System, Academic Year 2017/2018). Catholic University of Sacred Heart, Via Emilia Parmense, Piacenza, Italy, 2019. Available online: http://tesionline.unicatt.it/bitstream/10280/67851/1/tesiphd_completa_Minardi.pdf (accessed on 15 July 2021).
3. Bhagirath, C.; Kadambini, G. *The Development and Regulation of Bt Brinjal in India (Eggplant/Aubergine)*; International Service for the Acquisition of Agri-biotech Applications: Ithaca, NY, USA, 2009.
4. BBS. *Statistical Yearbook of Bangladesh. Bangladesh Bureau of Statistics*; Ministry of Planning, Government of the Peoples' Republic of Bangladesh: Dhaka, Bangladesh, 2020; p. 139.
5. Prabakaran, R.K.-S.; Balakrishnan, S.; RameshKumar, S.; Arumugam, T.; Anandakumar, C. Genetic diversity, trait relationship and path analysis in eggplant landraces. *Electron. J. Plant* **2015**, *6*, 831–837.
6. Van der Weerden, G.M.; Barendse, G.W. A web-based searchable database developed for the EGGNET project and applied to the Radboud University Solanaceae database. *Acta Hortic.* **2007**, *745*, 503–506. [CrossRef]
7. Boyaci, H.F.; Topcu, V.; Akin, T.; Yildirim, I.K.; Mehmet, O.; Aktas, A. Morphological and molecular characterization and relationships of Turkish local eggplant heirlooms. *Not. Bot. Horti Agrobot. Cluj Napoca* **2015**, *43*, 100–107. [CrossRef]
8. Mat Sulaiman, N.N.; Rafii, M.Y.; Duangjit, J.; Ramlee, S.I.; Phumichai, C.; Oladosu, Y.; Datta, D.R.; Musa, I. Genetic variability of eggplant germplasm evaluated under open field and glasshouse cropping conditions. *Agronomy* **2020**, *10*, 436. [CrossRef]
9. Muñoz-Falcón, J.; Prohens, J.; Vilanova, S.; Nuez, F. Diversity in commercial varieties and landraces of black eggplants and implications for broadening the breeders' gene pool. *Ann. Appl. Biol.* **2009**, *154*, 453–465. [CrossRef]
10. Gramazio, P.; Prohens, J.; Plazas, M.; Mangino, G.; Herraiz, F.J.; Vilanova, S. Development and genetic characterization of advanced backcross materials and an introgression line population of *Solanum incanum* in a *S. melongena* background. *Front. Plant Sci.* **2017**, *8*, 1477. [CrossRef]
11. Adeniji, O.; Aloyce, A. Farmer's knowledge of horticultural traits and participatory selection of African eggplant varieties (*Solanum aethiopicum*) in Tanzania. *Tropicultura* **2012**, *30*, 185–191.
12. Datta, D.R.; Rafii, M.Y.; Misran, A.; Jusoh, M.; Yusuff, O.; Sulaiman, N.M.; Momodu, J. Genetic diversity, heritability and genetic advance of *Solanum melongena* L. from three secondary centers of diversity. *Bangladesh J. Plant Taxonom.* **2021**, *28*, 155–169. [CrossRef]
13. Osei, M.; Oluoch, M.; Osei, C.; Banful, B. Morphological characterisafion of African Eggplant (*Solanum* spp.) Germplasm in some African countries. *Agric. Innov. Sustain. Dev.* **2010**. Available online: https://library.faraafrica.org/wp-content/uploads/2019/11/Agricultural-Innovations-for-Sustainable-Development.pdf#page=171 (accessed on 15 July 2021).

14. Plazas, M.; Vilanova, S.; Gramazio, P.; Rodríguez-Burruezo, A.; Fita, A.; Herraiz, F.J.; Ranil, R.; Fonseka, R.; Niran, L.; Fonseka, H. Interspecific hybridization between eggplant and wild relatives from different genepools. *J. Am. Soc. Hortic. Sci.* **2016**, *141*, 34–44. [CrossRef]

15. Polignano, G.; Uggenti, P.; Bisignano, V.; Della Gatta, C. Genetic divergence analysis in eggplant (*Solanum melongena* L.) and allied species. *Genet. Resour. Crop. Evol.* **2010**, *57*, 171–181. [CrossRef]

16. Cericola, F.; Portis, E.; Toppino, L.; Barchi, L.; Acciarri, N.; Ciriaci, T.; Sala, T.; Rotino, G.L.; Lanteri, S. The population structure and diversity of eggplant from Asia and the Mediterranean Basin. *PloS ONE* **2013**, *8*, e73702.

17. Syfert, M.M.; Castañeda-Álvarez, N.P.; Khoury, C.K.; Särkinen, T.; Sosa, C.C.; Achicanoy, H.A.; Bernau, V.; Prohens, J.; Daunay, M.C.; Knapp, S. Crop wild relatives of the brinjal eggplant (*Solanum melongena*): Poorly represented in genebanks and many species at risk of extinction. *Am. J. Bot.* **2016**, *103*, 635–651. [CrossRef] [PubMed]

18. Taher, D.; Solberg, S.Ø.; Prohens, J.; Chou, Y.-y.; Rakha, M.; Wu, T.-h. World vegetable center eggplant collection: Origin, composition, seed dissemination and utilization in breeding. *Front. Plant Sci.* **2017**, *8*, 1484. [CrossRef] [PubMed]

19. Collonnier, C.; Fock, I.; Kashyap, V.; Rotino, G.; Daunay, M.; Lian, Y.; Mariska, I.; Rajam, M.; Servaes, A.; Ducreux, G. Applications of biotechnology in eggplant. *Plant Cell Tissue Organ Cult.* **2001**, *65*, 91–107. [CrossRef]

20. Daunay, M.-C. Eggplant. In *Vegetables II*; Springer: Berlin/Heidelberg, Germany, 2008; pp. 163–220.

21. Mutegi, E.; Snow, A.A.; Rajkumar, M.; Pasquet, R.; Ponniah, H.; Daunay, M.C.; Davidar, P. Genetic diversity and population structure of wild/weedy eggplant (*Solanum insanum*, Solanaceae) in southern India: Implications for conservation. *Am. J. Bot.* **2015**, *102*, 140–148. [CrossRef]

22. Praneetha, K.J.D.S. Evaluation of brinjal (*Solanum melongena* L.) local types for yield and its quality characters. *IJCS* **2018**, *6*, 292–297.

23. Rahman, M.; Islam, K.; Jahan, M.; Uddin, M. Efficacy of some botanicals in controlling brinjal shoot and fruit borer, Leucinodes orbonalis. *Progress. Agric.* **2009**, *20*, 35–42. [CrossRef]

24. Sharmin, D.; Meah, M.; Moniruzzaman, M. Inheritance of resistance to phomopsis blight and fruit rot in brinjal. *J. Agrofor. Environ.* **2010**, *3*, 135–140.

25. Rocha, J.R.D.A.S.D.C.; Machado, J.C.; Carneiro, P.C.S. Multitrait index based on factor analysis and ideotype-design: Proposal and application on elephant grass breeding for bioenergy. *Gcb Bioenergy* **2018**, *10*, 52–60. [CrossRef]

26. Sellami, M.H.; Lavini, A.; Pulvento, C. Phenotypic and Quality Traits of Chickpea Genotypes under Rainfed Conditions in South Italy. *Agronomy* **2021**, *11*, 962. [CrossRef]

27. Olivoto, T.; Nardino, M. MGIDI: Toward an effective multivariate selection in biological experiments. *Bioinformatics* **2021**, *37*, 1383–1389. [CrossRef] [PubMed]

28. Banerjee, S.; Verma, A.; Bisht, Y.S.; Maurya, P.; Jamir, I.; Mondal, S.; Bhattacharjee, T.; Chattopadhyay, A. Genetic variability, correlation coefficient and path coefficient analysis in brinjal germplasm. *Int. J. Chem. Stud.* **2018**, *6*, 3069–3073.

29. Guide, F.F.R. *Bangladesh Agricultural Research Council (BARC)*; BARC: Dhaka, Bangladesh, 2012.

30. Withers, L.A.; Wheelans, S.K.; Williams, J.J.E. In vitro conservation of crop germplasm and the IBPGR databases. *Euphytica* **1990**, *45*, 9–22. [CrossRef]

31. Reynolds, M.; Balota, M.; Delgado, M.; Amani, I.; Fischer, R. Physiological and morphological traits associated with spring wheat yield under hot, irrigated conditions. *Funct. Plant Biol.* **1994**, *21*, 717–730. [CrossRef]

32. Radford, P. Growth analysis formulae-their use and abuse 1. *Crop. Sci.* **1967**, *7*, 171–175. [CrossRef]

33. Rocha, J.R.D.A.S.D.C.; Nunes, K.V.; Carneiro, A.L.N.; Marçal, T.D.S.; Salvador, F.V.; Carneiro, P.C.S.; Carneiro, J.E.S. Selection of superior inbred progenies toward the common bean ideotype. *Agron. J.* **2019**, *111*, 1181–1189.

34. Pinherio, J.; Bates, D.; DebRoy, S.; Deepayan, S.; Team, R.C. Linear and Nonlinear Mixed Effects Models. R Package Version 3.1-145. 2007, Volume 3, pp. 1–89. Available online: https://ftp.uni-bayreuth.de/math/statlib/R/CRAN/doc/packages/nlme.pdf (accessed on 15 July 2021).

35. Aravind, J.; Mukesh, S.; Wankhede, D.; Kaur, V. Augmented RCBD: Analysis of Augmented Randomised Complete Block Designs. R Package Version 0.1. 2020, Volume 2. Available online: https://aravind-j.github.io/augmentedRCBD/ (accessed on 15 July 2021).

36. Falconer, D.S. *Problems on Quantitative Genetics*; Longman: London, UK, 1983.

37. Burton, G.W. Quantitative inheritance in grasses. *6th Intl. Grassland Cong. Proc.* **1952**, *1*, 227–238.

38. Olivoto, T.; Nardino, M. MGIDI: A novel multi-trait index for genotype selection in plant breeding. *Bioinformatics* **2020**, 1–22. [CrossRef]

39. Wickham, H.; Chang, W.; Wickham, M.H. Package 'ggplot2'. Create elegant data visualisations using the grammar of graphics. *R Package Version* **2016**, *2*, 1–189.

40. Sarkar, M.; Kundagrami, S. Multivariate analysis in some genotypes of mungbean [*Vigna radiata* (L.) Wilczek] on the basis of agronomic traits of two consecutive growing cycles. *Legume Res. Int. J.* **2016**, *39*, 523–527. [CrossRef]

41. Carrillo-Perdomo, E.; Vidal, A.; Kreplak, J.; Duborjal, H.; Leveugle, M.; Duarte, J.; Desmetz, C.; Deulvot, C.; Raffiot, B.; Marget, P. Development of new genetic resources for faba bean (*Vicia faba* L.) breeding through the discovery of gene-based SNP markers and the construction of a high-density consensus map. *Sci. Rep.* **2020**, *10*, 1–14.

42. Islam, M.; Chhanda, R.; Pervin, N.; Hossain, M.; Chowdhury, R. Characterization and genetic diversity of brinjal germplasm. *Bangladesh J. Agric. Res.* **2018**, *43*, 499–512. [CrossRef]

43. Begum, F.; Islam, A.A.; Rasul, M.G.; Mian, M.K.; Hossain, M.M. Morphological diversity of eggplant (Solanum melongena) in Bangladesh. *Emir. J. Food Agric.* **2013**, *25*, 45–51. [CrossRef]

44. Solaiman, A.H.M.; Nishizawa, T.; Khatun, M.; Ahmad, S. Morphological characterization and genetic diversity studies of promising brinjal genotypes for hybridization program in Bangladesh. *J. Adv. Agric.* **2014**, *3*, 218–228. [CrossRef]

45. Parida, H.; Mandal, J.; Mohanta, S. A note on morphological characterization of brinjal (*Solanum melongena* L.) genotypes. *J. Crop. Weed* **2020**, *16*, 250–255. [CrossRef]

46. Shekar, K.C.; Ashok, P.; Sasikala, K. Characterization, character association, and path coefficient analyses in eggplant. *Int. J. Veg. Sci.* **2013**, *19*, 45–57. [CrossRef]

47. Kumar, G.; Meena, B.; Kar, R.; Tiwari, S.K.; Gangopadhyay, K.; Bisht, I.; Mahajan, R. Morphological diversity in brinjal (*Solanum melongena* L.) germplasm accessions. *Plant Genet. Resour.* **2008**, *6*, 232–236. [CrossRef]

48. Sunseri, F.; Polignano, G.B.; Alba, V.; Lotti, C.; Bisignano, V.; Mennella, G.; Drsquo, A.; Bacchi, M.; Riccardi, P.; Fiore, M.C.; et al. Genetic diversity and characterization of African eggplant germplasm collection. *Afr. J. Plant Sci.* **2010**, *4*, 231–241.

49. Dash, S.P.; Singh, J.; Sharma, D. Morphological characterization of brinjal (*Solanum melongena* L.) germplasm. *J. Pharmacogn. Phytochem.* **2019**, *8*, 1574–1578.

50. Tiwari, S.K.; Bisht, I.S.; Kumar, G.; Karihaloo, J.L. Diversity in brinjal (Solanum melongena L.) landraces for morphological traits of evolutionary significance. *Veg. Sci.* **2016**, *43*, 106–111.

51. Solaimana, A.H.M.; Nishizawa, T.; Khatun, M.; Ahmad, S. Physio-morphological characterization genetic variability and correlation studies in brinjal genotypes of Bangladesh. *Comput. Math. Biol.* **2015**, *4*, 1–36.

52. Begum, M.N.S.; Shirazy, B.J.; Mahbub, M.M.; Siddikee, M.A. Performance of Brinjal (Solanum Melongena) Genotypes through Genetic Variability Analysis. *Am. J. Plant Biol.* **2017**, *43*, 499–512.

53. Shinde, K.; Birajdar, U.; Bhalekar, M.; Patil, B. Genetic divergence in brinjal (*Solanum melongena* L.). *Veg. Sci.* **2012**, *39*, 103–104.

54. Singh, S.; Khemariya, P.; Rai, A.; Rai, A.C.; Koley, T.K.; Singh, B. Carnauba wax-based edible coating enhances shelf-life and retain quality of eggplant (*Solanum melongena*) fruits. *LWT* **2016**, *74*, 420–426. [CrossRef]

55. Hassan, I.; Jatoi, S.A.; Arif, M.; Siddiqui, S.U. *Genetic Variability in Eggplant for Agro-Morphological Traits*; World Vegetable Center: Tainan, Taiwan, 2015.

56. Premabati Devi, C.; Munshi, A.D.; Behera, T.K.; Choudhary, H.; Saha, P. Characterisation of cultivated breeding lines of eggplant (*Solanum melongena* L.) and related wild Solanum species from India. *J. Hortic. Sci. Biotechnol.* **2016**, *91*, 87–92. [CrossRef]

57. Koundinya, A.; Das, A.; Layek, S.; Chowdhury, R.; Pandit, M.e. Genetic variability, characters association and path analysis for yield and fruit quality components in Brinjal. *J. Appl. Nat. Sci.* **2017**, *9*, 1343–1349. [CrossRef]

58. Kumar, S.R.; Arumugam, T.; An, C.; Premalakshmi, V. Genetic variability for quantitative and qualitative characters in Brinjal (*Solanum melongena* L.). *Afr. J. Agric. Res.* **2013**, *8*, 4956–4959.

59. Sherly, J.; Shanthi, A. Variability, heritability and genetic advance in brinjal (*Solanum melongena* L.). *Res. Crop.* **2009**, *10*, 105–108.

60. Singh, O.; Kumar, J. Variability, heritability and genetic advance in brinjal. *Indian J. Hortic.* **2005**, *62*, 265–267.

61. Verma, P.; Kushwaha, M.; Panchbhaiya, A. Studies on variability, heritability and genetic advance for yield attributing traits in brinjal (*Solanum melongena* L.) for two different seasons. *Int. J. Curr. Microbiol. Appl. Sci.* **2018**, *7*, 1543–1552. [CrossRef]

62. Chattopadhyay, A.; Dutta, S.; Hazra, P. Characterization of genetic resources and identification of selection indices of brinjal (*Solanum melongena* L.) grown in eastern India. *Veg. Crop. Res. Bull.* **2011**, *74*, 39. [CrossRef]

63. Shende, R.; Desai, S.; Lachyan, T. Genetic variability and response to selection in brinjal (*Solanum melangena* L.). *Int. J. Curr. Res.* **2015**, *7*, 21545–21547.

64. Calus, M.P.; Veerkamp, R.F. Accuracy of multi-trait genomic selection using different methods. *Genet. Sel. Evol.* **2011**, *43*, 1–14. [CrossRef]

65. Konyak, W.; Kanaujia, S.; Jha, A.; Chaturvedi, H.; Ananda, A. Genetic variability, correlation and path coefficient analysis of brinjal. *SAARC J. Agric.* **2020**, *18*, 13–21. [CrossRef]

66. Saha, S.; Haq, M.E.; Parveen, S.; Mahmud, F.; Chowdhury, S.R.; Harun-Ur-Rashid, M. Variability, correlation and path coefficient analysis: Principle tools to explore genotypes of brinjal (*Solanum melongena* L.). *Asian J. Biotechnol. Genet. Eng.* **2019**, *2*, 1–9.

67. Dugard, P.; Todman, J.; Staines, H. *Approaching Multivariate Analysis: A Practical Introduction*; Routledge/Taylor & Francis Group: Abingdon, UK, 2010.

68. Otieno, O.V.; Owuor, O.N. Multivariate genotype and genotype by environment interaction biplot analysis of sugarcane breeding data using R. *Int. J. Stat. Distrib. Appl.* **2019**, *5*, 22.

69. Yan, W.; Frégeau-Reid, J. Breeding line selection based on multiple traits. *Crop. Sci.* **2008**, *48*, 417–423. [CrossRef]

70. Khan, M.M.H.; Rafii, M.Y.; Ramlee, S.I.; Jusoh, M.; Al Mamun, M. Genetic analysis and selection of Bambara groundnut (*Vigna Subterranea* [L.] Verdc.) Landraces for high yield revealed by qualitative and quantitative traits. *Sci. Rep.* **2021**, *11*, 1–21.

71. Meyer, R.S.; Bamshad, M.; Fuller, D.Q.; Litt, A. Comparing medicinal uses of eggplant and related Solanaceae in China, India, and the Philippines suggests the independent development of uses, cultural diffusion, and recent species substitutions. *Econ. Bot.* **2014**, *68*, 137–152. [CrossRef]

72. Virga, G.; Licata, M.; Consentino, B.B.; Tuttolomondo, T.; Sabatino, L.; Leto, C.; La Bella, S. Agro-morphological characterization of sicilian chili pepper accessions for ornamental purposes. *Plants* **2020**, *9*, 1400. [CrossRef] [PubMed]
73. Bhandari, H.; Bhanu, A.; Srivastava, K.; Singh, M.; Shreya, H.A. Assessment of genetic diversity in crop plants-an overview. *Adv. Plants Agric. Res.* **2017**, *7*, 00255.
74. Olivoto, T.; Diel, M.I.; Schmidt, D.; Lúcio, A.D.C. Multivariate analysis of strawberry experiments: Where are we now and where can we go? *BioRxiv* **2021**, 1–10. [CrossRef]

Article

Comparative Transcriptome Analysis Reveals Different Low-Nitrogen-Responsive Genes in Pepper Cultivars

Chunping Wang [1,†], Yifei Li [2,†], Wenqin Bai [1], Xiaomiao Yang [2], Hong Wu [1], Kairong Lei [1], Renzhong Huang [2], Shicai Zhang [2,*], Qizhong Huang [2,*] and Qing Lin [1,*]

[1] Chongqing Key Laboratory of Adversity Agriculture, Biotechnology Research Institute, Chongqing Academy of Agricultural Sciences, Chongqing 401329, China; wcp49930000@sina.com (C.W.); bwqbc82@126.com (W.B.); wvvhong-1968@163.com (H.W.); leikairong@126.com (K.L.)

[2] Vegetable and Flower Research Institute, Chongqing Academy of Agricultural Sciences, Chongqing 401329, China; liyifei201@163.com (Y.L.); xiaomiaoyangy@sina.com (X.Y.); Hrz23@aliyun.com (R.H.)

[*] Correspondence: shicaiz@126.com (S.Z.); hjshqz@126.com (Q.H.); linqingcq@126.com (Q.L.)

[†] Chunping Wang and Yifei Li contributed equally to this article.

Abstract: The molecular mechanisms underlying the variation in N-use efficiency (NUE) in pepper (*Capsicum annuum* L.) genotypes are poorly understood. In this work, two genotypes (750-1, low-N tolerant; ZCFB, low-N sensitive) with contrasting low-N tolerance were selected from 100 pepper cultivars on the basis of their relative leaf areas, shoot dry weights, root dry weights, and plant dry weights at the seedling stage. Subsequently, using RNA-Seq, the transcriptome of these two pepper genotypes under N starvation for 28 days was analyzed. We detected 2621/2470 and 3936/4218 different expressed genes (DEGs) in the leaves/roots of 750-1 and ZCFB, respectively. The changes in the expression of basic N metabolism genes were similar between 750-1 and ZCFB. However, different DEGs not directly involved in N metabolism were identified between the 750-1 and ZCFB cultivars. In 750-1, 110 unique DEGs were detected in the leaves, of which 103 were down-regulated, including genes associated with protein metabolism, photosynthesis, secondary metabolism, cell wall metabolism, stress response, and disease resistance. In ZCFB, 142 unique DEGs were detected in the roots, of which 117 were up-regulated, resulting in enhancement of processes such as protein degradation, secondary metabolites synthesis, lipid metabolism, endocytosis, the tricarboxylic acid cycle (TCA), transcriptional regulation, stress response, and disease resistance. Our results not only facilitate an understanding of the different regulatory process in low-N-tolerant and low-N-sensitive pepper cultivars, but also provide abundant candidate genes for improving the low-N tolerance of pepper cultivars.

Keywords: *Capcicum annuum* L.; RNA-Seq; nitrogen-use efficiency; low-nitrogen tolerance

Citation: Wang, C.; Li, Y.; Bai, W.; Yang, X.; Wu, H.; Lei, K.; Huang, R.; Zhang, S.; Huang, Q.; Lin, Q. Comparative Transcriptome Analysis Reveals Different Low-Nitrogen-Responsive Genes in Pepper Cultivars. *Horticulturae* **2021**, *7*, 110. https://doi.org/10.3390/horticulturae7050110

Academic Editor: Yuyang Zhang

Received: 1 April 2021
Accepted: 11 May 2021
Published: 13 May 2021

Publisher's Note: MDPI stays neutral with regard to jurisdictional claims in published maps and institutional affiliations.

Copyright: © 2021 by the authors. Licensee MDPI, Basel, Switzerland. This article is an open access article distributed under the terms and conditions of the Creative Commons Attribution (CC BY) license (https://creativecommons.org/licenses/by/4.0/).

1. Introduction

Nitrogen (N) is one of the most important macronutrients for plants, and it can be absorbed and assimilated by the roots in various forms, including nitrate, ammonium, and amino acids. Nitrate and ammonium are the most common forms used by plants, with nitrate being the dominant form [1]. N metabolism can be divided into three processes: uptake, assimilation, and remobilization. In the uptake stage, at least six transporters participate in nitrate uptake in *Arabidopsis* (NPF6.3/NRT1.1, NPF4.6/NRT1.2, NRT2.1, NRT2.2, NRT2.4, and NRT2.5) [2,3]. In the assimilation process, nitrate is reduced to ammonium by nitrate and nitrite reductases, and then the ammonium is assimilated into amino acids by glutamine synthetase (GS), glutamine antinotransferase (GOGAT), and asparagine synthetase (AS) [4]. The N remobilization process comprises protein degradation and amino acid transport. Several genes involved in the ubiquitin-26S proteasome pathway [5] and encoding amino acid transporters [6,7] have also been shown to be associated with these processes. These studies have provided important clues for understanding the mechanism of N metabolism. However, present knowledge about the complicated N regulatory network remains incomplete, and the molecular basis governing the genetic variation of N-use efficiency (NUE) among crop cultivars remains unclear.

Pepper (*Capsicum annuum* L.) is one of the most important vegetable crops of the Solanaceae family and is grown worldwide as a food, medicine source and ornamental plant [8]. It is well documented that, within a certain range, there is a positive correlation between the relative growth rate of pepper and the N concentration in the soil. However, inappropriate N utilization can lead to undesirable growth, yield, and quality [9,10]. In recent years, in areas with fertile soil, farmers have been using a higher amount of N fertilizer than is required, which has resulted in low NUE and serious environmental pollution [11]. By contrast, in some areas where the soil quality is poor, the need for a high input of N fertilizer represents a great economic burden on pepper growers [12]. Therefore, improving NUE is urgently needed for the development of sustainable pepper production. Genetic variation in NUE has been reported for different crops such as rice [13], barley [14], wheat [15], rapeseed [16], maize [17], and cotton [18], but molecular knowledge about the genetic variation of NUE is still very poor. Several genes responsible for improved NUE have been identified in rice, including *DEP1* [19], *OsNRT1.1B* [20], *OsNRT2.3b* [21], *ARE1* [22], *OsNRT1.1A* [23], and *OsNPF6.1* [24]. Unfortunately, studies on the molecular regulation mechanism of N metabolism in pepper are scarce, let alone those studies focusing on NUE variation.

The next-generation high-throughput RNA sequencing technology (RNA-Seq) is a powerful tool for revealing genome-wide changes under biotic/abiotic stresses and can provide system level information regarding the N metabolism network. RNA-Seq analysis has been applied to the transcriptome analysis of low-N response of a single plant genotype, such as those of cucumber [25], maize [26], wheat [27], physic nut [28], and rice [29]. A large number of candidate genes involved in low-N response were detected [25,27–29]. Furthermore, the potential regulatory roles of lncRNAs in response to N stress have also been investigated [26]. However, it is difficult to reveal variations in NUE using only one genotype. Therefore, comparative transcriptome analysis of genotypes with different low N tolerances has become more recognized as a tool for understanding NUE [30–35]. A high abundance of transcripts related to high affinity nitrate transporters (NRT2.2, NRT2.3, NRT2.5, and NRT2.6) in the N-stress tolerant sorghum genotypes [31] and an energy-saving assimilation pattern in N-stress tolerant Tibetan wild barley genotype have been revealed [32].

The focus of this study was to identify low-N-responsive genes that were differentially expressed between low-N-tolerant and low-N-sensitive genotypes after long-term low-N stress, with the aim of providing more information on NUE variation. First, we examined the low-N tolerance of 100 pepper cultivars. Subsequently, we analyzed the genome-wide gene expression changes of two pepper cultivars with contrasting low-N tolerance under low-N stress. Lastly, the different low-N-responsive genes between the two cultivars were intensively analyzed using RNA-Seq and various bioinformatics methods, revealing potential new candidate genes for improving the low-N tolerance of pepper cultivars.

2. Materials and Methods

2.1. Screening of Pepper Cultivars with Different Low-N Tolerance

The low-N tolerance of 100 pepper cultivars was analyzed hydroponically in a greenhouse at the Chongqing Key Laboratory of Adversity Agriculture. Yamazaki nutrient solution for pepper was used as a sufficient-N solution containing 1.50 mM $Ca(NO_3)_2 \cdot 4H_2O$, 6.00 mM KNO_3, 0.83 mM $NH_4H_2PO_4$, 0.75 mM $MgSO_4 \cdot 7H_2O$, 0.10 mM $FeSO_4 \cdot 7H_2O$, 0.10 mM Na_2EDTA, 0.05 mM H_3BO_3, 0.0008 mM $ZnSO_4 \cdot 7H_2O$, 0.01 mM $MnSO_4 \cdot 4H_2O$, 0.0003 mM $CuSO_4 \cdot 5H_2O$, and 0.00002 mM $(NH_4)_6Mo_7O_{24} \cdot 4H_2O$. Pepper seeds were soaked in warm water (55 °C) for 30 min, and then germinated in the dark on moist filter papers at 25–28 °C. Small buds were then transferred to foam boards floating in black plastic boxes filled with distilled water. The foam boards consisted of 15 uniform holes in which 15 plants grew. After seven days, the distilled water was changed to one-half-strength sufficient-N solution. A week later, the solution of half of the seedlings was changed to low-N solution, in which the concentrations of $Ca(NO_3)_2 \cdot 4H_2O$, KNO_3, and $NH_4H_2PO_4$ were 20% that of the sufficient-N solution, and the other nutrients were the same as the sufficient-N solution. Furthermore, K_2SO_4, KH_2PO_4, and $CaCl_2$ were added in moderation to the low-N solution to avoid K, P, and Ca deficiency. The rest of the seedlings grew under N-sufficient condition as controls. Each cultivar had three replicates (15 plants/replicate) for each N condition (sufficient-N or low-N). The culture solution was refreshed every seven days at pH 6.2–6.4. All plants were grown in a greenhouse at 28 °C (day) and 25 °C (night) with a relative humidity of 70–80% and a photoperiod of 14/10 h, 300 μmol m^{-2} s^{-1}. The leaf areas, shoot dry weights, root dry weights, and plant dry weights were evaluated 28 days later.

2.2. RNA Preparation and Sequencing

Seeds from the 750-1 (low-N-tolerant) and ZCFB (low-N-sensitive) pepper cultivars were germinated, and the seedlings were cultured under two different N conditions (sufficient-N and low-N) as described above. Four weeks later, the youngest fully expanded leaves and roots of the seedlings grown in sufficient-N solution and low-N solution were collected separately at 10:00–11:00 A.M., flash frozen in liquid N, and then stored at −80 °C. Samples were collected from 45 independent plants from three replicates (15 plants/replicates). Total RNA was extracted using an RNA Plant Kit (Aidlab Biotech, Beijing, China). The quantity and quality of total RNA was examined on a NanoDrop 2000 spectrophotometer (Thermo Fisher Scientific, Madison, WI, USA) and an Agilent 2100 Bioanalyzer (Waldbronn, Germany), respectively. cDNA libraries were constructed and sequenced on a BGISEQ-500 platform at BGI (Shenzhen, China).

2.3. Transcriptome De Novo Assembly, Gene Functional Annotation, and Differentially Expressed Genes (DEGs) Analyses

The raw reads of transcriptome sequencing were filtered using SOAPnuke v.15.2 (BIG, Shenzhen, China), and a set of clean reads was obtained. After de novo assembling, the clean reads were mapped to the pepper reference genome [36]. Genes were functionally annotated based on the NCBI non-redundant (Nr) [37], Gene Ontology (GO) [38], and the Kyoto Encyclopedia of Genes and Genomes (KEGG) databases [39]. For gene expression analysis, the numbers of matched reads were calculated and then normalized to RPKM by RSEM v.1.2.12. Significant differential expression genes (DEGs) were identified as those with a fold-change $\geq$ 2.0 and an FDR $\leq$ 0.001. DEGs were clustered using GO-Term Finder software, and pathway enrichment analysis was performed based on terms from the KEGG database.

2.4. Quantitative Real-Time PCR (qRT-PCR) Analysis

To verify the reliability of RNA-Seq data, total RNA was extracted from the leaves and roots of biological triplicates from the two sequenced genotypes, and the template cDNA samples were prepared using an iScrip First Strand Synthesis System Kit (Bio-Rad Laboratories, Hercules, CA, USA). Primers for each PCR reaction were designed to have a melting temperature of 58–62 °C and to produce a PCR product between 100 and 200 bp (Table S1). The pepper actin gene (*AY572427*) was used as an internal control. qRT-PCR reactions were performed using a CFX96^{TM} Real-Time System (Bio-Rad Laboratories, Hercules, CA, USA). The reaction conditions were as follows: 95 °C for 30 s, 40 cycles of 95 °C for 5 s, and 60 °C for 30 s. The 2$^{-\Delta\Delta Ct}$ method was used to calculate the expression levels of genes.

3. Results

3.1. Screening for Pepper Cultivars with Contrasting Low-N Tolerance

To identify pepper cultivars with contrasting low-N tolerance, 100 pepper cultivars were screened based on various metrics. To avoid the effects of natural variation in individual embryos and/or endosperm size on initial growth from different genotypes, the relative value (low-N treated sample/control) of each character was used in the following analyses. Our results showed that peppers from these 100 cultivars had a wide range of relative values of leaf areas, shoots, roots, and plant dry weights, and the coefficient of variation (CV) of all of the relative values was above 15%, demonstrating the high variation in tolerance to low-N among cultivars (Table 1 and Table S2).

Table 1. Range of relative values of growth characteristics in pepper seedlings.

Index	Mean	Range	CV (%)
Relative value of leaf area	0.63	0.38–0.95	18.9
Relative value of shoot dry weight	0.68	0.41–0.99	16.8
Relative value of root dry weight	1.08	0.66–1.61	15.6
Relative value of plant dry weight	0.73	0.44–1.03	15.4

Cluster analysis was then carried out based on the relative values of leaf areas, shoot dry weights, root dry weights, and plant dry weights. The 100 pepper cultivars could be grouped into three clusters, consisting of five (Group I), six (Group II), and 89 (Group III) cultivars (Figure 1). There were significant differences in each characteristic between the three groups (Table 2). The mean relative values of each characteristic in Group I were significantly higher than those of Group II, and the mean relative values of each characteristic in Group III were between those of Groups I and II, indicating that Group I was a low-N-tolerant group, while Group II was a low-N-sensitive group.

Table 2. The low-nitrogen tolerance of each cluster group.

Cluster	Relative Value of Leaf Area		Relative Value of Shoot Dry Weight		Relative Value of Root Dry Weight		Relative Value of Plant Dry Weight	
	Mean	Range	Mean	Range	Mean	Range	Mean	Range
Group I	0.86a	0.75–0.95	0.90a	0.81–0.99	1.46a	1.31–1.61	0.97a	0.89–1.03
Group II	0.44c	0.38–0.48	0.47c	0.41–0.52	0.85c	0.66–1.03	0.52c	0.44–0.57
Group III	0.63b	0.41–0.91	0.68b	0.52–0.94	1.08b	0.67–1.44	0.73b	0.57–0.99

Note: Values followed by different letters within the same column are significantly different at the 0.05 level.

750-1 had the highest values for the relative leaf area as well as shoot and plant dry weights (Table S2) and was classified into Group I. Thus, it was considered low-N tolerant. ZCFB had the lowest relative values of shoot, root and plant dry weights (Table S2) and was classified into Group II. This cultivar was therefore considered low-N sensitive.

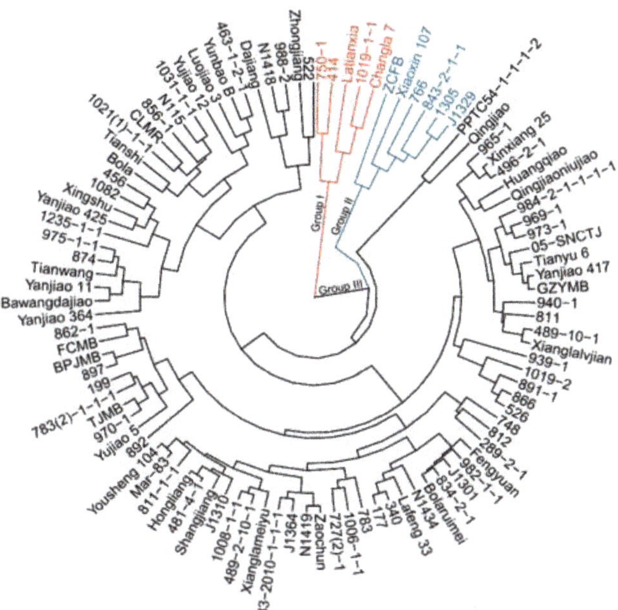

Figure 1. Cluster analysis of the low-nitrogen tolerance from 100 pepper cultivars. Cluster analysis was carried out based on the relative values of leaf area and shoot, root, and plant dry weights. Agglomerative hierarchical clustering was performed using the hclust algorithm in R, with dst = dist (dat1, method = "euclidean") and hclust (dst, method = "average"). Group I, low-N-tolerant group; Group II, low-N-sensitive group; the low-nitrogen tolerance of Group III was between that of Groups I and II.

3.2. RNA-Seq Analysis of Two Pepper Cultivars with Contrasting Low-N Tolerances

To better understand the molecular mechanism governing the genetic variation of N-use efficiency (NUE) in pepper cultivars, the global gene expression profile of two pepper genotypes (750-1 and ZCFB) with contrasting low-N tolerance was next analyzed using RNA-Seq. Samples were taken 28 days after low-N treatment, and the following samples were used for sequencing: TCL1 (750-1-leaf control), TCR2 (750-1-root control), TTL3 (750-1-leaf low-N treatment), TTR4 (750-1-root low-N treatment), SCL5 (ZCFB-leaf control), SCR6 (ZCFB-root control), STL7 (ZCFB-leaf low-N treatment), and STR8 (ZCFB-root low-N treatment). Sequence data are available from the NCBI SRA database with accession number PRJNA700470. Over 65 million clean reads were generated from each library. A total of 91.75–93.55% of these clean reads could be well aligned to the pepper reference genome, at least 69% could be uniquely mapped to the reference genome (71.88% on average). The number of genes with at least one mapped transcript were 29,167, 29,817, 28,970, 29,981, 29,330, 29,814, 29,078, and 30,335 for libraries TCL1, TCR2, TTL3, TTR4, SCL5, SCR6, STL7, and STR8, respectively (Table 3).

Table 3. Summary of genome mapping.

Sample	Total Clean Reads	Total Mapping Ratio	Uniquely Mapping Ratio	Total Gene Number
TCL1	66,496,250	92.38%	70.18%	29,167
TCR2	66,186,556	92.03%	72.58%	29,817
TTL3	65,608,344	92.54%	69.65%	28,970
TTR4	65,388,728	92.26%	73.41%	29,981
SCL5	65,498,816	91.75%	69.00%	29,330
SCR6	65,052,664	91.53%	72.37%	29,814
STL7	66,211,026	93.55%	72.20%	29,078
STR8	66,062,774	93.09%	75.68%	30,335

3.3. Identification of DEGs and Validation of RNA-Seq by qRT-PCR

To identify low-N-responsive genes, comparisons were made between the low-N treated samples and controls (Figure 2; Tables S3–S6). We found that 548/2073 differentially expressed genes (DEGs) (TCL1 vs. TTL3) were up/down-regulated in leaves, while 1385/1085 DEGs (TCR2 vs. TTR4) were up/down-regulated in the roots of 750-1. Additionally, 1239/2697 DEGs (TCL5 vs. TTL7) were up/down-regulated in leaves, and 2956/1262 DEGs (TCR6 vs. TTR8) were up/down-regulated in the roots of ZCFB.

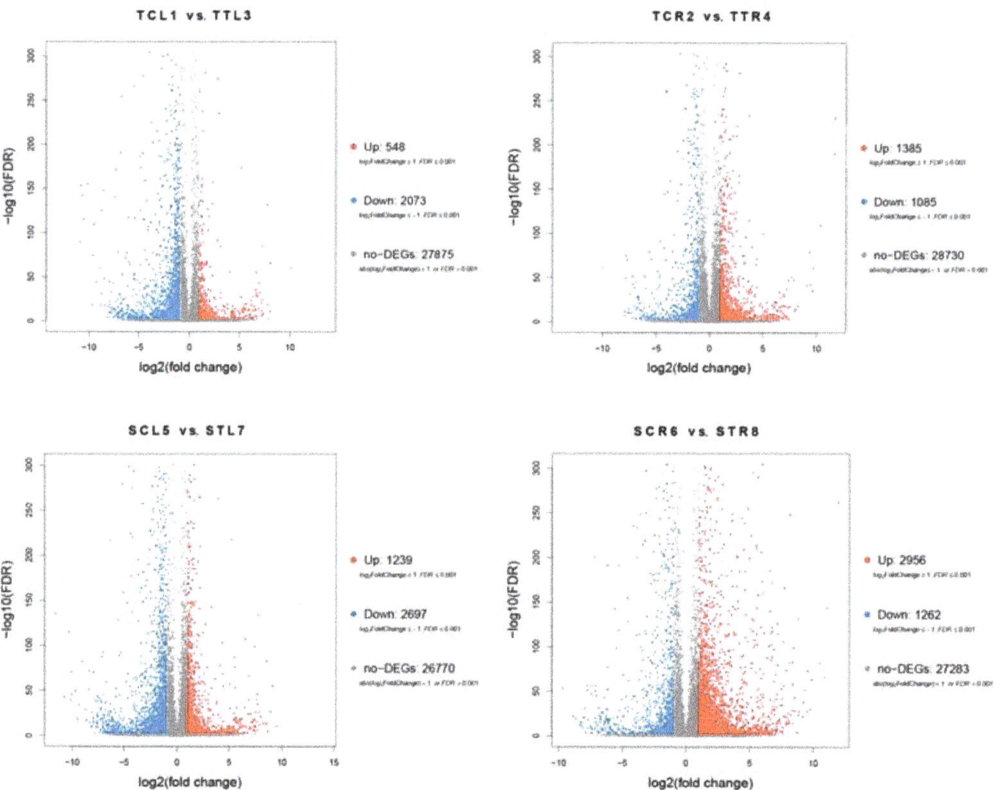

Figure 2. Volcano plot of DEGs. The X-axis represents the value of the difference multiple after log2 conversion, and the Y-axis indicates the significance value after log10 conversation. The three colors red, blue, and gray, represent up-regulated DEGs, down-regulated DEGs, and unchanged genes, respectively.

To validate our RNA-Seq data, the expression levels of eight genes (*LOC107841845, LOC107855963, LOC107858815, LOC107867964, LOC107870519, LOC107875600, LOC107875602,* and *LOC107875603*) in the leaves and roots of 750-1 and ZCFB were next confirmed by qRT-PCR. The expression trends of the eight genes were consistent with our results from RNA-Seq in the eight libraries (Figure 3).

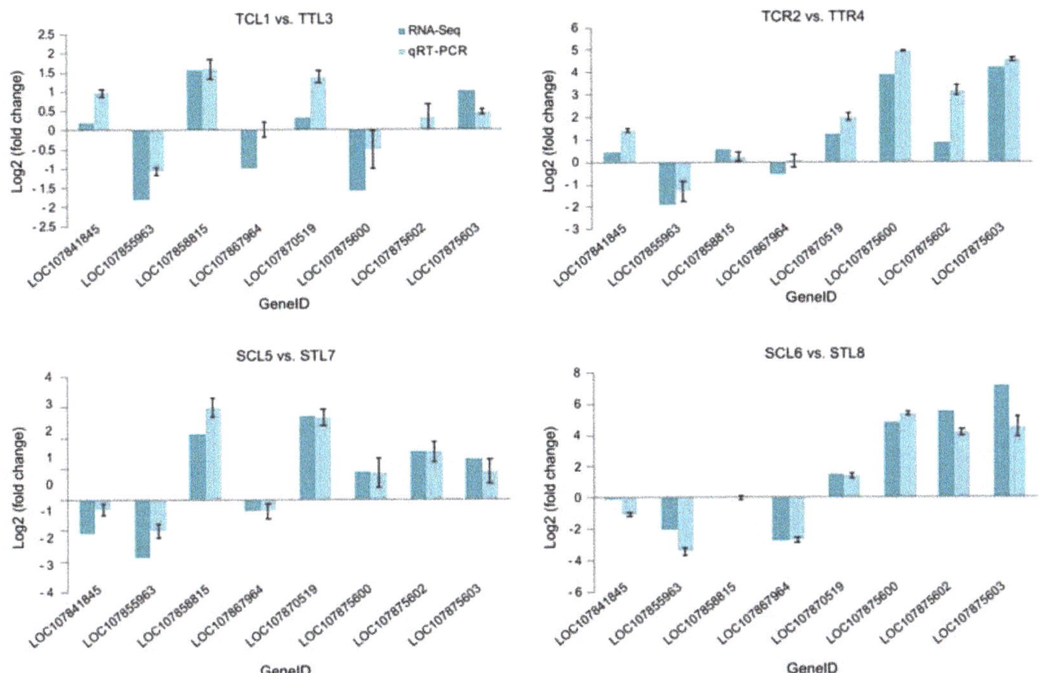

Figure 3. RNA-Seq results confirmed by qRT-PCR. A total of eight genes (*LOC107841845, LOC107855963, LOC107858815, LOC107867964, LOC107870519, LOC107875600, LOC107875602,* and *LOC107875603*) were examined by qRT-PCR. The pepper actin gene (AY572427) was used as an internal control.

3.4. GO and KEGG Enrichment Analysis of DEGs

Next, GO analysis was carried out using the identified DEGs in the three main GO categories: biological process, cellular component, and molecular function (Figure 4A). KEGG pathway analysis showed that the DEGs involved in N metabolism were enriched in all of the four libraries (Figure 4B). The DEGs involved in nitrogen metabolism were then intensively analyzed. The DEGs associated with nitrate transportation, assimilation, and remobilization processes were detected in each genotype (Figure 5 and Figure S1A–D). The genes of the nitrate transporter family were up-regulated, especially in the roots, while N assimilation systems were suppressed. The genes participating in N remobilization were also enhanced. Overall, the expression patterns of the DEGs involved in N metabolism were similar in both genotypes. Thus, further analysis is needed to find clues governing different NUEs between the two genotypes examined.

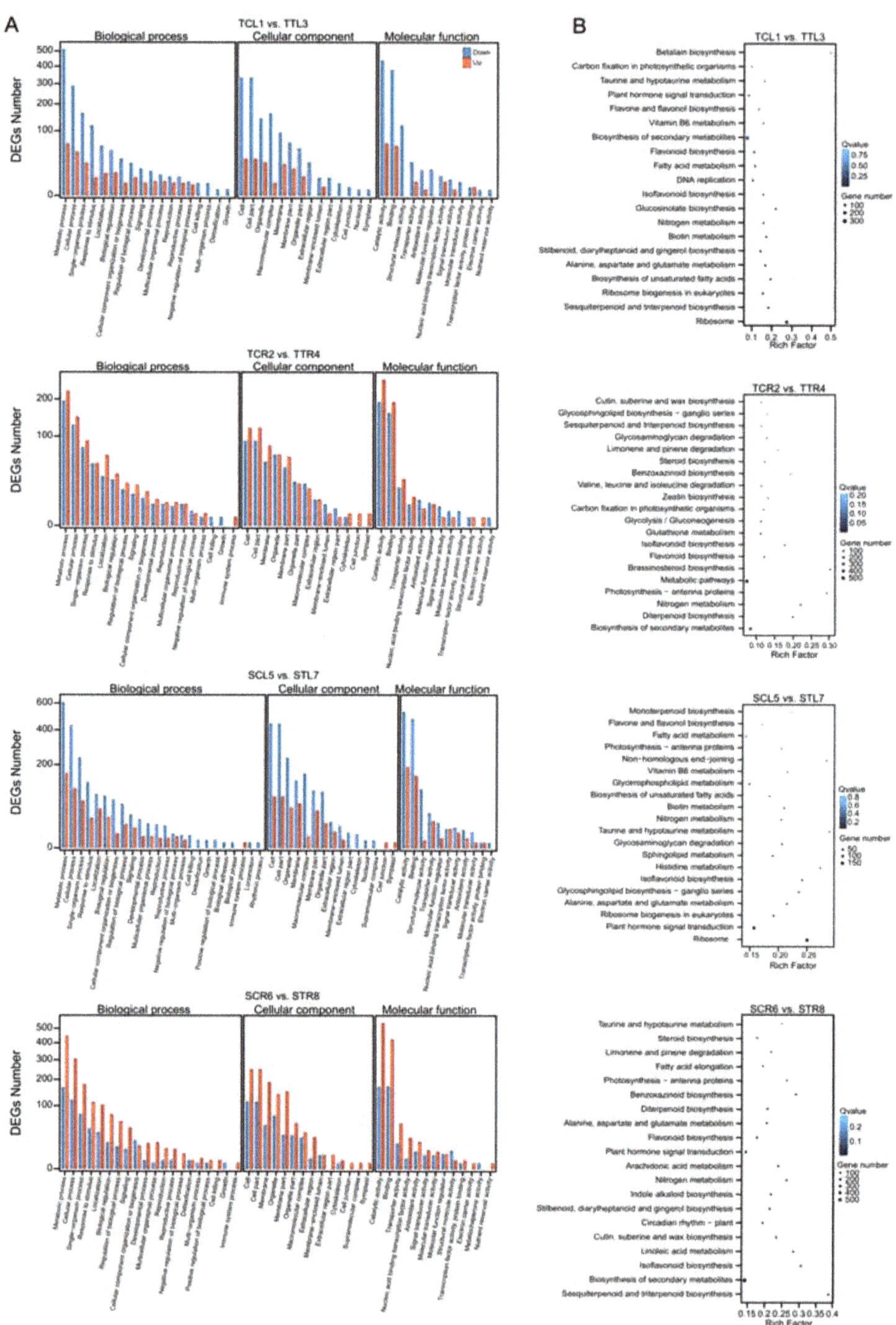

Figure 4. Function classifications of DEGs. (**A**): GO functional enrichment analysis of DEGs. The X-axis represents GO functional classification, and the Y-axis represents the number of up-regulated and down-regulated genes in each corresponding GO term. (**B**): Pathway enrichment analysis of DEGs. The X-axis represents enrichment factor values, and Y-axis represents pathway names.

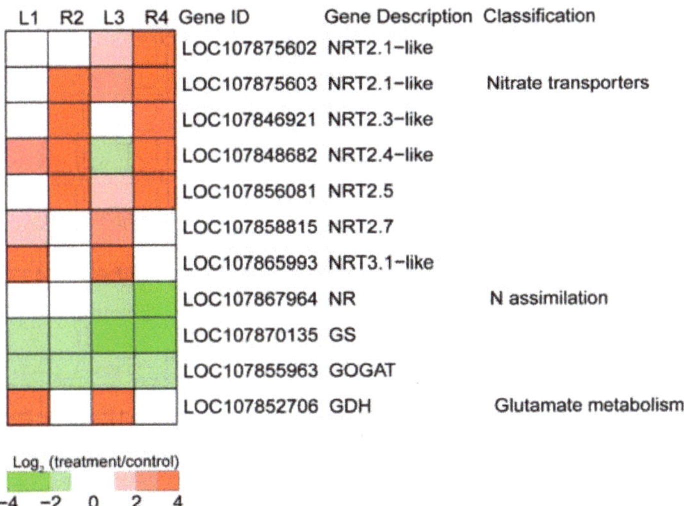

Figure 5. Heatmap of the DEGs involved in N metabolism. L1, TCL1 vs. TTL3; R2, TCR2 vs. TTR4; L3, TCL5 vs. TTL7; R4, TCR6 vs. TTR8.

3.5. Different DEGs between Two Cultivars

To examine the different DEGs between the two cultivars, comparisons were made between the DEGs of 750-1 and ZCFB (Figure 6A,B). The absolute values of log2 fold-changes were set as equal to or greater than five. Furthermore, the DEGs involved in N metabolism, which were analyzed previously, were excluded from these analyses. The results indicated that 110 unique DEGs from 750-1 could be detected in leaves, and most were down-regulated (Figure 6A), while 142 unique DEGs from ZCFB were detected in roots, and most were up-regulated (Figure 6B). Most of the unique DEGs are described in the following section (Figure 7A–J).

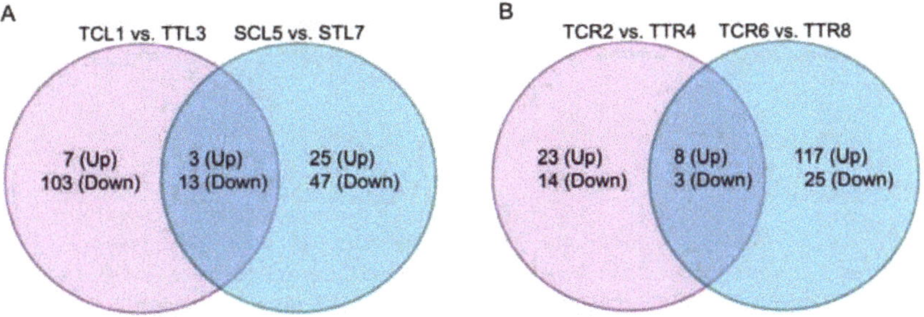

Figure 6. Venn diagram of unique DGEs in each genotype. (**A**): Unique DEGs in leaves. (**B**): Unique DEGs in roots. The absolute value of log2 fold change was equal to or greater than 5.

Figure 7. Classification of unique DEGs. (**A**): Protein metabolism. (**B**): Carbon metabolism. (**C**): Lipid metabolism. (**D**): Secondary metabolism. (**E**): Cell wall synthesis and structure. (**F**): Transport. (**G**): Signal transduction. (**H**): Transcription factor. (**I**): Stress response. (**J**): Disease resistance. L1, TCL1 vs. TTL3; R2, TCR2 vs. TTR4; L3, SCL5 vs. SCL7; R4, SCR6 vs. STR8.

3.5.1. Protein Metabolism

In 750-1, other than S-adenosylmethionine decarboxylase proenzyme (*LOC107855485*), all the unique DEGs required for protein metabolism were downregulated in the leaves (Figure 7A). In ZCFB, more than half of the genes associated with protein degradation were up-regulated, especially the members of ubiquitin-proteosome pathway (*LOC107853739, LOC107862077, LOC107851773, LOC107877293, LOC107869401, LOC10789245*, and *LOC107856197*).

3.5.2. Carbon and Lipid Metabolism

In the carbon metabolism process, genes associated with photosynthesis were downregulated in the leaves of 750-1 (Figure 7B). Two genes (*LOC107843574* and *LOC107857137*) of the tricarboxylic acid cycle (TCA) were upregulated in the leaves and roots of ZCFB. In the lipid metabolism process, DEGs involved in lipid synthesis and degradation were inhibited in the leaves of 750-1, while those in the roots of ZCFB were significantly enhanced, including members of the 3-ketoacyl-CoA synthase (*LOC107845580, LOC107847631*, and *LOC107868466*), fatty acid desaturase (*LOC107851810*), and phospholipase (*LOC107859298*) families (Figure 7C).

3.5.3. Secondary Metabolism

The genes responsible for the biosynthesis of phenylpropanoid, sesquiterpenoid, diterpenoid, flavonoid, anthocyanin, and other secondary metabolites showed altered expression under N starvation in both genotypes, including members of the cytochrome P450 family (*LOC107844024, LOC107860971, LOC107849698, LOC107871074, LOC107867086, LOC107845094, LOC107877698, LOC107856388,* and *LOC107874613*) and anthocyanin biosynthesis (*LOC107845761* and *LOC107866550*) (Figure 7D). The flavonoid biosynthesis process consists of the largest number of DEGs followed by the phenylpropannoid biosynthesis process. All the unique DEGs in the leaves of 750-1 were repressed, while most of the unique DEGs in the roots of ZCFB were up-regulated.

3.5.4. Cell Wall Synthesis and Structure

In the leaves, all of the unique DEGs in this category were repressed in both of the genotypes, including *xyloglucan 6-xylosyltransferase 2* (*LOC107842350*), which is responsible for cell wall synthesis in ZCFB, as well as *Glucan edo-1,3-beta-glucosidase* (*LOC107879143*) and *beta-D-xylosidase* (*LOC107854300*), which functions in cell wall degradation in 750-1 (Figure 7E). Most of these genes in the roots of the two genotypes were up-regulated, including *probable pectate lyase 5* (*LOC107848186* and *LOC107848183*) in 750-1 and *putative pectinesterase 11* (*LOC107869412*) and *polygalacturonase-like* (*LOC107856362* and *LOC107869773*) in ZCFB.

3.5.5. Transport

In 750-1, all of the unique DEGs engaged in transport were down-regulated in the leaves, including genes associated with osmosis (*LOC107865680*) and stoma movement (*LOC107844000*), while in the roots, except for the gene encoding sodium-dependent phosphate transporter protein 1 (*LOC107857112*), all of the genes were up-regulated, including metal (*LOC107869634*) and sugar transport (*LOC107873879*) (Figure 7F). In ZCFB, most of the genes were up-regulated in both of the leaves and roots, including genes associated with channel (*LOC107855209*), endocytosis (*LOC107848190*), sodium/metabolite and urea transport (*LOC107871496* and *LOC107870432*) in the leaves, and gene encoding channel proteins (*LOC107840638*) and proteins involved in endocytosis (*LOC107848190, LOC107840435,* and *LOC107845831*) in the roots.

3.5.6. Signal Transduction and Transcription Factor

For hormone synthesis and regulation, all of the genes were inhibited in the leaves of both genotypes (Figure 7G), including those involved in ethylene synthesis (*LOC107840369*) and response (*LOC107865651*) in 750-1 and those associated with the auxin response (*LOC107840512* and *LOC107853594*) in ZCFB. In 750-1, the genes of the LRR receptor-like serine/threonine-protein kinase family (*LOC107869144, LOC107869171,* and *LOC107869351*) and those involved in Ca2+ signaling (*LOC107847049* and *LOC107844087*) were up-regulated in the leaves and roots, respectively. The genes of transcription factors (TFs) that showed changes in expression mainly fell into the *MYB, bHLH, NAC, WRKY, BEE,* and *ERF* families (Figure 7H).

3.5.7. Stress Response and Disease Resistance

In 750-1, most of the unique DEGs related to stress response and disease resistance were observable in the leaves, and except for *protein SRG1-like* (*LOC107859182*) and *NBS-LRR root-knot nematode resistance protein* (*LOC107864520*), all the DEGs of the two pathways were down-regulated (Figure 7I,J). Meanwhile, in ZCFB, DEGs involved in stress response and disease resistance were detected in both the leaves and roots, and most of the genes in the leaves were down-regulated, in contrast to in the roots.

4. Discussion

In this study, a large number of low-N-responsive genes were identified from low-N-tolerant (750-1) and low-N-sensitive (ZCFB) pepper genotypes after long-term N deficiency, and the number of low-N-responsive genes in ZCFB was far greater than those in 750-1 (Figure 2). The expression profiles of the genes associated with nitrate transport, assimilation, and remobilization were similar between the 750-1 and ZCFB (Figure 5). However, many unique DEGs were detected between the two pepper genotypes after long-term N deficiency, implying different molecular mechanisms of resistance to N deficiency in these two genotypes.

4.1. Unique DEGs in the Low-N-Tolerant Genotype

In 750-1, the unique DEGs involved in primary metabolism, secondary metabolism, stress response, and disease resistances were down-regulated in the leaves after long-term N deficiency. However, the relative biomass of 750-1 was significantly higher than that of the other cultivars (Table S2), indicating that the growth of 750-1 was repressed less than other cultivars under N starvation. The relative values of root dry weights and leaf areas of 750-1 were much higher than those of other cultivars (Table S2), implying a larger absorption area of N and stronger photosynthesis in 750-1. Interestingly, genes of *somatic embryogenesis receptor kinase 2 (SERK2)-like (LOC107857123)* and *LRR receptor-like serine/threonine-protein kinase (LOC107869144 and LOC107869171)* were up-regulated in the leaves of 750-1 (Figure 7G). Both SERK2 and LRR receptor-like serine/threonine-protein kinase belong to the receptor-like kinase (RLK) family. SERK2 is essential for male microsporogenesis in *Arabidopsis* [40]. It has been postulated that at an early low-N stress stage, processes such as absorption, transportation, and assimilation of N might be highly enhanced in 750-1 under low-N conditions, as well as primary metabolism, which could result in improved biomass accumulation in 750-1 relative to other cultivars. After long-term N limitation, 750-1 might slowly decrease vegetative growth and enter reproductive stage earlier to accelerate its life cycle and seed handing down.

4.2. Unique DEGs in the Low-N Sensitive Genotype

Lipids are a major subcellular component, the biosynthesis and composition of which are influenced by N [41,42]. As essential components of the membrane, lipids play an important role in endocytosis, in which plasma membrane lipids and associated proteins are internalized in vesicles that fuse with the endosomal system [43]. In our study, a number of DEGs involved in lipid metabolism were up-regulated in the roots of ZCFB (Figure 7C). Meanwhile, several genes associated with endocytosis were also enhanced (Figure 7F). In plants, endocytosis plays an important role not only in transporting membrane proteins, lipids, and extracellular molecules into the cell but also in nutrient delivery, toxin avoidance, and pathogen defense [44,45]. To our knowledge, few studies have detected the involvement of endocytosis in N starvation. Here, the detection of endocytosis-related genes being involved in N starvation could provide new insights into N metabolism in plants.

Secondary metabolites have no direct role in plant growth, but their roles in stress defense have garnered much interest in recent years [46,47]. In the present study, genes involved in the synthesis of secondary metabolites, such as phenylpropanoid and flavonoid, were significantly up-regulated after low-N stress in ZCFB, as well as those encoding ATP-binding cassette (ABC) transporters that are involved in the transport of plant secondary metabolites (Figure 7F) [48,49]. We also found that genes involved in the oxidation/antioxidation response (Figure 7H) and disease resistance (Figure 7I) were up-regulated in ZCFB, as well as the expression of genes belonging to the *NAC, ERF, MYB,* and *bHLH* families (Figure 7H), which play essential roles in stress response. Considering the low relative biomass of ZCFB (Table S2), enhanced secondary metabolism and stress response could suggest that there was a diversion of materials from primary metabolism to the defense system under N deficiency. Indeed, previous studies have shown that N limitation results in the reduction of plant growth; however, some secondary metabolites were also shown to accumulate [25,50]. Moreover, genes involved in the tricarboxylic acid cycle (TCA) were up-regulated in ZCFB (Figure 7B), implying that to survive N deficiency, ZCFB may require more energy to support enhanced biological processes, such as endocytosis, secondary metabolism, and transcriptional regulation.

4.3. P and N Crosstalk

Phosphorus (P) is acquired by plants primarily in the form of inorganic phosphate (Pi) by Pi transporters (PHT). The *Nitrogen limitation adaptation* (*NLA*) gene is involved in adaptive responses to low-N stress in *Arabidopsis* [51] and is an important regulator of PHTs under N deficiency [52–55]. Over-expression of the rice Pi transporter gene OsPT2 enhances tolerance to Pi deficiency, as well as N2 fixation and ammonium assimilation [56,57]. These studies indicate the intimate crosstalk between the N limitation and P deficiency response pathways. In the present study, *Sodium-dependent phosphate transport protein 1* (*LOC107857112*) was repressed in the roots of 750-1, and *PHT1;2* (*LOC107866602*) and *PHT1;9* (*LOC107866766*) were up-regulated in the roots of ZCFB under low-N stress (Figure 7F), providing new information regarding the low-N induced crosstalk between N and P.

5. Conclusions

Improving NUE is urgently needed for the development of sustainable pepper production. However, present knowledge about the molecular basis governing the genetic variation of N-use efficiency (NUE) among pepper cultivars remains unclear. In this study, transcriptome of two pepper genotypes with contrary low-N tolerance (750-1, tolerant; ZCFB, sensitive) were compared using RNA-Seq and various bioinformatics methods after N starvation for 28 days. The results showed that the transcriptomic responses to low-N stress differed considerably between 750-1 and ZCFB, especially in genes that were not directly involved in N metabolism. The unique low-N-responsive genes between the two genotypes provide new insights for a comprehensive understanding of genotypic variation in NUE.

Supplementary Materials: The following are available online at https://www.mdpi.com/article/10.3390/horticulturae7050110/s1, Figure S1: DEGs involved in nitrogen metabolism. Table S1: Primers for qRT-PCR. Table S2: Relative values of each character. Table S3: DEGs of TCL1vs.TTL3. Table S4: DEGs of TCR2 vs. TTR4. Table S5: DEGs of SCL5 vs. STL7. Table S6: DEGs of SCR vs. STR8.

Author Contributions: Data curation, C.W., Y.L. and W.B.; formal analysis, C.W., K.L., R.H., Q.H. and Q.L.; investigation, C.W., S.Z., W.B., X.Y., H.W., Y.L.; methodology, C.W., Q.H. and Q.L.; project administration, R.H., S.Z., Q.H. and Q.L.; resources, S.Z., X.Y., R.H. and Q.H.; software, C.W. and W.B.; writing—original draft, C.W. and Y.L.; writing—review & editing, C.W., Y.L., K.L., R.H., Q.H. and Q.L. All authors have read and agreed to the published version of the manuscript.

Funding: This work was supported by grants from the China Agriculture Research System (CARS-24-A-04) and the Chongqing Science and Technology Bureau of China (cstc2019jscx-gksbX0123; cstc2020jscx-msxmX0054).

Institutional Review Board Statement: Not applicable.

Informed Consent Statement: Not applicable.

Data Availability Statement: The data supporting the findings of this study are available from the corresponding author upon reasonable request.

Conflicts of Interest: The authors declare no competing interest.

References

1. Liu, Q.; Chen, X.; Wu, K.; Fu, X. Nitrogen signaling and use efficiency in plants: What's new? *Curr. Opin. Plant Biol.* **2015**, *27*, 192–198. [CrossRef] [PubMed]
2. Krapp, A.; David, L.C.; Chardin, C.; Girin, T.; Marmagne, A.; Leprince, A.S.; Chaillou, S.; Ferrario-Méry, S.; Meyer, C.; Daniel-Vedele, F. Nitrate transport and signalling in Arabidopsis. *J. Exp. Bot.* **2014**, *65*, 789–798. [CrossRef] [PubMed]
3. Tegeder, M.; Masclaux-Daubresse, C. Source and sink mechanisms of nitrogen transport and use. *New Phytol.* **2018**, *217*, 35–53. [CrossRef] [PubMed]

4. Xu, G.; Fan, X.; Miller, A.J. Plant nitrogen assimilation and use efficiency. *Annu. Rev. Plant Biol.* **2012**, *63*, 153–182. [CrossRef] [PubMed]

5. Peng, M.; Bi, Y.M.; Zhu, T.; Rothstein, S.J. Genome-wide analysis of Arabidopsis responsive transcriptome to nitrogen limitation and its regulation by the ubiquitin ligase gene NLA. *Plant Mol. Biol.* **2007**, *65*, 775–797. [CrossRef] [PubMed]

6. Feng, L.; Yang, T.; Zhang, Z.; Li, F.; Chen, Q.; Sun, J.; Shi, C.; Deng, W.; Tao, M.; Tai, Y.; et al. Identification and characterization of cationic amino acid transporters (CATs) in tea plant (*Camellia sinensis*). *Plant Growth Regul.* **2018**, *84*, 57–69. [CrossRef]

7. Liu, S.; Wang, D.; Mei, Y.; Xia, T.; Xu, W.; Zhang, Y.; You, X.; Zhang, X.; Li, L.; Wang, N.N. Overexpression of *GmAAP6a* enhances tolerance to low nitrogen and improves seed nitrogen status by optimizing amino acid partitioning in soybean. *Plant Biotechnol. J.* **2020**, *18*, 1749–1762. [CrossRef] [PubMed]

8. Qin, C.; Yu, C.; Shen, Y.; Fang, X.; Chen, L.; Min, J.; Zhang, Z. Whole-genome sequencing of cultivated and wild peppers provides insights into *Capsicum* domestication and specialization. *Proc. Natl. Acad. Sci. USA* **2014**, *111*, 5135–5140. [CrossRef] [PubMed]

9. Ayodele, O.J.; Alabi, E.O.; Aluko, M. Nitrogen fertilizer effects on growth, yield and chemical composition of hot pepper (rodo). *Int. J. Agric. Sci.* **2015**, *8*, 666–673.

10. Wahocho, N.A.; Zeshan Ahmed, S.; Jogi, Q.; Talpur, K.H.; Leghari, S.J. Growth and productivity of chilli (*Capsicum annuum* L.) under various nitrogen levels. *Sci. Int. Lahore* **2016**, *28*, 1321–1326.

11. Zhu, J.H.; Li, X.L.; Christie, P.; Li, J.L. Environmental implications of low nitrogen use efficiency in excessively fertilized hot pepper (*Capsicum frutescens* L.) cropping systems. *Agric. Ecosyst. Environ.* **2005**, *111*, 70–80. [CrossRef]

12. Amare, T.; Nigussie, D.; Kebede, W.T. Performance of hot pepper (*Capsicum annuum*) varieties as influenced by nitrogen and phosphorus fertilizers at Bure, Upper Watershed of the Blue Nile in Northwestern Ethiopia. *Int. J. Agric. Sci.* **2013**, *3*, 599–608.

13. Sachiko, N.; Toriyama, K.; Yoshimichi, F. Genetic variations in dry matter production and physiological nitrogen use efficiency in rice (*Oryza sativa* L.) varieties. *Breed. Sci.* **2009**, *59*, 269–276.

14. Anbessa, Y.; Juskiw, P.; Good, A.; Nyachiro, J.; Helm, J. Genetic variability in nitrogen use efficiency of spring barley. *Crop Sci.* **2009**, *49*, 1259–1269. [CrossRef]

15. Górny, A.G.; Banaszak, Z.; Ługowska, B.; Ratajczak, D. Inheritance of the efficiency of nitrogen uptake and utilization in winter wheat (*Triticum aestivum* L.) under diverse nutrition levels. *Euphytica* **2010**, *177*, 191–206. [CrossRef]

16. Bouchet, A.-S.; Laperche, A.; Bissuel-Belaygue, C.; Snowdon, R.; Nesi, N.; Stahl, A. Nitrogen use efficiency in rapeseed. *Agron. Sustain. Dev.* **2016**, *36*, 38. [CrossRef]

17. Rodrigues, M.C.; Rezende, W.M.; Silva, M.E.J.; Faria, S.V.; Zuffo, L.T.; Galvão, J.C.C.; DeLima, R.O. Genotypic variation and relationships among nitrogen-use efficiency and agronomic traits in tropical maize inbred lines. *Genet. Mol. Res.* **2017**, *16*, gmr16039757. [CrossRef]

18. Zhang, H.; Fu, X.; Wang, X.; Gui, H.; Dong, Q.; Pang, N.; Wang, Z.; Zhang, X.; Song, M. Identification and screening of nitrogen-efficient cotton genotypes under low and normal nitrogen environments at the seedling stage. *J. Cotton. Res.* **2018**, *1*, 6. [CrossRef]

19. Sun, H.; Qian, Q.; Wu, K.; Luo, J.; Wang, S.; Zhang, C.; Fu, X. Heterotrimeric G proteins regulate nitrogen-use efficiency in rice. *Nat. Genet.* **2014**, *46*, 652–656. [CrossRef]

20. Hu, B.; Wang, W.; Ou, S.; Tang, J.; Li, H.; Che, R.; Chu, C. Variation in *NRT1.1B* contributes to nitrate-use divergence between rice subspecies. *Nat. Genet.* **2015**, *47*, 834–838. [CrossRef]

21. Fan, X.; Tang, Z.; Tan, Y.; Zhang, Y.; Luo, B.; Yang, M.; Lian, X.; Shen, Q.; Miller, A.; Xu, G. Overexpression of a pH-sensitive nitrate transporter in rice increases crop yields. *Proc. Natl. Acad. Sci. USA* **2016**, *113*, 7118–7123. [CrossRef]

22. Wang, Q.; Nian, J.; Xie, X.; Yu, H.; Zhang, J.; Bai, J.; Zuo, J. Genetic variations in *ARE1* mediate grain yield by modulating nitrogen utilization in rice. *Nat. Commun.* **2018**, *9*, 735. [CrossRef]

23. Wang, W.; Hu, B.; Yuan, D.; Liu, Y.; Che, R.; Hu, Y.; Chu, C. Expression of the nitrate transporter gene *OsNRT1.1A/OsNPF6.3* confers high yield and early maturation in rice. *Plant Cell* **2018**, *30*, 638–651. [CrossRef]

24. Tang, W.; Ye, J.; Yao, X.; Zhao, P.; Xuan, W.; Tian, Y.; Wan, J. Genome-wide associated study identifies NAC42-activated nitrate transporter conferring high nitrogen use efficiency in rice. *Nat. Commun.* **2019**, *10*, 5279. [CrossRef]

25. Zhao, W.; Yang, X.; Yu, H.; Jiang, W.; Sun, N.; Liu, X.; Gu, X. RNA-Seq-based transcriptome profiling of early nitrogen deficiency response in cucumber seedlings provides new insight into the putative nitrogen regulatory network. *Plant Cell Physiol.* **2015**, *56*, 455–467. [CrossRef]

26. Lv, Y.; Liang, Z.; Ge, M.; Qi, W.; Zhang, T.; Lin, F.; Peng, Z.; Zhao, H. Genome-wide identification and functional prediction of nitrogen-responsive intergenic and intronic long non-coding RNAs in maize (*Zea mays* L.). *BMC Genom.* **2016**, *17*, 350. [CrossRef]

27. Curci, P.L.; Cigliano, R.A.; Zuluaga, D.L.; Janni, M.; Sanseverino, W.; Sonnante, G. Transcriptomic response of durum wheat to nitrogen starvation. *Sci. Rep.* **2017**, *7*, 1176. [CrossRef]

28. Kuang, Q.; Zhang, S.; Wu, P.; Chen, Y.; Li, M.; Jiang, H.; Wu, G. Global gene expression analysis of the response of physic nut (*Jatropha curcas* L.) to medium- and long-term nitrogen deficiency. *PLoS ONE* **2017**, *12*, e0182700.

29. Xin, W.; Zhang, L.; Zhang, W.; Gao, J.; Yi, J.; Zhen, X.; Li, Z.; Zhao, Y.; Peng, C.; Zhao, C. An integrated analysis of the rice transcriptome and metabolome reveals differential regulation of carbon and nitrogen metabolism in response to nitrogen availability. *Int. J. Mol. Sci.* **2019**, *20*, 2349. [CrossRef]

30. Hao, Q.N.; Zhou, X.A.; Sha, A.H.; Wang, C.; Zhou, R.; Chen, S.L. Identification of genes associated with nitrogen use efficiency by genome-wide transcriptional analysis of two soybean genotypes. *BMC Genom.* **2011**, *12*, 525. [CrossRef]

31. Gelli, M.; Duo, Y.D.; Konda, A.R.; Zhang, C.; Holding, D.; Dweikat, I. Identification of differentially expressed genes between sorghum genotypes with contrasting nitrogen stress tolerance by genome-wide transcriptional profiling. *BMC Genom.* **2014**, *15*, 179–194. [CrossRef] [PubMed]

32. Quan, X.; Zeng, J.; Ye, L.; Chen, G.; Han, Z.; Shah, J.; Zhang, G. Transcriptome profiling analysis for two Tibetan wild barley genotypes in responses to low nitrogen. *BMC Plant Biol.* **2016**, *16*, 30. [CrossRef] [PubMed]

33. Li, W.; Xiang, F.; Zhong, M.; Zhou, L.; Liu, H.; Li, S.; Wang, X. Transcriptome and metabolite analysis identifies nitrogen utilization genes in tea plant (*Camellia sinensis*). *Sci. Rep.* **2017**, *7*, 1693.

34. Sinha, S.K.; Sevanthi, V.A.M.; Chaudhary, S.; Tyagi, P.; Venkadesan, S.; Rani, M.; Mandal, P.K. Transcriptome analysis of two rice varieties contrasting for nitrogen use efficiency under chronic N starvation reveals differences in chloroplast and starch metabolism-related genes. *Genes* **2018**, *9*, 206. [CrossRef]

35. Iqbal, A.; Dong, Q.; Wang, X.R.; Gui, H.P.; Zhang, H.H.; Zhang, X.L.; Song, M.Z. Transcriptome analysis reveals differences in key genes and pathways regulating carbon and nitrogen metabolism in cotton genotypes under N starvation and resupply. *Int. J. Mol. Sci.* **2020**, *21*, 1500.

36. Available online: https://www.ncbi.nlm.nih.gov/genome/10896?genome_assembly_id=227510 (accessed on 3 August 2017).

37. Available online: ftp://ftp.ncbi.nih.gov/blast/db (accessed on 3 August 2017).

38. Available online: http://www.geneontology.org/ (accessed on 3 August 2017).

39. Available online: http://www.genome.jp/kegg/ (accessed on 3 August 2017).

40. Albrech, C.; Russinova, E.; Kemmerling, B.; Kwaaitaal, M.; de Vries, S.C. Arabidopsis somatic embryogenesis receptor kinase proteins serve brassinosteroid-dependent and -independent signaling pathways. *Plant Physiol.* **2008**, *148*, 611–619. [CrossRef]

41. Yang, Y.; Yu, X.; Song, L.; An, C. ABI4 activates *DGAT1* expression in Arabidopsis seedlings during nitrogen deficiency. *Plant Physiol.* **2011**, *156*, 873–883. [CrossRef]

42. Liu, M.Y.; Burgos, A.; Ma, L.; Zhang, Q.; Tang, D.; Ruan, J. Lipidomics analysis unravels the effect of nitrogen fertilization on lipid metabolism in tea plant (*Camellia sinensis* L.). *BMC Plant Biol.* **2017**, *17*, 165. [CrossRef]

43. D'Hondt, K.; Heese-Peck, A.; Riezman, H. Protein and lipid requirements for endocytosis. *Annu. Rev. Genet.* **2000**, *34*, 255–295. [CrossRef]

44. Chen, X.; Irani, N.G.; Friml, J. Clathrin-mediated endocytosis: The gateway into plant cells. *Curr. Opin. Plant Biol.* **2011**, *14*, 674–682. [CrossRef]

45. Fan, L.; Li, R.; Pan, J.; Ding, Z.; Lin, J. Endocytosis and its regulation in plants. *Trends Plant Sci.* **2015**, *20*, 388–397. [CrossRef]

46. Isah, T. Stress and defense responses in plant secondary metabolites production. *Biol. Res.* **2019**, *52*, 39. [CrossRef]

47. Khare, S.; Singh, N.B.; Singh, A.; Hussain, I.; Niharika, K.; Yadav, V.; Bano, C.; Yadav, R.K.; Amist, N. Plant secondary metabolites synthesis and their regulations under biotic and abiotic constraints. *J. Plant Biol.* **2020**, *63*, 203–216. [CrossRef]

48. Yazaki, K. ABC transporters involved in the transport of plant secondary metabolites. *FEBS Lett.* **2006**, *580*, 1183–1191. [CrossRef]

49. Lv, H.; Li, J.; Wu, Y.; Garyali, S.; Wang, Y. Transporter and its engineering for secondary metabolites. *Appl. Microbiol. Biotechnol.* **2016**, *100*, 6119–6130. [CrossRef]

50. Larbat, R.; Robin, C.; Lillo, C.; Drengstig, T.; Ruoff, P. Modeling the diversion of primary carbon flux into secondary metabolism under variable nitrate and light/dark conditions. *J. Theor. Biol.* **2016**, *402*, 144–157. [CrossRef]

51. Peng, M.; Hannam, C.; Gu, H.; Bi, Y.M.; Bothstein, S.J. A mutation in *NLA*, wich encodes a RING-type ubiquitin ligase, disrupts the adaptability of *Arabidopsis* to nitrogen limitation. *Plant J.* **2007**, *50*, 320–337. [CrossRef]

52. Kant, S.; Peng, M.; Rothstein, S.J. Genetic regulation by NLA and microRNA827 for maintaining nitrate-dependent phosphate homeostasis in *Arabidopsis*. *PLoS Genet.* **2011**, *7*, e1002021. [CrossRef]

53. Lin, W.Y.; Huang, T.K.; Chiou, T.J. Nitrogen limitation adaptation, a target of microRNA827, mediates degradation of plasma membrane-localized phosphate transporters to maintain phosphate homeostasis in *Arabidopsis*. *Plant Cell* **2013**, *25*, 4061–4074. [CrossRef]

54. Park, B.S.; Seo, J.S.; Chua, N.H. Nitrogen limitation adaptation recruits phosphate2 to target the phosphate transporter PT2 for degradation during the regulation of *Arabidopsis* phosphate homeostasis. *Plant Cell* **2014**, *26*, 454–464. [CrossRef]

55. Yue, W.; Ying, Y.; Wang, C.; Zhao, Y.; Dong, C.; Whelan, J.; Shou, H. OsNLA1, a RING-type ubiquitin ligase, maintains phosphate homeostasis in *Oryza sativa* via degradation of phosphate transporters. *Plant J.* **2017**, *90*, 1040–1051. [CrossRef]

56. Chen, G.H.; Yan, W.; Yang, S.P.; Wang, A.; Gai, J.Y.; Zhu, Y.L. Overexpression of rice phosphate transporter gene OsPT2 enhances tolerance to low phosphorus stress in soybean. *J. Agric. Sci. Technol.* **2015**, *17*, 469–482.

57. Zhu, W.; Yang, L.; Yang, S.; Gai, J.; Zhu, Y. Overexpression of rice phosphate transporter gene OsPT2 enhances nitrogen fixation and ammonium assimilation in transgenic soybean under phosphorus deficiency. *J. Plant Biol.* **2016**, *59*, 172–181. [CrossRef]

Article

CaHSP18.1a, a Small Heat Shock Protein from Pepper (*Capsicum annuum* L.), Positively Responds to Heat, Drought, and Salt Tolerance

Yan-Li Liu [1,†], Shuai Liu [1,†], Jing-Jing Xiao [1], Guo-Xin Cheng [1], Haq Saeed ul [1,2] and Zhen-Hui Gong [1,*]

1   College of Horticulture, Northwest A&F University, Yangling 712100, China; 2018060114@nwafu.edu.cn (Y.-L.L.); liushuai5887@nwsuaf.edu.cn (S.L.); 2018050339@nwafu.edu.cn (J.-J.X.); 2014060108@nwsuaf.edu.cn (G.-X.C.); Saeed_ulhaq@nwsuaf.edu.cn (H.S.u.)
2   Department of Horticulture, University of Agriculture Peshawar, Peshawar 25120, Pakistan
*   Correspondence: zhgong@nwsuaf.edu.cn; Tel.: +86-029-8708-2102; Fax: +86-029-8708-2613
†   Co-first author.

Citation: Liu, Y.-L.; Liu, S.; Xiao, J.-J.; Cheng, G.-X.; ul, H.S.; Gong, Z.-H. CaHSP18.1a, a Small Heat Shock Protein from Pepper (*Capsicum annuum* L.), Positively Responds to Heat, Drought, and Salt Tolerance. *Horticulturae* **2021**, *7*, 117. https://doi.org/10.3390/horticulturae7050117

Academic Editor: Yuyang Zhang

Received: 1 April 2021
Accepted: 11 May 2021
Published: 18 May 2021

Publisher's Note: MDPI stays neutral with regard to jurisdictional claims in published maps and institutional affiliations.

Copyright: © 2021 by the authors. Licensee MDPI, Basel, Switzerland. This article is an open access article distributed under the terms and conditions of the Creative Commons Attribution (CC BY) license (https://creativecommons.org/licenses/by/4.0/).

Abstract: Pepper is a thermophilic crop, shallow-rooted plant that is often severely affected by abiotic stresses such as heat, salt, and drought. The growth and development of pepper is seriously affected by adverse stresses, resulting in decreases in the yield and quality of pepper crops. Small heat shock proteins (s HSPs) play a crucial role in protecting plant cells against various stresses. A previous study in our laboratory showed that the expression level of *CaHSP18.1a* was highly induced by heat stress, but the function and mechanism of CaHSP18.1a responding to abiotic stresses is not clear. In this study, we first analyzed the expression of *CaHSP18.1a* in the thermo-sensitive B6 line and thermo-tolerant R9 line and demonstrated that the transcription of *CaHSP18.1a* was strongly induced by heat stress, salt, and drought stress in both R9 and B6, and that the response is more intense and earlier in the R9 line. In the R9 line, the silencing of *CaHSP18.1a* decreased resistance to heat, drought, and salt stresses. The silencing of *CaHSP18.1a* resulted in significant increases in relative electrolyte leakage (REL) and malonaldehyde (MDA) contents, while total chlorophyll content decreased under heat, salt, and drought stresses. Overexpression analyses of *CaHSP18.1a* in transgenic *Arabidopsis* further confirmed that *CaHSP18.1a* functions positively in resistance to heat, drought, and salt stresses. The transgenic *Arabidopsis* had higherchlorophyll content and activities of superoxide dismutase, catalase, and ascorbate peroxidase than the wild type (WT). However, the relative conductivity and MDA content were decreased in transgenic Arabidopsis compared to the wild type (WT). We further showed that the CaHSP18.1a protein is localized to the cell membrane. These results indicate CaHSP18.1a may act as a positive regulator of responses to abiotic stresses.

Keywords: *CaHSP18.1a*; gene silencing; transgenic *Arabidopsis*; heat stress; pepper; gene expression

1. Introduction

Plants can tolerate considerable biotic and abiotic stresses in their complex and changing environments, including drought, high salt, extreme temperatures, and oxidation [1,2]. To mitigate stresses, plants have developed several protective mechanisms. Heat shock proteins (HSPs) can maintain protein homeostasis and prevent or repair the misfolding of proteins in abiotic stresses response. Moreover, HSPs are evolutionarily conserved molecular chaperones widely found among various plant taxa [3–5]. Plant HSPs also play critical roles in the folding, transport, degradation, and assembly of proteins under normal and stress conditions [6]. In response to high temperatures, plant cells dramatically increase the concentrations of HSPs to prevent heat-related damage and increase plant thermotolerance [7]. In addition, HSPs are also involved in plant growth and development under normal conditions, including the growth of flowers and seeds as well as fruit set, development [8], tuberization [9], and nutrient uptake [10]. HSPs are present in the cell membrane and cytoplasm, nucleus, and cell organelles such as the mitochondria, chloroplasts, and endoplasmic reticulum [11,12].

HSPs, based on their sequence homology and molecular weight, are generally grouped into the following different families: HSP20s, HSP60s, HSP70s, HSP90s, and HSP100s [13,14]. Of the five conserved families, HSP20s, are also called small heat shock proteins (s HSPs). The molecular weights of HSP20s range between 15 and 42 kDa [13,15,16]. Furthermore, one of the distinctive characteristics of HSP20s is their ability to bind to substrate proteins without ATP, and they also have a strong ability to bind to denatured substrates [15–18]. Thus, s HSPs are highly able to maintain the stability of foreign proteins in cells to prevent them from aggregating. Although there are many types of substrate proteins, s HSPs have a flexible N-terminus and α-crystallin domain (ACD) hydrophobic surface that can adapt to bind these different protein substrates. In addition, s HSPs can be combined with different substrates in different ways, which makes s HSPs able to bind to a wider variety of proteins and to provide more complicated mechanisms of action among HSPs [19].

Korotaeva et al. [20] and Nieto-Sotelo et al. [21] showed that different HSPs are differentially expressed in different species, and even among different genotypes of the same species. It has been reported that the overexpression of AtHSP17. 6A increased the penetration resistance of *Arabidopsis* [22]. *AtHSP21* improved the heat resistance of transgenic *Arabidopsis* and extended the memory time of plants subjected to heat resistance, such that *Arabidopsis* was more heat resistant when subjected to heat stress again [23]. Some studies have also reported that HSP gene expression positively regulated protective enzyme activities. For example, in *Arabidopsis*, overexpression of *AtHSP17.8* enhanced SOD activity [24]. Similarly, overexpression of *HSP16.9* in tobacco increased the activities of POD, CAT, and SOD [25].

Pepper (*Capsicum annuum* L.) is one of the most important economical and medicinal vegetable crops worldwide [26].Pepper is usually cultivated in warm regions under temperatures of 15–34 °C [27]. Salt, drought, and heat stress can limit pepper growth and development and severely damage pepper pollination and seed set, which can lead to flower and fruit abscission and thus lower pepper fruit yield and quality [28,29].HSP20s in pepper play a major role in environmental stress responses, and a total of 35 pepper HSP20s were identified by Guo et al. [30]. All HSP20s were named based on their molecular weights, and stress-related cis-elements were detected in the promoter regions, including heat shock elements (HSEs), TATA boxes, CCAAT motifs, and TC-rich repeats [26]. Many CaHSP20 genes are not expressed across different pepper tissues (i.e., root, stem, leaf, and flower tissues). In recent years, the functions of CaHSP22.4, CaHSP25.9, CaHSP16.4, CaHSP24.2, and CaHSP26 have been identified. CaHSP16.4 is localized to the cytoplasm and nucleus, while in *Arabidopsis* lines with *CaHSP16.4* overexpression, increased tolerance to heat stress has been observed [31]. Guo et al. [26] also found that overexpression of CaHSP22.4, which is located in the mitochondria and cytoplasm, increased heat tolerance in *Arabidopsis*, with the expression increasing when pepper plants were subjected to high temperature. Similarly, the CaHSP25.9 protein was localized to the cell membrane and cytoplasm, and positively regulates heat, salt, and drought stress tolerance in pepper (*Capsicum annuum* L.) [32]. Pepper CaHSP24.2 is localized to mitochondria, the cytoplasm, and chloroplasts, where *CaHSP24.2* enhances the thermo-tolerance of transgenic *Arabidopsis* plants and regulates the expression of heat stress-related genes [30]. He et al. [33] overexpressed *CaHSP26*, which enhanced the tolerance of heat stress in *Arabidopsis*. Interestingly, heat-tolerance and salt-tolerance decreased in *CaHSP22.0*-silenced pepper [34]. All these studies suggest that sHSP20s may participate in responses to heat stress [35] and contribute to the acquisition of pepper thermo-tolerance [30].

Among the 35 *CaHSP20s* examined, the expression level of *CaHSP18.1a* was increased in both the B6 and R9 lines under heat stress [30]. Moreover, sequence analysis showed that *CaHSP18.1a* contained an HSE, and some other stress-related elements were also identified [30]. Based on the above findings, we analyzed the subcellular localization and expression pattern of *CaHSP18.1a* in different pepper tissues, as well as its response to salt, drought, and heat stresses. Virus-induced gene silencing (VIGS) was preliminarily used to analyze the functions of *CaHSP8.1a* in response to stress in pepper plants. In addition, overexpression (OE) in transgenic *Arabidopsis thaliana* indicated that *CaHSP18.1a* plays a positive regulatory role in the responses to heat, salt, and drought stress. Our results provide a basis for further functional studies of *CaHSP18.1a* in other important crop species and in its role in stress tolerance.

2. Materials and Methods

2.1. Plant Materials and Growth Conditions

The thermo-tolerant pepper line R9 (a sweet pepper from the World/Asia Vegetable Research and Development Center, PP0042-51) and the thermo-sensitive pepper line B6 (selected by the Pepper Research Group, College of Horticulture, Northwest A&F University, Yangling, China) were used in this study. Pepper seedlings were cultivated in a growth chamber under the following growth conditions prior to various treatments: daily 16 h light/8 h dark cycles and 65% relative humidity until the 6–8 true leaves stage. The temperature was changed throughout the course of the experiment. R9 peppers were grown under 25/20 °C day/night temperatures to enable analyzing gene expression [30,36,37]. However, the growing conditions for use of virus-induced gene silencing (VIGS) in the R9 pepper line were 22/18 °C day/night temperatures [38]. *Arabidopsis* ecotype Col-0 variety seedlings were incubated at 65% relative humidity, 22/18 °C (day/night), and 16 h/8 h (light/dark) photoperiod conditions [37,38].

2.2. RNA Extraction and Real-Time Fluorescent Quantitative PCR qRT-PCR Analysis

Total RNA was extracted using the Trizol method [28]. Synthesis of cDNA was conducted with the PrimeScript™ kit (Takara, Dalian, China) according to the manufacturer's instructions. First, we downloaded the amino acid sequence of *CaHSP18.1a* from the Pepper Genomics Database (accessed date on 1 January 2020, http://peppergenome.snu.ac.kr/: Accession number: CA08g17060). Primer Premier 5.0 was used to design primers, and primer specificity was detected using NCBI Primer BLAST (https://www.ncbi.nlm.nih.gov/tools/primer-blast/, accessed on 5 January 2020) (Supplementary Table S1). The pepper ubiquitin binding gene *CaUbi3* (Accession number AY486137) was used as a reference gene [39]. qRT-PCR was performed using the iQ5.0 Bio-Rad iCycler thermal cycler (Bio-Rad, Hercules, CA, USA). The SYBR Green Super mix (Takara, Dalian, China) was used in the qRT-PCR reaction system following the manufacturer's instructions. The relative expression levels of the gene were analyzed using the $2^{-\Delta\Delta CT}$ method [40].

2.3. Subcellular Localization of CaHSP18.1a Protein

The ORF (open reading frame) of *CaHSP18.1a* without a termination codon was PCR-amplified using a specific primer pair (Supplementary Table S1). The resulting *CaHSP18.1a* fragment was cloned into the pVBG2307: GFP vector with *Xba*I and *Kpn*I restriction sites. The pVBG2307:*CaHSP18.1a*: GFP fusion protein transient expression vector and the control vector pVBG2307: GFP, after having been successfully constructed, were transformed into *Agrobacterium tumefaciens* strain GV3101, which was then injected into tobacco (*Nicotiana tabacum*) leaves to induce transient expression. After dark cultivation for approximately 36 h, epidermis samples of tobacco leaves were photographed under a fully automatic upright fluorescence microscope on the public platform of the College of Horticulture, Northwest A&F University, and the fluorescence patterns in the cells were observed; we specifically used the method described by Yu et al. [41].

2.4. Virus-Induced Gene Silencing of CaHSP18.1a

A 256-bp fragment of the *CaHSP18.1a* ORF was PCR-amplified using a specific primer pair (Supplementary Table S1). The underlined sequences are restriction enzyme cleavage sites (for *Xba*I and *Kpn*I). The resulting *CaHSP18.1a* fragment was inserted into TRV2:00 vectors, with the empty vector TRV2:00 and TRV2: *CaPDS* (phytoene desaturase gene) used as negative and positive controls. When R9 plants reached the two true leaves stage, we followed the method of Wang et al [38], which involved mixing the pTRV1 bacterial culture with an equal volume of the TRV2:00, TRV2: *CaPDS*, and TRV2: *CaHSP18.1a* cultures; this solution was injected into the leaves of R9 plants. After incubation in the dark at 18 °C for 2 days, plants were transferred to incubators under preset normal conditions. After 35 days, when most of the leaves of the TRV2-*CaPDS* pepper plants had become bleached, total RNA was extracted from the leaves of the silenced TRV2: *CaHSP18.1a* plants and the negative control TRV2:00 plants, and qRT-PCR was used to detect the *CaHSP18.1a* expression level, which was used to calculate silencing efficiency.

2.5. Generation of CaHSP18.1a-Overexpression Arabidopsis Lines

The entire coding regions of *CaHSP18.1a* were cloned into the pVBG2307 vector between the *XbaI* and *KpnI* restriction sites to yield the final plasmid pVBG2307: *CaHSP18.1a* used for genetic transformation (the primers used for this experiment are given in Supplementary Table S1). The recombinant fusion vector was transformed into *Agrobacterium* strain GV3101 and transformed into *Arabidopsis thaliana* as described by Clough and Bent [42]. Transformed strains of pVBG2307 expression vector were screened with kanamycin and confirmed by PCR verification. We extracted DNA to detect the correctness of the target band. First, the fragment lengths of the bands were compared with the target band, obtaining, respectively, the OE1, OE2, OE3, OE4, and OE5 lines. Next, we performed real-time qRT-PCR quantitative analysis and detected the transcript from the inserted construct (Supplementary Figure S2B). *CaHSP18.1a* was thus determined to be expressed in large quantities in the OE3, OE2, and OE1 strains, but the wild-type gene was not detected. Both the target band and qRT-PCR results indicated that the *CaHSP18.1a* gene was successfully transferred into *Arabidopsis thaliana*, and the obtained T3-generation *Arabidopsis thaliana* could thus be used for further experiments.

2.6. Experimental Treatments and Sample Collection

The roots, stems, and leaves of R9 and B6 pepper seedlings (at the 4-to-6-leaf stage) grown under normal conditions were sampled in order to analyze expression of *CaHSP18.1a* in different tissues. For the thermotolerance treatment, R9 and B6 pepper seedlings (again, at the 4-to-6-leaf stage) were grown at 42 °C for 24 h, and root, stem, and leaf samples were collected from stress-treated seedlings at 0, 0.5, 1, 3, and 6 h post-treatment. For the drought stress treatment, the roots of R9 seedlings were soaked in 300 mM mannitol, and root, stem, and leaf samples were collected from stress-treated seedlings at 0, 3, 6, 12, and 24 h after treatment.

To analyze the function of *CaHSP18.1a* in response to pepper abiotic stress, silenced pepper seedlings and TRV2:00 pepper seedlings were grown at 42 °C for 24 h. For the drought and salt stresses, seedlings were treated with 300 mM mannitol and 300 mM NaCl for 24 h. Samples were collected and malondialdehyde (MDA) content, total chlorophyll content, and relative electrolyte leakage (REL) were determined. To identify the tolerance of *CaHSP18.1a*-overexpression in *Arabidopsis thaliana* in response to heat, salt, and drought stress, T3-generation *Arabidopsis thaliana* and wild-type lines were treated as described.

For heat stress, 2-week-old OE3 seedlings were treated at 42 °C for 24 h. For drought stress, water was withheld from 3-week-old transgenic *Arabidopsis* seedlings for 10 d. Samples were collected to measure the total chlorophyll contents, MDA content, REL, the activity levels of CAT, SOD, and ascorbic acid peroxidase (APX). For salt stress, the seeds of WT and transgenic lines were sown on MS medium with 0, 100, and 150 mM NaCl, and the roots lengths were measured after 10 days of treatment. The germination rate was determined after 6 d. The 3-week-old WT and transgenic plants were irrigated with 200 mM NaCl solution for 7 days, once every 2 days.

2.7. Measurement of Physiological Indicators

REL was estimated using the thiobarbituric acid reaction [43]. Total chlorophyll content was determined in the leaves according to methods previously described by Arkus et al [44]. Lipid peroxidation was determined by measuring the MDA content following the method of Campos et al [45]. POD and SOD activity levels were measured following the methods of Guo et al [46]. APX activity was measured using the methods of Nakano and Asada [47]. CAT activity was determined following AebiH [48].

2.8. Statistical Analyses

The experimental data were analyzed using Origin (Origin Lab, Northampton, MA, USA) and SPPS (SPSS Inc., Chicago, IL, USA). Significance tests for differences between control and stress treatments were assessed at a $p \leq 0.05$ level of significance. All experiments were performed and analyzed separately based on three biological replicates.

3. Results

3.1. Expression of the CaHSP18.1a in Pepper Plants under Abiotic Stress

To confirm whether heat, drought, and salt have an effect on the expression of *CaHSP18.1a*, R9 and B6 pepper lines were used to analyze the expression of *CaHSP18.1a* under heat, drought, and salt stress. Under heat stress (Figure 1A,B), the expression levels of *CaHSP18.1a* were significantly upregulated (samples IV, V, and VI) at 3 h in the roots, stems, and leaves of R9 plants and peaked at 6 h (sample Point V) in B6 plants. However, during the 22 °C recovery stage, the recovery times of *CaHSP18.1a* in the R9 and B6 strains differed. In roots, the expression levels of *CaHSP18.1a* in R9 and B6 plants returned to baseline after a 6 h recovery at the normal temperature (sample VII point) (Figure 1B). In stems, the expression level of *CaHSP18.1a* returned to a normal level after 3 h under the 22 °C recovery conditions for both the R9 and B6 plants (point VI in samples) (Figure 1C). In leaves, the expression level of *CaHSP18.1a* returned to normal levels at 24 h (sample VIII point) (Figure 1D). We also analyzed the expression pattern of *CaHSP18.1a* under salt and drought stresses (Figure 1E–H). After 6-h NaCl treatments at different concentrations, the expression of *CaHSP18.1a* in R9 and B6 leaves and roots was highest under the 150 mM NaCl, 100 mM NaCl treatments, respectively. The transcription of *CaHSP18.1a* was higher in R9 under different concentrations of NaCl treatment (Figure 1E, F). After 6-h treatments with different concentrations of mannitol, the expression of *CaHSP18.1a* in R9 and B6 leaves was the highest after the 150 mM mannitol treatment; the highest expression of *CaHSP18.1a* was observed in R9 and B6 roots subjected to the 50mM mannitol treatment (Figure 1G, H). In addition, the transcription of *CaHSP18.1a* was higher in R9 under different concentrations of mannitol. This analysis showed that the expression of *CaHSP18.1a* in pepper was induced by heat, salt, and drought. The response times of *CaHSP18.1a* in different organs of different pepper lines differed, and the response was more intense and more early in the R9 line, which suggests that *CaHSP18.1a* plays a substantial role in plant responses to heat stress.

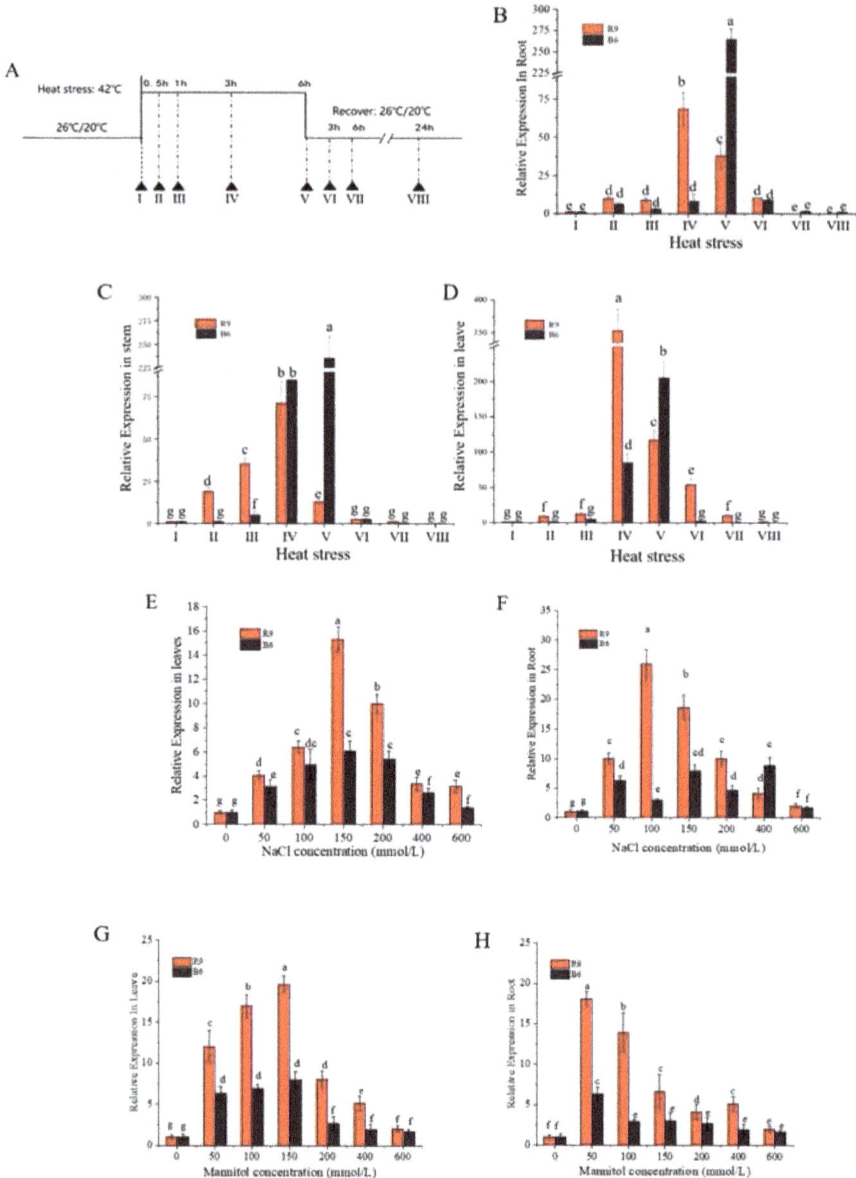

Figure 1. The expression characteristics of *CaHSP18.1a* in peppers in response to heat stress. (**A**) Time course of heat stress treatment and normal temperature recovery; the sampling time points are represented by triangles (pepper sample points I–VIII); (**B–D**) The expression levels of *CaHSP18.1a* in roots, stems, and leaves of R9 and B6 plants at each sampling time point; the expression levels of B6 and R9 plants were based on the reference level of their samples, and *CaUBI-3* was selected as the reference gene.(**E–H**) The expression levels of *CaHSP18.1a* following salt and drought treatment in R9 and B6 leaves and roots. The data presented are means with standard deviations of three biological replicates. Different letters denote statistical significance ($p \leq 0.05$).

3.2. Subcellular Localization of CaHSP18.1a Protein

To explore the subcellular localization of CaHSP18.1a, we constructed the pVBG2307: CaHSP18.1a: GFP fusion expression vector. Both pVBG2307: GFP and pVBG2307:CaHSP 18.1a: GFP fusion plasmids were introduced into *Nicotiana tabacum* leaves, and fluorescence was confirmed in the transformed tobacco cells with a microscope (Figure 2). We found that the green fluorescence signal of pVBG2307: CaHSP18.1a: GFP was detected in the cell membrane (Figure 2A), while the fluorescence of the empty pVBG2307: GFP vector was distributed throughout the cell (Figure 2B), indicating that CaHSP18.1a is localized to the cell membrane.

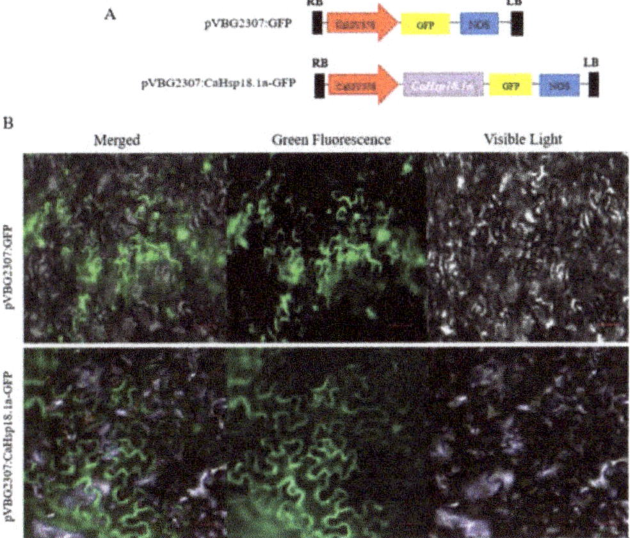

Figure 2. Transient expression of CaHSP18.1a in tobacco. (**A**) Schematic diagram of the CaHSP18.1a subcellular localization expression vector. (**B**) Subcellular localization of the CaHSP18.1a protein in tobacco leaves, with pVBG2307: GFP as control. Scale bar = 50 μm.

3.3. CaHSP18.1a-Silenced Plants Sensitive to Abiotic Stress

Confirming the VIGS procedure, after about 40–45 days, plants injected with the positive control TRV2: *CaPDS* showed a large area of typical white leaves (Supplementary Figure S1A), while under normal conditions there was no difference between *CaHSP18.1a*-silenced (TRV2:*CaHSP18.1a*) and negative control (TRV2:00) pepper plants. The silencing efficiency of *CaHSP18.1a*-silenced and TRV2:00 plants was assessed using q RT-PCR. As shown in Supplementary Figure S1A, the expression level of *CaHSP18.1a* in the silenced pepper plants decreased to less than 20% of that observed in the negative control plants. Thus, the silencing efficiency for *CaHSP18.1a*-silenced plants reached more than 80% (Supplementary Figure S1B). Therefore, control plants (TRV2:00) and silenced plants (TRV2:*CaHSP18.1a*) were used for the follow-up investigation.

HS (42 °C) was applied to *CaHSP18.1a*-silenced and control pepper plants for 3 h, and the silenced plants and the control group began to show different degrees of wilting. The heat-stress treatment (42 °C) induced significantly different symptoms after 24 h, such that the new growth of *CaHSP18.1a*-silenced plants was seriously wilted with curled leaves and shed lower leaves, while the leaves of the control plants were only slightly curled (Figure 3A). In addition, the MDA content and REL was lower in the control plants compared to *CaHSP18.1a*-silenced pepper plants (Figure 3B,C); however, the total chlorophyll content was higher in the control than in the *CaHSP18.1a*-silenced plants (Supplementary Figure S1C).

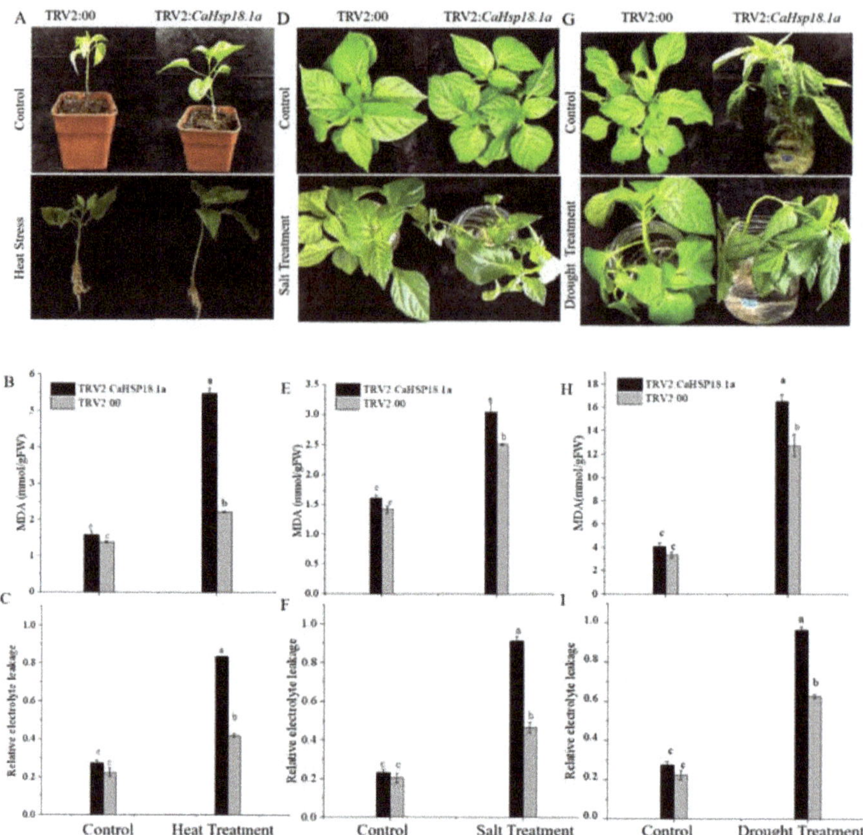

Figure 3. TRV2:*CaHSP18.1a* and TRV2:00 plant phenotypes under heat, drought, and salt treatments, respectively. (**A–C**) TRV2:*CaHSP18.1a* and TRV2:00 plant phenotypes, malonaldehyde (MDA) content, and relative electrolyte leakage (REL) under the 42 °C heat treatment for 24 h; (**D–F**) TRV2:*CaHSP18.1a* and TRV2:00 plant phenotypes, MDA content, and relative electrolyte leakage (REL) following salt stress by being soaked in 300 mM NaCl solution for 24 h; (**G–I**) TRV2:*CaHSP18.1a* and TRV2:00 plant phenotypes, MDA content, and REL following drought stress by being soaked in 300 mM mannitol solution for 24 h. Data are means with standard deviations of three biological replicates. Different letters denote statistical significance ($p \leq 0.05$).

To study the salt-tolerance of silenced and control plants, we washed their roots and soaked them in 300 mM NaCl solution for 24 h. The leaves of silenced plants showed symptoms of wilting, shriveling, and serious yellowing, with lower leaves that had begun to absciss, while the leaves of the control plants only showed some yellowing and did not exhibit obvious wilting. The leaves of the control plants showed only yellowing and no apparent wilting (Figure 3D). The MDA content of both plants increased significantly, but that of silenced plants was higher than that of control plants (Figure 3E). Relative electrolyte leakage (REL) was higher in silenced plants compared to control plants (0.96 versus 0.65) (Figure 3F). To study the effects of *CaHSP18.1a*-silencing on drought tolerance, the silenced and control plants were soaked in 300 mM mannitol solution for 36 h. The *CaHSP18.1a*-silenced pepper showed severe loss of water and wilting, while control plants showed no obvious change (Figure 3G). Furthermore, the MDA content and REL both exhibited a similar increase in the silenced pepper plants (Figure 3H,I). This indicated that silencing of *CaHSP18.1a* reduced the drought tolerance of pepper plants.

3.4. Effect of CaHSP18.1a Overexpression on Transgenic Arabidopsis

3.4.1. Overexpression of CaHSP18.1a Enhances Plant Tolerance of Heat Stress

First, we transformed pVBG2307:*CaHSP18.1a* into *Agrobacterium tumefaciens* strain GV3101, which was used to transfect *Arabidopsis thaliana* using the dipping method; successful transformants were identified through resistance gene screening and molecular level detection until homozygous T3 lines were obtained (Supplementary Figure S2A). The wild-type (WT) line and five transgenic lines were cultured on Murashige and Skoog (MS) medium for 10 days, and the lengths of their roots were measured. The survival rate of WT plants was lower than those of the OE1, OE2, and OE3 seedlings (Supplementary Figure S2C,D). Transgenic plants and WT plants were cultured under normal growth conditions for 48 days, and the growth rates of the OE1, OE2, and OE3 lines exceeded those of WT plants (Supplementary Figure S2D). Thus, the OE3, OE2, and OE1 lines were selected for follow-up experiments.

The obtained transgenic lines and WT plants were heat treated (42 °C for 24 h) at the 3-week stage. After heat-stress treatment, the WT plants showed wilting symptoms. Notably, restorable wilt or indistinct-symptoms were observed among the *CaHSP18.1a*-OE seedlings (Figure 4A), indicating that *CaHSP18.1a* plays an active role in increasing the thermotolerance of transgenic *Arabidopsis*. In addition, REL and MDA content increased significantly in both the OE and WT lines after heat treatment, while the MDA content was notably lower in the transgenic lines relative to the WT plants (Figure 4B). In addition, the SOD and POD activities of *CaHSP18.1a*-OE seedlings were clearly higher than those of the WT plants (Figure 4C,D). However, the catalase (CAT) activity did not significantly differ between the WT and transgenic lines (Figure 4E). *CaHSP18.1a* played a role in transgenic *Arabidopsis*, probably by regulating the expression of endogenous genes.

In the present study, among 18 stress-related genes, 12 were up-regulated in transgenic lines, while the other 6 did not change much (Figure 5). Among the up-regulated genes, *AtHSPC30, AtAPX3, AtCAT, AtHSP70,* and *AtRab1* were more prominently expressed in the transgenic OE3 line. Moreover, the expression of the 18 stress-related genes was markedly increased in both transgenic *Arabidopsis* and WT plants under heat stress. However, the expression of these genes in WT seedlings was lower than that in transgenic seedlings (Figure 5).

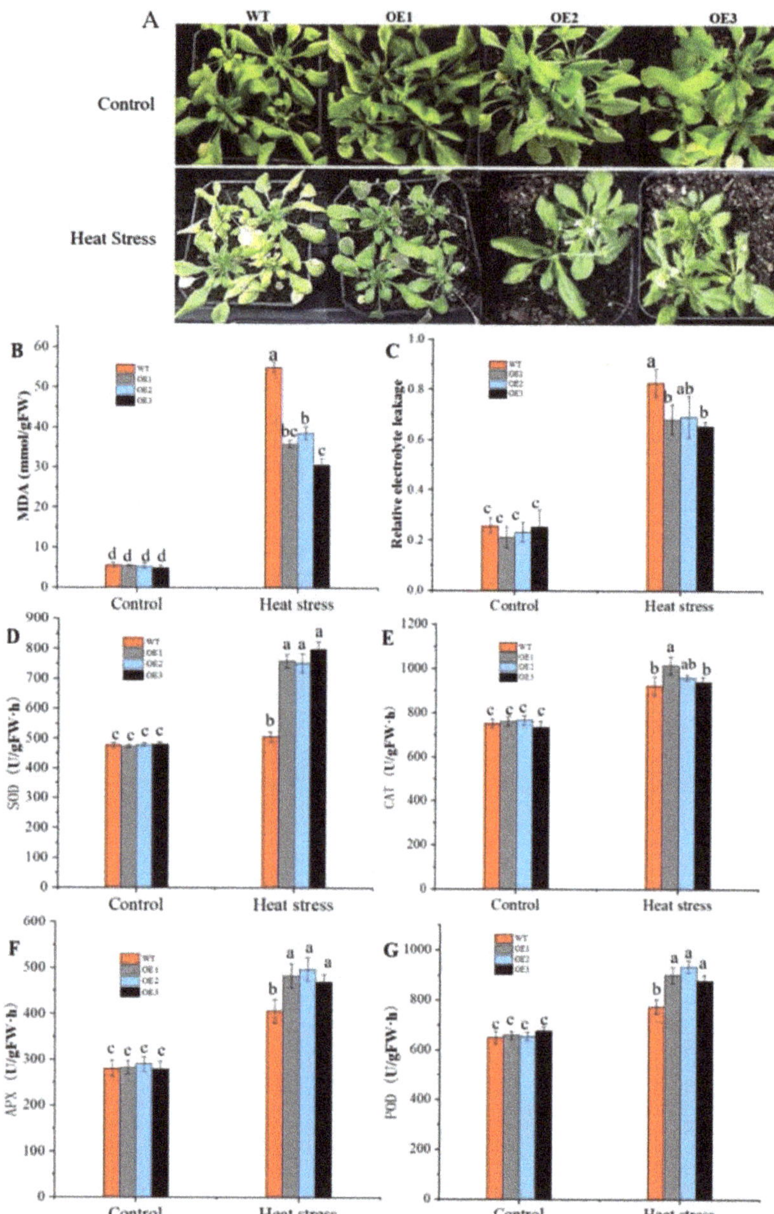

Figure 4. Heat resistance of transgenic *CaHSP18.1a*-OE *Arabidopsis* plants. (**A**) Phenotypes of 42 °C-treated wild-type (WT) and transgenic *Arabidopsis*; (**B–C**) Malonaldehyde (MDA) and relative electrolyte leakage (REL) of WT and transgenic *Arabidopsis*; (**D–G**) Superoxide dismutase (SOD), catalase (CAT), peroxidase (POD), and ascorbic acid peroxidase (APX) activity of WT and transgenic *Arabidopsis*. Data are means with standard deviations of three biological replicates. Different letters denote statistical significance ($p \leq 0.05$).

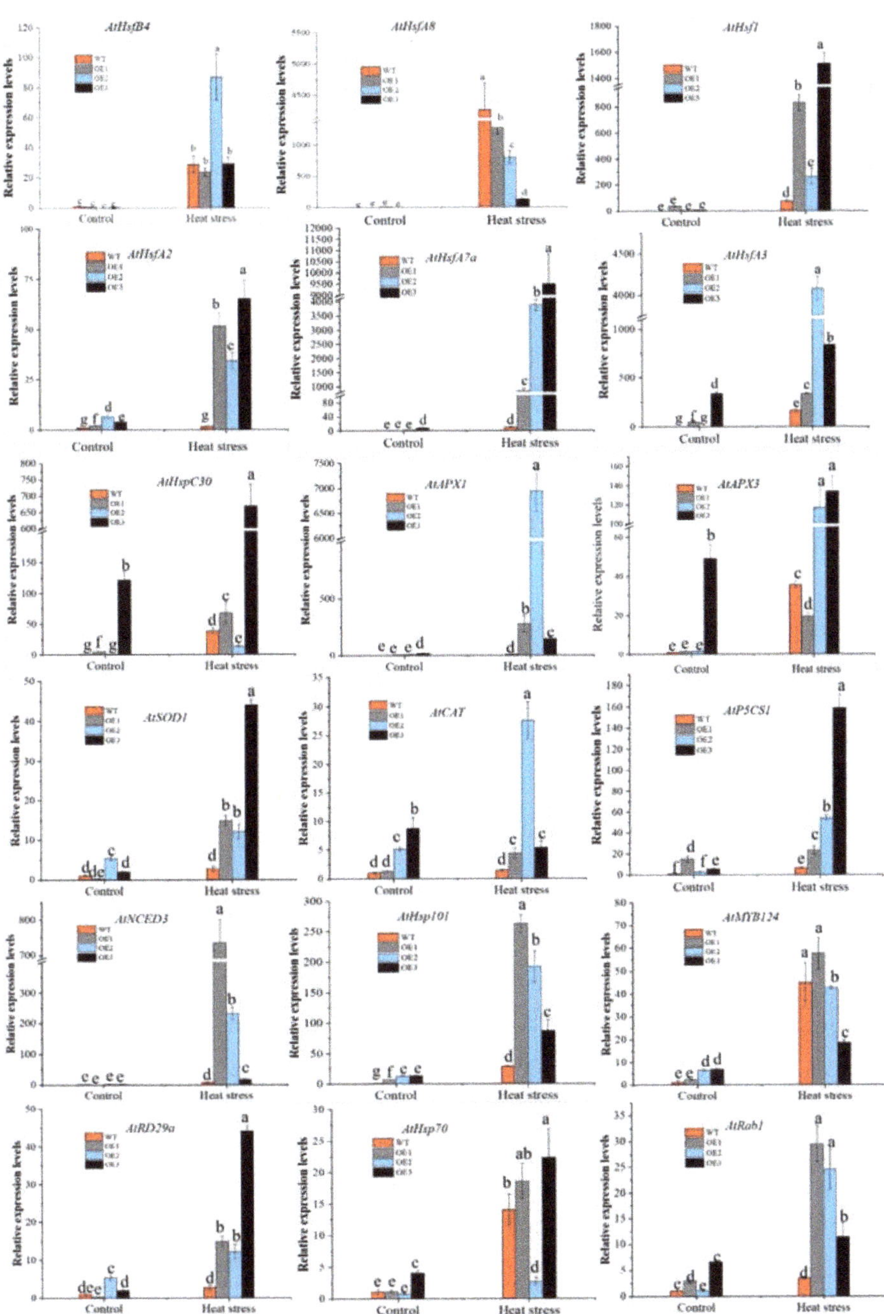

Figure 5. Expression patterns of stress–response genes in wild-type (WT), OE1, OE2, and OE3 lines before and after 42 °C heat treatment for 24 h. Data are means with standard deviations of three biological replicates. Different letters denote statistical significance ($p \leq 0.05$).

3.4.2. Overexpression of CaHSP18.1a Enhances Plant Tolerance to Drought Stress

CaHSP18.1a is a molecular chaperone, though its response to drought stress is still unclear. To further study its function under drought and salt stress, *CaHSP18.1a* transgenic *Arabidopsis thaliana* and WT seedlings were drought treated (Figure 6A). After water control treatment was conducted for 10 d on 3-week-old plants with consistent growth, WT plants showed severe wilting; the leaves turned yellow, while overexpression plants grew better than WT plants (Figure 6A). These results indicated that *CaHSP18.1a* increases drought tolerance of transgenic *Arabidopsis*. In addition, the MDA content and REL were increased in both the WT and OE lines, whereas the MDA content of transgenic seedlings was obviously lower than that of WT plants (Figure 6B,C). Thus, the degree of damage in OE plants was lower than that in WT plants. The SOD, CAT, and peroxidase (POD) activity showed an upward trend in both the WT and OE lines, but the activity level of SOD, CAT, and POD in transgenic seedlings was notably higher than that in WT plants (Figure 6D,F,G). While ascorbic acid peroxidase (APX) activity increased, there was, however, no visible difference between *CaHSP18.1a*-OE and WT lines (Figure 6E). The expression levels of the 18 stress-related genes were induced to varying degrees by drought stress. However, the expression of *AtHsfA2*, *AtHSPC30*, and *AtAPX1* exhibited almost no change in WT seedlings, and all of them were strongly increased in *CaHSP18.1a*-OE lines after drought stress. In addition, the expression levels of other genes were higher in transgenic lines compared to the WT plants after drought stress (Figure 7). Thus, the 18 stress-related genes examined may be involved at different levels in the response of *CaHSP18.1a*-OE lines to drought stress.

3.4.3. Overexpression of CaHSP18.1a Enhances Plant Tolerance to Salt Stress

To study the role of *CaHSP18.1a* in salt stress, transgenic *Arabidopsis thaliana* was subjected to salt stress. First, the germination rate of transgenic seeds under salt stress was observed (Supplementary Figure S3A). OE2, OE1, and WT seeds exhibited normal germination on MS plates without NaCl; the germination rate and seedling growth of transgenic lines were almost unaffected by 100 mM NaCl MS plates. However, the seed germination rate differed when the NaCl concentration was increased to 150 mM, and the germination rate of WT seeds was slightly lower than that of transgenic seeds. The salt tolerance of *CaHSP18.1a* transgenic *Arabidopsis* seeds increased. After the germinated seedlings were moved to MS plates with NaCl concentrations of 0, 100, and 150 mM, compared with the untreated seedlings, the growth of OE2, OE3, OE1, and WT seedlings after 6 d under 100 and 150 mM NaCl treatments was worse (Supplementary Figure S3A,B); root elongation was significantly inhibited, and the root length of transgenic lines seedlings was greater than that of WT plants (Figure 8A).

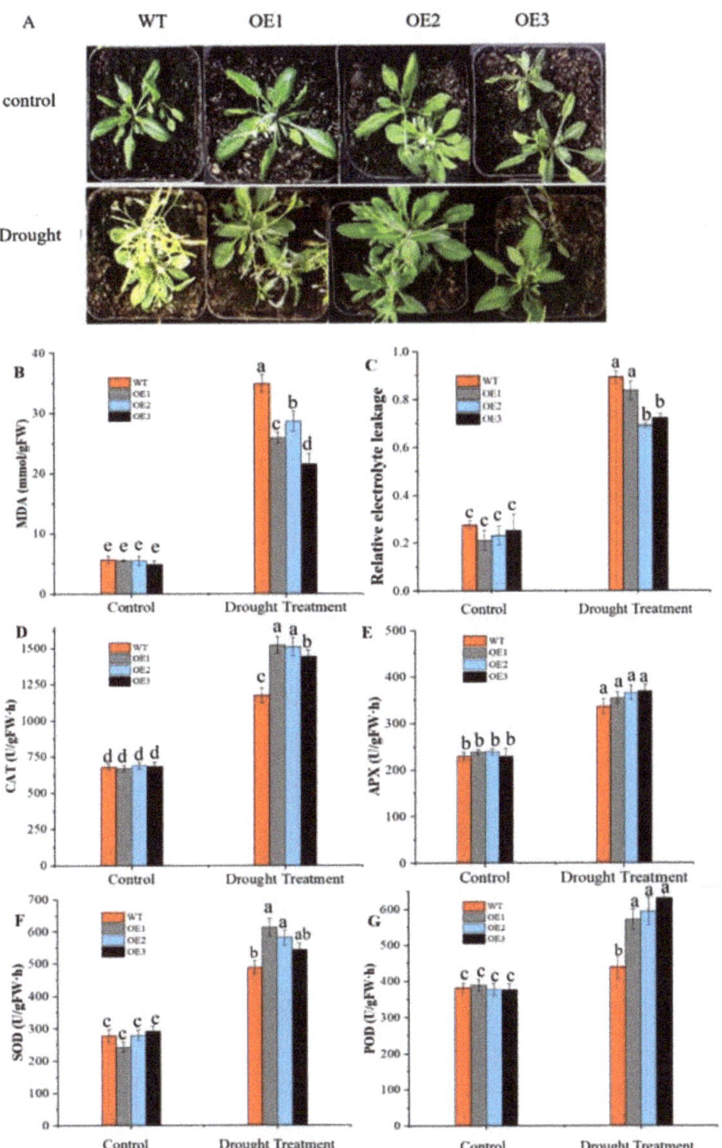

Figure 6. Drought resistance of transgenic *CaHSP18.1a*-OE *Arabidopsis* plants. (**A**) Phenotypes of wild-type (WT) and *CaHSP18.1a*-OE *Arabidopsis*; (**B,C**) Malonaldehyde (MDA) content and relative electrolyte leakage (REL) of WT and *CaHSP18.1a*-OE *Arabidopsis*; (**D–G**) Superoxide dismutase (SOD), catalase (CAT), peroxidase (POD), and ascorbic acid peroxidase (APX) activity of WT and *CaHSP18.1a*-OE *Arabidopsis* without watering for 10 days. Data are means with standard deviations of three biological replicates. Different letters denote statistical significance ($p \leq 0.05$).

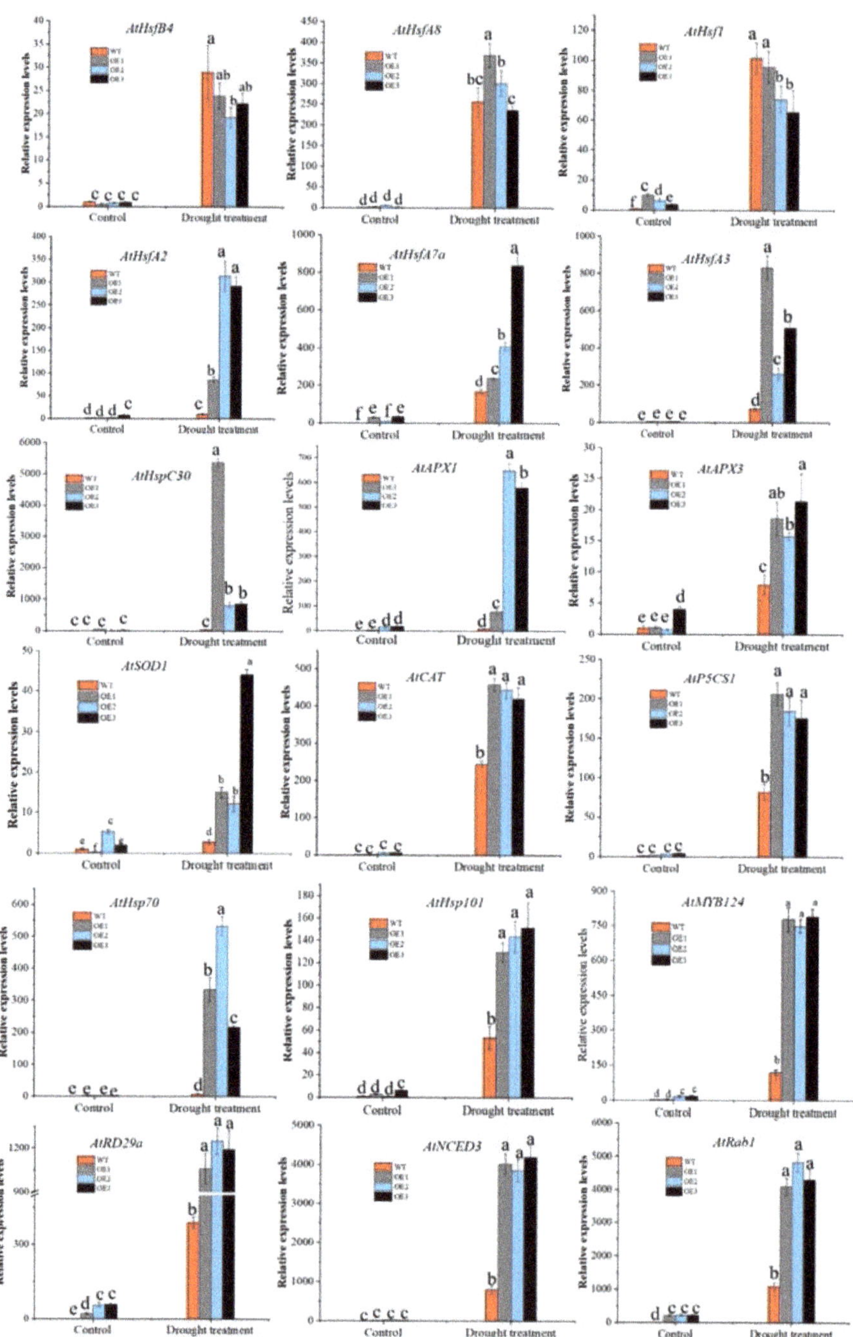

Figure 7. Expression pattern of stress–response genes in wild-type (WT), OE1, OE2, and OE3 lines before and after 10 days of drought treatment. Data are means with standard deviations of three biological replicates. Different letters denote statistical significance ($p \leq 0.05$).

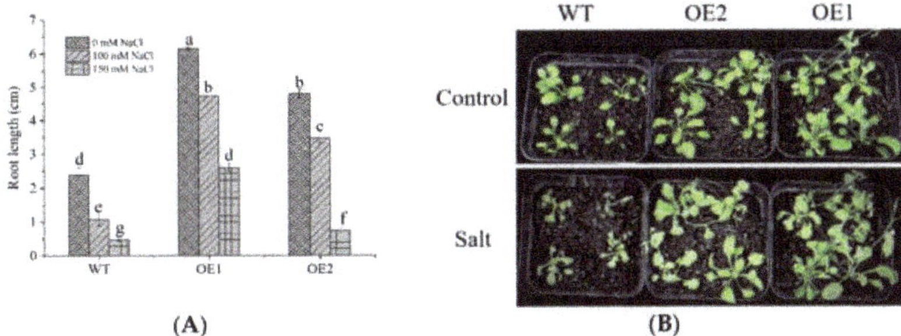

Figure 8. Overexpression of *CaHSP18.1a* enhanced tolerance to salt stress. (**A**) Root length of wild-type (WT) and *CaHSP18.1a*-OE *Arabidopsis* lines grown for 10 d on Murashige and Skoog (MS) medium containing 0, 100, and 150 mM NaCl. (**B**) Seedling growth for WT and transgenic plants exposed to NaCl. Three-week-old seedlings were watered with 200 mM NaCl once every 1–2 days. The images were taken after 7 days.

To explore the effect of *CaHSP18.1a* on the salt tolerance of *Arabidopsis thaliana*, 3-week-old WT and transgenic plants were irrigated with 200 mM NaCl solution for 7 days. As shown in Figure 8B, WT plants showed dehydration and wilting with weak growth. However, the transgenic *Arabidopsis thaliana* showed no significant difference from the control plants except for a slight yellowing phenotype in their leaves.

4. Discussion

Plants are inevitably subjected to various extreme environmental conditions, such as heat, drought, oxidation and salt damage [36]. Under such adverse conditions, sHsP20s make a valuable protective contribution [19]. Plant HSPs are linked to heat tolerance and have been confirmed in many species [31,49]. Guo et al. [30] identified sHsP20s in pepper, and showed that *CaHSP18.1a* was induced in different tissues of pepper plants under heat stress, but the function of *CaHSP18.1a* under heat, salt, and drought stress has not been further studied. In this study, we identified that *CaHSP18.1a* is positively involved in plant tolerance to heat and salt, drought stress.

CaHSP18.1a was responsive to heat stress in both R9 and B6 plants and strongly induced [30]. In accordance with the results, we also found that the expression level of *CaHSP18.1a* was strongly induced in both the R9 and B6 lines after heat stress treatment (Figure 1B–D). However, under heat stress, the expression of *CaHSP18.1a* in R9 plants was higher than that in B6 plants (Figure 1B–D). This may be because R9 is a thermo-tolerant cultivar, it has better thermo-tolerance and adaptability than B6 under heat stress. The heat tolerance of plants is related to the dynamic expression patterns of heat stress-related genes [50]. Under heat stress, other HSP20s or HSPs in R9 are also strongly and rapidly induced in the early stage of heat stress (0.5–1 h). It had also been reported that the expression level of *CaHSP25.8* and *CaHSP30.1* in R9 was higher than B6, but with the extension of heat stress treatment time, the expression level of these two genes in B6 were higher than R9 [30]. However, these results also showed that the expression of *CaHSP18.1a* was lower at V in R9 than in B6. The expression of pepper HSP20s is regulated by many transcription factors, such as HSFs [51]. Under heat stress, it is because HSFs that regulate the expression of *CaHSP18.1a* in pepper variety R9 and B6 are different, or the expression of the same HSFs that regulate the expression of *CaHSP18.1a* is different in R9 and B6, causing the differential expression of *CaHSP18.1a* in B6 and R9 [51,52]. Therefore, the difference in the expression of *CaHSP18.1a* between heat-resistant and heat-sensitive varieties is due to the above reasons. However, the relationship between the function of this gene and the heat-resistance mechanism of pepper still needs further research.

In addition, *CaHSP18.1a* was induced under salt and drought stress (Figure 1E–H). The expression of *CaHSP18.1a* in R9 leaves and roots was highest under the 150 mM, 100 mM NaCl treatments, respectively (Figure 1E,F). The expression of *CaHSP18.1a* in R9 leaves was the highest after the 150 mM mannitol treatment; the highest expression of *CaHSP18.1a* was observed in R9 roots subjected to the 50 mM mannitol treatment (Figure 1G,H). However, the expression of the *CaHSP18.1a* decreased at higher NaCl and mannitol concentrations. The response pattern of *CaHSP18.1a* that rapidly and sharply responded to salt and drought stress in a short time, and then had slight variations, was similar to quite a few HSP20s such as *TaHSP23.9* [53] and *ClHSP22.8* [54]. Thus, *CaHSP18.1a* may play a role in pepper which rapidly adapts to drought and salt stress.

VIGS technology is an important method used to study gene function under adverse environments [55]. In the R9 line, silencing of *CaHSP16.4* reduces heat tolerance and drought resistance of pepper plants [31]; *CaHSP22.0*-silenced peppers showed more sensitivity to salt and heat stress, which was mainly reflected in decreased antioxidant enzyme activity, increased leaf conductivity, and increased superoxide anion and MDA contents [34]. MDA content and REL are products of cell membrane lipid peroxidation, which damages the integrity of plasma membranes under salt or heat stress [35] and may sensitize plants to subsequent stress [56]. MDA content, total chlorophyll content, and REL are widely used to determine the degree to which plants have been damaged by abiotic stress [4]. It has also been reported that proline content, MDA content, and POD and SOD activity of pepper were significantly related to the variation in heat tolerance and temperature stress time, which can be used as an index for heat resistance identification [57].

In this study, after treatments with high temperature, salt, and drought stress, the content of MDA and REL in peppers that had been silenced for *CaHSP18.1a* was higher than that of the controls, indicating that the damage to cell membranes increased in *CaHSP18.1a*-silenced plants (Figure 3A–C). It was also found that the *CaHSP18.1a*-silenced plants had lower total chlorophyll content when exposed to heat stress (Figure 3B and Supplementary Figure S1C). These results demonstrated that silencing of *CaHSP18.1a* reduced pepper stress tolerance (Figure 3). In contrast, overexpression of *CaHSP18.1a* in *Arabidopsis thaliana* transgenic lines was associated with minimal injury symptoms, increased REL, and decreased MDA content compared with WT plants (Figures 4 and 6). These results showed that *CaHSP18.1a* increases plant tolerance to heat, salt, and drought stresses.

HSP20s are widely distributed in plants, and their location may be related to their function, as exemplified by *AtHsP21* being localized to chloroplasts [23]. *CaHSP18.1a* was predicted to have cytoplasm localization [30]. Subcellular localization of *CaHSP18.1a* is shown in Figure 2, which confirmed that it is localized to the cell membrane.

Studies have shown that HSP20 is a molecular chaperone that can also participate in antioxidant mechanisms of plants [6,58]. HSP20scan cooperate with the plant's antioxidant scavenging system to protect plants from secondary damage [59,60]. For example, overexpression of *AtHSP17.6* can increase CAT enzyme activity and further regulate abiotic stress responses [61]. Furthermore, the over-expression of *ZmHSP16.9* in tobacco can increase the activities of POD, CAT and SOD, and enhance oxidative stress tolerance [62]. In this study, SOD, CAT, POD, and APX enzyme activities in overexpression of *CaHSP18.1a* plants were significantly enhanced under heat and drought stress. This is similar to the results obtained with *Arabidopsis* transformed with *CaHSP25.9*; that is, by increasing the activities of ROS-scavenging related antioxidant enzymes, the heat, salt, and drought tolerance of plants can be increased [31,63]. It has been reported that plants have built defense mechanisms that scavenge excess reactive oxygen species (ROS) throughout their long evolutionary histories [64–66], such as ROS-scavenging non-enzymatic antioxidants (e.g., ascorbic acid (AsA), glutathione, and proline) [67] and antioxidant enzymes (e.g., peroxidase (POD), catalase (CAT), superoxide dismutase (SOD), and glutathione peroxidase (GPX)) that prevent secondary oxidative stress caused by abiotic stress [68–70]. Moreover, the expression levels of *AtSOD1*, *AtAPX1*, *AtAPX3*, and *AtCAT1* were also increased by heat and drought stress, and were higher in the *CaHSP18.1a* transgenic *Arabidopsis* than in the WT. This indicated that *CaHSP18.1a* may improve stress resistance through the ROS-scavenging system, but the specific mechanism needs further study.

Many stress-related genes are involved in plant responses to heat, salt, and drought stresses. It has been reported that *AtHsfA2* is a heat shock transcription factor that enables prolonged acquired thermo-tolerance, and it can enhance tolerance to salt and osmotic stresses [71–73]. s HSPs are downstream target genes of HsfA2 [74]. Burke [75,76] have also shown that *AtHSA32* and *AtHSP101* expression can be induced by high temperature and participate in the acquired thermo-tolerance of plants. *AtMYB44* can be induced by salt, drought, and other stresses to participate in the abscisic acid (ABA) signaling pathway; Refs. [77,78] found that the ABA signaling response gene *AtDREB2A* can be induced by low temperature stress. The drought responsive gene *AtRD29A* was up-regulated under heat, salt, and drought stresses [79], while the molecular chaperone HSP70 participates in drought and heat stress responses [28]. *NCED3* is related to biological metabolism and also participates in defense responses to drought stress [80]. Notably, HSP20s can regulate many of these stress-related genes [31,36]. For example, *CaHSP16.4* and *OsMSR-4* can increase the expression of these genes in transgenic seedlings, thereby enhancing stress resistance [31,81]. In this study, we assessed the expression levels of 13 stress-related genes in WT and transgenic plants. *CaHSP18.1a* enhanced heat tolerance in transgenic *Arabidopsis*, which may be closely related to its regulation of the expression of many heat-stress-related genes in *Arabidopsis*. Overexpressed genotypes compared with wild type under normal conditions also showed higher values of stress-related genes expression, higher expression of *AtP5CS*, higher expression of *AtNCED*, higher expression of *AtMYB*, and higher expression of *AtRD29, AtHsfA2, AtRab1* and *AtHSP30*. These results showed that *CaHSP18.1a* may play an important role in regulation of these genes. Under heat stress, the expressions levels of *AtHsfB4, AtHSFA8, AtHSFA2, AtHSFA7a, AtHSPC30, AtHSFA3, AtHSP70,* and *AtHSP101* in transgenic plants were significantly higher than WT plants (Figure 5). In particular, the *AtHSPC30* and *AtHSP70* transcripts were present at levels nearly 3-fold higher in transgenic seedings than in WT plants; the transcript level of *AtHsfA2* was also up-regulated in transgenic plants. The expression levels of *AtHSP70, AtHSP101, AtDREB2A, AtMYB124, AtNCED3, AtRD29A,* and *AtRAB1* were higher (Figure 7) in the *CaHSP18.1a* transgenic *Arabidopsis* than in WT plants under drought stress. Similar results were also reported by Feng and Huang [31,32]. Thus, *CaHSP18.1a* may respond to heat and drought stress through its complex regulatory network.

5. Conclusions

In this study, we first analyzed the expression of *CaHSP18.1a* in R9 and B6 pepper lines and demonstrated that *CaHSP18.1a* was expressed when induced by abiotic stress factors such as high temperature, drought, and high salinity. *CaHSP18.1a* silencing decreased the resistance of pepper plants to heat, drought, and salt stresses through different molecular and physiological mechanisms. Overexpression analyses of *CaHSP18.1a* in transgenic *Arabidopsis* further confirmed that *CaHSP18.1a* functions positively in responses to heat, drought, and salt stresses. The expression levels of other stress-related genes were also measured, and some were determined to be significantly affected by *CaHSP18.1a* overexpression. We further confirmed that *CaHSP18.1a* protein was localized in the cell membrane. Collectively, these results show *CaHSP18.1a* likely acts as a positive regulator of the response to abiotic stresses in pepper.

Supplementary Materials: The following are available online at https://www.mdpi.com/article/10.3390/horticulturae7050117/s1, Figure S1: Detection of silencing efficiency of CaHsp18.1a gene mediated by TRV2; Figure S2: Validation and acquisition of homozygous strain of T3 generation of Arabidopsis with overexpression of CaHsp18.1a; Figure S3: Germination of the transgenic Arabidopsis under salt stress; Table S1: The main primers sequence used in this research.

Author Contributions: S.L., Y.-L.L. and Z.-H.G. conceived and designed the research; S.L. and Y.-L.L. conducted the experiments and wrote the manuscript; G.-X.C. and Y.-L.L. analyzed the data; H.S.u., S.L. and J.-J.X. critically revised the manuscript; Z.-H.G. contributed reagents and funded the project. All authors have read and agreed to the published version of the manuscript.

Funding: We highly appreciate the financial support of the funding from the National Natural Science Foundation of China (No. U1603102, No. 31772309).

Institutional Review Board Statement: Not applicable.

Informed Consent Statement: Not applicable.

Data Availability Statement: Data is contained within the article.

Conflicts of Interest: The authors declare no conflict of interest.

Abbreviations

ACD	alpha-crystallin domain
HS	heat stress
HSP	heat shock proteins
sHSPs	small heat shock proteins
REL	relative electrolyte leakage
MDA	Malondialdehyde
OE	Overexpression
OE1	No. 1 Arabidopsis line with overexpressed CaHSP18.1a
OE2	No. 2 Arabidopsis line with overexpressed CaHSP18.1a
OE3	No. 3 Arabidopsis line with overexpressed CaHSP18.1a
R9	a thermo-tolerant line
qRT-PCR	real-time fluorescence quantitative PCR
VIGS	virus-induced gene silencing
PDS	phytoene desaturase
TRV	tobacco rattle virus
ROS	reactive oxygen species
APX	ascorbate peroxidase
CAT	catalase
SOD	superoxide dismutase
POD	peroxidase

References

1. Ahuja, I.; de Vos, R.C.H.; Bones, A.M.; Hall, R.D. Plant molecular stress responses face climate change. *Trends Plant Sci.* **2010**, *15*, 664–674. [CrossRef] [PubMed]
2. Wang, W.; Vinocur, B.; Altman, A. Plant responses to drought, salinity and extreme temperatures: Towards genetic engineering for stress tolerance. *Planta* **2003**, *218*, 1–14. [CrossRef] [PubMed]
3. McLoughlin, F.; Basha, E.; Fowler, M.E.; Kim, M.; Bordowitz, J.; Katiyar-Agarwal, S.; Vierling, E. Class I and II Small Heat Shock Proteins Together with HSP101 Protect Protein Translation Factors during Heat Stress. *Plant Physiol.* **2016**, *172*, 1221–1236. [PubMed]
4. Muthusamy, S.K.; Dalal, M.; Chinnusamy, V.; Bansal, K.C. Genome-wide identification and analysis of biotic and abiotic stress regulation of small heat shock protein (HSP20) family genes in bread wheat. *J. Plant Physiol.* **2017**, *211*, 100–113. [CrossRef] [PubMed]
5. Xiang, J.; Chen, X.; Hui, W.; Xiang, Y.; Yan, M.; Wang, J. Overexpressing heat-shock protein OsHSP50.2 improves drought tolerance in rice. *Plant Cell Rep.* **2018**, *37*, 1585–1595. [CrossRef]
6. Wang, W.; Vinocur, B.; Shoseyov, O.; Altman, A. Role of plant heat-shock proteins and molecular chaperones in the abiotic stress response. *Trends Plant Sci.* **2004**, *9*, 244–252. [CrossRef]
7. Zhang, L.; Liu, J. Research Progress of Heat Shock Protein. *Guangdong Seric.* **2006**, *40*, 39–42.
8. Beck, E.H.; Fettig, S.; Knake, C.; Hartig, K.; Bhattarai, T. Specific and unspecific responses of plants to cold and drought stress. *J. Biosci.* **2007**, *32*, 501–510. [CrossRef]
9. Agrawal, L.; Narula, K.; Basu, S.; Shekhar, S.; Ghosh, S.; Datta, A.; Chakraborty, S. Comparative Proteomics Reveals a Role for Seed Storage Protein AmA1 in Cellular Growth, Development, and Nutrient Accumulation. *J. Proteome Res.* **2013**, *12*, 4904–4930. [CrossRef]
10. Shekhar, S.; Mishra, D.; Gayali, S.; Buragohain, A.K.; Chakraborty, S.; Chakraborty, N. Comparison of proteomic and metabolomic profiles of two contrasting ecotypes of sweetpotato (*Ipomoea batata* L.). *J. Proteom.* **2016**, *143*, 306–317. [CrossRef]
11. Waters, E.R.; Lee, G.J.; Vierling, E. Evolution, structure and function of the small heat shock proteins in plants. *J. Exp. Bot.* **1996**, *47*, 325–338. [CrossRef]

12. Boston, R.S.; Viitanen, P.V.; Vierling, E. Molecular chaperones and protein folding in plants. *Plant Mol. Biol.* **1996**, *32*, 191–222. [CrossRef]
13. Hu, W.; Hu, G.; Han, B. Genome-wide survey and expression profiling of heat shock proteins and heat shock factors revealed overlapped and stress specific response under abiotic stresses in rice. *Plant Sci.* **2009**, *176*, 583–590. [CrossRef]
14. Li, J.; Zhang, J.; Jia, H.; Li, Y.; Xu, X.; Wang, L.; Lu, M. The Populus trichocarpa PtHSP17.8 involved in heat and salt stress tolerances. *Plant Cell Rep.* **2016**, *35*, 1587–1599. [CrossRef]
15. Eyles, S.J.; Gierasch, L.M. Nature's molecular sponges: Small heat shock proteins grow into their chaperone roles. *Proc. Natl. Acad. Sci. USA* **2010**, *107*, 2727–2728. [CrossRef]
16. Martin, H. A first line of stress defense: Small heat shock proteins and their function in protein homeostasis. *J. Mol. Biol.* **2015**, *427*, 1537–1548.
17. Tyedmers, J.; Mogk, A.; Bukau, B. Cellular strategies for controlling protein aggregation. *Nat. Rev. Mol. Cell Biol.* **2010**, *11*, 777–788. [CrossRef]
18. Waters, E.R. The evolution, function, structure, and expression of the plant sHSPs. *J. Exp. Bot.* **2013**, *64*, 391–403. [CrossRef]
19. Lambert, W.; Koeck, P.J.B.; Ahrman, E.; Purhonen, P.; Cheng, K.; Elmlund, D.; Hebert, H.; Emanuelsson, C. Subunit arrangement in the dodecameric chloroplast small heat shock protein HSP21. *Protein Sci.* **2011**, *20*, 291–301. [CrossRef]
20. Korotaeva, N.E.; Antipina, A.I.; Grabelnykh, O.I.; Varakina, N.N.; Borovskii, G.B.; Voinikov, V.K. Mitochondrial Low-Molecular-Weight Heat-Shock Proteins and the Tolerance of Cereal Mitochondria to Hyperthermia. *Russ. J. Plant Physiol.* **2001**, *48*, 798–803. [CrossRef]
21. Nieto-Sotelo, J.; Martínez, L.M.; Ponce, G.; Cassab, G.I.; Alagón, A.; Meeley, R.B.; Ribaut, J.-M.; Yang, R. Maize HSP101 Plays Important Roles in Both Induced and Basal Thermotolerance and Primary Root Growth. *Plant Cell Online* **2002**, *14*, 1621–1633. [CrossRef]
22. Sun, W.; Bernard, C.; Cotte, B.V.D.; Montagu, M.V.; Verbruggen, N. At-HSP17.6A, encoding a small heat-shock protein in Arabidopsis, can enhance osmotolerance upon overexpression. *Plant J.* **2010**, *27*, 407–415. [CrossRef]
23. Sedaghatmehr, M.; MuellerRoeber, B.; Balazadeh, S. The plastid metalloprotease FtsH6 and small heat shock protein HSP21 jointly regulate thermomemory in Arabidopsis. *Nat. Commun.* **2016**, *7*, 12439. [CrossRef]
24. Kim, D.H.; Xu, Z.; Hwang, I. AtHSP17.8 overexpression in transgenic lettuce gives rise to dehydration and salt stress resistance phenotypes through modulation of ABA-mediated signaling. *Plant Cell Rep.* **2013**, *32*, 1953–1963. [CrossRef]
25. Nicky, D.; Xu, J.; Peters, J.L.; Sunghun, P.; Ivo, R. Multi-Level Interactions Between Heat Shock Factors, Heat Shock Proteins, and the Redox System Regulate Acclimation to Heat. *Front. Plant Sci.* **2015**, *6*, 999.
26. Guo, M.; Liu, J.H.; Ma, X.; Zhai, Y.F.; Gong, Z.H.; Lu, M.H. Genome-wide analysis of the HSP70 family genes in pepper (*Capsicum annuum* L.) and functional identification of CaHSP70-2 involvement in heat stress. *Plant Sci.* **2016**, *252*, 246–256. [CrossRef]
27. Zou, X. Cultivation Seasons and Cropping patterns of Capsicum in China. *J. China Capsicum* **2002**, *3*, 1672–4542.
28. Guo, M.; Zhai, Y.-F.; Lu, J.-P.; Chai, L.; Chai, W.-G.; Gong, Z.-H.; Lu, M.-H. Characterization of CaHSP70-1, a Pepper Heat-Shock Protein Gene in Response to Heat Stress and Some Regulation Exogenous Substances in *Capsicum annuum* L. *Int. J. Mol. Sci.* **2014**, *15*, 19741–19759. [CrossRef]
29. Pagamas, P.; Nawata, E. Sensitive stages of fruit and seed development of chili pepper (*Capsicum annuum* L. var. Shishito) exposed to high-temperature stress. *Sci. Hortic.* **2008**, *117*, 21–25. [CrossRef]
30. Guo, M.; Liu, J.H.; Lu, J.P.; Zhai, Y.F.; Wang, H.; Gong, Z.H.; Wang, S.B.; Lu, M.H. Genome-wide analysis of the CaHSP20 gene family in pepper: Comprehensive sequence and expression profile analysis under heat stress. *Front. Plant Sci.* **2015**, *6*, 806. [CrossRef]
31. Huang, L.; Cheng, G.; Khan, A.; Wei, A.; Gong, Z. CaHSP16.4, a small heat shock protein gene in pepper, is involved in heat and drought tolerance. *Protoplasma* **2019**, *256*, 39–51. [CrossRef] [PubMed]
32. Feng, X.; Zhang, H.; Ali, M.; Gai, W.; Gong, Z. A small heat shock protein CaHSP25.9 positively regulates heat, salt, and drought stress tolerance in pepper (*Capsicum annuum* L.). *Plant Physiol. Biochem.* **2019**, *142*, 151–162. [CrossRef] [PubMed]
33. He, L.; Gao, N.; Shan, Z.; Guo, S. Pepper Chloroplast Small Heat Shock Proteins CaHSP26 Enhanced the Ability of Anti-high Temperature Stress in Arabidopsis. *North. Hortic.* **2011**, *6*, 140–143.
34. Sun, J.; Cheng, G.; Huang, L.; Liu, S.; Ali, M.; Khan, A.; Yu, Q.; Yang, S.; Luo, D.; Gong, Z. Modified expression of a heat shock protein gene, CaHSP22.0, results in high sensitivity to heat and salt stress in pepper (*Capsicum annuum* L.). *Sci. Hortic.* **2019**, *249*, 364–373. [CrossRef]
35. Yu, J.; Cheng, Y.; Feng, K.; Ruan, M.; Ye, Q.; Wang, R.; Li, Z.; Zhou, G.; Yao, Z.; Yang, Y.; et al. Genome-Wide Identification and Expression Profiling of Tomato HSP20 Gene Family in Response to Biotic and Abiotic Stresses. *Front. Plant Sci.* **2016**, *7*, 1215. [CrossRef]
36. Wang, M.; Zou, Z.; Li, Q.; Sun, K.; Chen, X.; Li, X. The CsHSP17.2 molecular chaperone is essential for thermotolerance in Camellia sinensis. *Sci. Rep.* **2017**, *7*, 1237. [CrossRef]
37. Zhai, Y.; Wang, H.; Liang, M.; Lu, M. Both silencing- and over-expression of pepper CaATG8c gene compromise plant tolerance to heat and salt stress. *Environ. Exp. Bot.* **2017**, *141*, 10–18. [CrossRef]
38. Wang, J.E.; Liu, K.K.; Li, D.W.; Zhang, Y.L.; Zhao, Q.; He, Y.M.; Gong, Z.H. A novel peroxidase CaPOD gene of pepper is involved in defense responses to Phytophthora capsicum infection as well as abiotic stress toleranc. *Int. J. Mol. Sci.* **2013**, *14*, 3158–3177. [CrossRef]

39. Wan, H.; Yuan, W.; Ruan, M.; Ye, Q.; Wang, R.; Li, Z.; Zhou, G.; Yao, Z.; Zhao, J.; Liu, S.; et al. Identification of reference genes for reverse transcription quantitative real-time PCR normalization in pepper (*Capsicum annuum* L.). *Biochem. Biophys. Res. Commun.* **2011**, *416*, 24–30. [CrossRef]

40. Livaka, K.J.; Schmittgen, T.D. Analysis of relative gene expression data using real-time quantitative PCR and the $2^{-\Delta\Delta C_T}$ Method. *J. Methods* **2001**, *25*, 402–408. [CrossRef]

41. Yu, C.; Zhan, Y.; Feng, X.; Huang, Z.A.; Sun, C. Identification and Expression Profiling of the Auxin Response Factors in *Capsicum annuum* L. under Abiotic Stress and Hormone Treatments. *Int. J. Mol. Sci.* **2017**, *18*, 2719. [CrossRef]

42. Clough, S.J.; Bent, A. Floral dip: A simplified method for Agrobacterium -mediated transformation of Arabidopsis thaliana. *Plant J.* **1998**, *16*, 735–743. [CrossRef] [PubMed]

43. Yin, Y.X.; Guo, W.L.; Zhang, Y.L.; Ji, J.J.; Xiao, H.J.; Yan, F.; Zhao, Y.Y.; Zhu, W.C.; Chen, R.G.; Chai, W.G. Cloning and characterisation of a pepper aquaporin, CaAQP, which reduces chilling stress in transgenic tobacco plants. *Plant Cell Tissue Organ Cult.* **2014**, *118*, 431–444. [CrossRef]

44. Arkus, K.A.J.; Cahoon, E.B.; Jez, J.M. Mechanistic analysis of wheat chlorophyllase. *Arch. Biochem. Biophys.* **2005**, *438*, 146–155. [CrossRef]

45. Campos, P.S.; nia Quartin, V.; chicho Ramalho, J.; Nunes, M.A. Electrolyte leakage and lipid degradation account for cold sensitivity in leaves ofCoffea sp. plants. *J. Plant Physiol.* **2003**, *160*, 283–292. [CrossRef] [PubMed]

46. Guo, W.L.; Chen, R.G.; Gong, Z.H.; Yin, Y.X.; Ahmed, S.S.; He, Y.M. Exogenous abscisic acid increases antioxidant enzymes and related gene expression in pepper (*Capsicum annuum*) leaves subjected to chilling stress. *Genet. Mol. Res. GMR* **2012**, *11*, 4063–4080. [CrossRef] [PubMed]

47. Nakano, Y.; Asada, K. Hydrogen Peroxide is Scavenged by Ascorbate-specific Peroxidase in Spinach Chloroplasts. *Plant Cell Physiol.* **1981**, *22*, 867–880.

48. Aebi, H. [13] Catalase in vitro. *Methods Enzymol.* **1984**, *105*, 121–126.

49. Jiang, C.; Xu, J.; Zhang, H.A.; Zhang, X.; Shi, J.; Li, M.I.; Ming, F. A cytosolic class I small heat shock protein, RcHSP17.8, of Rosa chinensis confers resistance to a variety of stresses to Escherichia coli, yeast and Arabidopsis thaliana. *Plant Cell Environ.* **2009**, *32*, 1046–1059. [CrossRef]

50. Lopes-Caitar, V.S.; de Carvalho, M.C.; Darben, L.M.; Kuwahara, M.K.; Nepomuceno, A.L.; Dias, W.P.; Abdelnoor, R.V.; Marcelino-Guimarães, F.C. Genome-wide analysis of the HSP20 gene family in soybean: Comprehensive sequence, genomic organization and expression profile analysis under abiotic and biotic stresses. *BMC Genom.* **2013**, *14*, 1471–2164. [CrossRef]

51. Wang, X.; Huang, W.; Yang, Z.; Liu, J.; Huang, B. Transcriptional regulation of heat shock proteins and ascorbate peroxidase by CtHsfA2b from African bermudagrass conferring heat tolerance in Arabidopsis. *Sci. Rep.* **2016**, *6*, 28021. [CrossRef]

52. Schramm, F.; Ganguli, A.; Kiehlmann, E.; Englich, G.; Walch, D.; von Koskull-Döring, P. The Heat Stress Transcription Factor HsfA2 Serves as a Regulatory Amplifier of a Subset of Genes in the Heat Stress Response in Arabidopsis. *Plant Mol. Biol.* **2006**, *60*, 759–772. [CrossRef] [PubMed]

53. Wang, J.; Gao, X.; Dong, J.; Tian, X.; Wang, J.; Palta, J.A.; Xu, S.; Fang, Y.; Wang, Z. Over-Expression of the Heat-Responsive Wheat Gene TaHSP23.9 in Transgenic Arabidopsis Conferred Tolerance to Heat and Salt Stress. *Front Plant Sci.* **2020**, *11*, 243. [CrossRef]

54. He, Y.; Yao, Y.; Li, L.; Li, Y.; Gao, J.; Fan, M. A heat-shock 20 protein isolated from watermelon (ClHSP22.8) negatively regulates the response of Arabidopsis to salt stress via multiple signaling pathways. *PeerJ.* **2021**, *9*, e10524. [CrossRef] [PubMed]

55. Yao, B.; Yue, S.; Sun, L.; Ma, Z.; Su, J.; Liu, X. Development of Virus Induced Gene Silencing and Application of VIGS in Plant Abiotic Stresses. *Chin. Agric. Sci. Bull.* **2016**, *32*, 131–136.

56. Ammar, J.B.; Lanoiselle, J.-L.; Lebovka, N.I.; Hecke, E.V.; Vorobiev, E.J.J. Impact of a Pulsed Electric Field on Damage of Plant Tissues: Effects of Cell Size and Tissue Electrical Conductivity. *J. Food Sci.* **2011**, *76*, E90–E97. [CrossRef]

57. Ma, B.; Lu, M.; Gong, Z.H. Responses of growth and physiology of pepper (*Capsicum annuum* L.) seedlings to high temperature stress. *J. Northwest A F Univ.* **2013**, *41*, 112–118.

58. Shang, H.; Cao, S.; Yang, Z.; Cai, Y.; Zheng, Y. Effect of Exogenous γ-Aminobutyric Acid Treatment on Proline Accumulation and Chilling Injury in Peach Fruit after Long-Term Cold Storage. *J. Agric. Food Chem.* **2011**, *59*, 1264–1268. [CrossRef]

59. Rossel, J.B.; Wilson, I.W.; Pogson, B.J. Global Changes in Gene Expression in Response to High Light in Arabidopsis. *Plant Physiol.* **2002**, *130*, 1109–1120. [CrossRef] [PubMed]

60. Aghdam, M.S.; Sevillano, L.; Flores, F.B.; Bodbodak, S. Heat shock proteins as biochemical markers for postharvest chilling stress in fruits and vegetables. *Sci. Hortic.* **2013**, *160*, 54–64. [CrossRef]

61. Li, G.; Li, J.; Hao, R.; Guo, Y. Activation of catalase activity by a peroxisome-localized small heat shock protein HSP17.6CII. *J. Genet. Genom.* **2017**, *44*, 395–404. [CrossRef] [PubMed]

62. Sun, L.; Liu, Y.; Kong, X.; Zhang, D.; Pan, J.; Zhou, Y.; Wang, L.; Li, D.; Yang, X. ZmHSP16.9, a cytosolic class I small heat shock protein in maize (Zea mays), confers heat tolerance in transgenic tobacco. *Plant Cell Rep.* **2012**, *31*, 1473–1484. [CrossRef] [PubMed]

63. Meifang, L.; Xiujian, W.; Shangjing, G. Obtain of Transgenic Sweet Pepper Plants with Gene from Endoplasmic Reticulum Small Heat Shock Protein CaHSP22.5. *Mol. Plant Breed.* **2018**, *7*, 2212–2218.

64. Zhao, J.; Liang, J.; Wu, X.; Liu, X.; Li, H.; Zhu, S. Expression Profiling of Rice ROS Scavenging System Related Genes under Salt or Low Temperature Stress. *Acta Bot. Boreali-Occident. Sin.* **2015**, *35*, 872–883.

65. Kim, J.M.; Woo, D.H.; Kim, S.H.; Lee, S.Y.; Park, H.Y.; Seok, H.Y.; Chung, W.S.; Moon, Y.H. Arabidopsis MKKK20 is involved in osmotic stress response via regulation of MPK6 activity. *Plant Cell Rep.* **2012**, *31*, 217–224. [CrossRef]

66. Tuteja, N.; Banu, M.S.A.; Huda, K.M.K.; Gill, S.S.; Jain, P.; Xuan, H.P.; Tuteja, R. Pea p68, a DEAD-Box Helicase, Provides Salinity Stress Tolerance in Transgenic Tobacco by Reducing Oxidative Stress and Improving Photosynthesis Machinery. *PLoS ONE* **2014**, *9*, e98287. [CrossRef]

67. Ron, M.; Sandy, V.; Martin, G.; Frank, V.B. Reactive oxygen gene network of plants. *Trends Plant Sci.* **2004**, *9*, 490–498.

68. Cao, S.; Yang, Z.; Cai, Y.; Zheng, Y. Fatty acid composition and antioxidant system in relation to susceptibility of loquat fruit to chilling injury. *Food Chem.* **2011**, *127*, 1777–1783. [CrossRef]

69. Foyer, C.H.; Noctor, G. Redox homeostasis and antioxidant signaling: A metabolic interface between stress perception and physiological responses. *Plant Cell Online* **2005**, *17*, 1866–1875. [CrossRef]

70. Moller, I.M. Plant Mitochondria and Oxidative Stress: Electron Transport, Nadph Turnover, and Metabolism of Reactive Oxygen Species. *Annu. Rev. Plant Physiol. Plant Mol. Biol.* **2001**, *52*, 561–591. [CrossRef]

71. Charng, Y.; Liu, H.; Liu, N.; Chi, W.; Wang, C.; Chang, S.; Wang, T. A Heat-Inducible Transcription Factor, HsfA2, Is Required for Extension of Acquired Thermotolerance in Arabidopsis. *Plant Physiol.* **2007**, *143*, 251–262. [CrossRef] [PubMed]

72. Heerklotz, D.; Döring, P.; Bonzelius, F.; Winkelhaus, S.; Nover, L. The balance of nuclear import and export determines the intracellular distribution and function of tomato heat stress transcription factor HsfA2. *Mol. Cell. Biol.* **2001**, *21*, 1759–1768. [CrossRef] [PubMed]

73. Ogawa, D.; Kazuo, Y.; Nishiuchi, T. High-level overexpression of the Arabidopsis HsfA2 gene confers not only increased themotolerance but also salt/osmotic stress tolerance and enhanced callus growth. *J. Exp. Bot.* **2007**, *58*, 3373–3383. [CrossRef] [PubMed]

74. Pv, K.D.; Scharf, K.D.; Nover, L. The diversity of plant heat stress transcription factors. *Trends Plant Sci.* **2007**, *12*, 452–457.

75. Burke, J.J.; Chen, J. Enhancement of reproductive heat tolerance in plants. *PLoS ONE* **2015**, *10*, e0122933. [CrossRef]

76. Charng, Y.Y.; Liu, H.C.; Liu, N.Y.; Hsu, F.C.; Ko, S.S. Arabidopsis Hsa32, a novel heat shock protein, is essential for acquired thermotolerance during long recovery after acclimation. *Plant Physiol.* **2006**, *4*, 1297–1305. [CrossRef]

77. Nguyen, N.H.; Nguyen, C.T.T.; Jung, C.; Cheong, J.-J. AtMYB44 suppresses transcription of the late embryogenesis abundant protein gene AtLEA4-5. *Biochem. Biophys. Res. Commun.* **2019**, *511*, 931–934. [CrossRef]

78. Nakashima, K.; Shinwari, Z.K.; Sakuma, Y.; Seki, M.; Miura, S.; Shinozaki, K.; Yamaguchi-Shinozaki, K. Organization and expression of two Arabidopsis DREB2 genes encoding DRE-binding proteins involved in dehydration- and high-salinity-responsive gene expression. *Plant. Mol. Biol.* **2000**, *42*, 657–665. [CrossRef]

79. Verbruggen, N.; Hermans, C. Proline accumulation in plants: A review. *Amino Acids* **2008**, *35*, 753–759. [CrossRef]

80. Rymaszewski, W.; Vile, D.; Bediee, A.; Dauzat, M.; Luchaire, N.; Kamrowska, D.; Granier, C.; Hennig, J. Stress-Related Gene Expression Reflects Morphophysiological Responses to Water Deficit. *Plant. Physiol.* **2017**, *174*, 1913–1930. [CrossRef]

81. Yin, X.; Huang, L.; Zhang, X.; Wang, M.; Xu, G.; Xia, X. Expression of rice geneOsMSR4confers decreased ABA sensitivity and improved drought tolerance inArabidopsis thaliana. *Plant. Growth Regul.* **2015**, *75*, 549–556. [CrossRef]

Article

Genome Wide Characterization, Comparative and Genetic Diversity Analysis of Simple Sequence Repeats in Cucurbita Species

Lei Zhu [1,2], Huayu Zhu [1], Yanman Li [1], Yong Wang [1], Xiangbin Wu [2], Jintao Li [2], Zhenli Zhang [2], Yanjiao Wang [2], Jianbin Hu [1,2], Sen Yang [1], Luming Yang [1,2,*] and Shouru Sun [1,2,*]

1   Henan Key Laboratory of Fruit and Cucurbit Biology, Henan Agricultural University, Zhengzhou 450002, China; zhulei2018@henau.edu.cn (L.Z.); hyzhu@henau.edu.cn (H.Z.); liyanman2004@163.com (Y.L.); yongwang@henau.edu.cn (Y.W.); jianbinhu@henau.edu.cn (J.H.); senyang1989@henau.edu.cn (S.Y.)
2   International Joint Laboratory of Horticultural Biology, College of Horticulture, Henan Agricultural University, 95 Wenhua Road, Zhengzhou 450002, China; wxb15993635378@163.com (X.W.); 118739921624@163.com (J.L.); Zhangzl15225961525@163.com (Z.Z.); jiaoblue123@163.com (Y.W.)
*   Correspondence: lumingyang@henau.edu.cn (L.Y.); ssr365@henau.edu.cn (S.S.)

Citation: Zhu, L.; Zhu, H.; Li, Y.; Wang, Y.; Wu, X.; Li, J.; Zhang, Z.; Wang, Y.; Hu, J.; Yang, S.; et al. Genome Wide Characterization, Comparative and Genetic Diversity Analysis of Simple Sequence Repeats in Cucurbita Species. *Horticulturae* **2021**, *7*, 143. https://doi.org/10.3390/horticulturae7060143

Academic Editor: Yuyang Zhang

Received: 19 May 2021
Accepted: 5 June 2021
Published: 8 June 2021

Publisher's Note: MDPI stays neutral with regard to jurisdictional claims in published maps and institutional affiliations.

Copyright: © 2021 by the authors. Licensee MDPI, Basel, Switzerland. This article is an open access article distributed under the terms and conditions of the Creative Commons Attribution (CC BY) license (https://creativecommons.org/licenses/by/4.0/).

Abstract: Simple sequence repeats (SSRs) are widely used in mapping constructions and comparative and genetic diversity analyses. Here, 103,056 SSR loci were found in *Cucurbita* species by in silico PCR. In general, the frequency of these SSRs decreased with the increase in the motif length, and di-nucleotide motifs were the most common type. For the same repeat types, the SSR frequency decreased sharply with the increase in the repeat number. The majority of the SSR loci were suitable for marker development (84.75% in *Cucurbita moschata*, 94.53% in *Cucurbita maxima*, and 95.09% in *Cucurbita pepo*). Using these markers, the cross-species transferable SSR markers between *C. pepo* and other Cucurbitaceae species were developed, and the complicated mosaic relationships among them were analyzed. Especially, the main syntenic relationships between *C. pepo* and *C. moschata* or *C. maxima* indicated that the chromosomes in the *Cucurbita* genomes were highly conserved during evolution. Furthermore, 66 core SSR markers were selected to measure the genetic diversity in 61 *C. pepo* germplasms, and they were divided into two groups by structure and unweighted pair group method with arithmetic analysis. These results will promote the utilization of SSRs in basic and applied research of *Cucurbita* species.

Keywords: pumpkin; simple sequence repeat (SSR); syntenic relationship; cross-species markers; population structure

1. Introduction

The *Cucurbita* genus (2n = 2x = 40), belonging to the Cucurbitaceae family, contains more than 13 species [1]. Most *Cucurbita* species are wild resources, and only three domesticated species, *Cucurbita maxima*, *Cucurbita moschata*, and *Cucurbita pepo*, are widely cultivated and have become important food crops globally [2]. At present, Asia has the largest pumpkin cultivation area, and China is the main producer of pumpkins. In 2012, the planting area of pumpkin was approximately 3.8×10^4 Hm^2 in China, and the total output reached 7.0×10^6 tons (http://www.fao.org/faostat/zh/#data/QC/visualize, 2020). Due to the fact of their long history of cultivation and domestication, *Cucurbita* species show a greater diversity in fruit shape, size, and color than other Cucurbitaceae species [3]. Furthermore, *Cucurbita* species have strong roots and exhibit good adaption to different biotic and abiotic stresses, such as cold, viruses, and salinity, and so they are widely used as rootstocks in grafting [4,5]. Although they are a common global crop, fundamental genetic research on *Cucurbita* is lacking, and few studies have been conducted to improve the cultivation and breeding of this genus.

Simple sequence repeats (SSRs) are widely used in genetic mapping constructions, genetic diversity analyses, and genome-wide association studies due to the fact of their relative abundance, multi-allelism, co-dominance, and low cost [6,7]. In the Cucurbitaceae family, the whole-genome sequencing of *Cucumis sativus*, *Cucumis melon*, and *Citrullus lanatus* has been completed [8–10], and genome-wide SSR markers have been characterized and developed in these crops, which has greatly promoted their application in gene and quantitative trait locus (QTL) mapping as well as in comparative genomics [11–13]. A rough syntenic relationship between melon (2n = 2x = 24) and cucumber (2n = 2x = 14) chromosomes was revealed by comparative mapping using 199 SSR markers developed from cucumber [14]. Later, Yang et al. (2014) developed a higher density map of *Cucumis hystrix* containing 416 SSR markers, and 151 and 50 markers were derived from cucumber and melon, respectively. With these shared markers among the three *Cucumis* species, the chromosome-level syntenic relationships were well established, which was further confirmed by fluorescence in situ hybridization (FISH) [15]. Ninety-one syntenic blocks were divided between cucumber and melon, and 53 syntenic blocks were identified between cucumber and *Cucumis hystrix*. Furthermore, the genome-wide SSR markers developed from melon and watermelon have made it possible to more clearly define chromosomal syntenic relationships, and the complicated mosaic patterns of chromosome synteny between melon, watermelon, and cucumber have been well established based on cross-species SSR markers [12,15]. However, the syntenic relationships and chromosomal rearrangements between *Cucurbita* species and other Cucurbitaceae crops are still largely fragmented and incomplete.

Based on the conserved sequences among species or genera, some amplified fragment length polymorphism (AFLP), random amplified polymorphic DNA (RAPD), and SSR markers were developed in previous studies [16–20]. However, these restricted markers are insufficient for research on genetic diversity, genetic mapping, and comparative genomics. Esteras et al. (2012) constructed the first genetic map in pumpkin using 304 single nucleotide polymorphisms (SNPs) and 11 SSR markers and found that the linkage groups of pumpkin were partially homoeologous to cucumber chromosomes. The applications of these expressed sequence tag (EST)-SNP markers are still greatly limited due to the small numbers of markers, the high cost of enzymes, and the complicated operating procedure [21]. Due to the lack of genome-wide coverage and polymorphic markers, in-depth application and comparative analysis still cannot be conducted. With the development of high-throughput sequencing technology, there has been an increase in studies on *Cucurbita*, and the whole-genome sequences of three important cucurbit crops have become available. Based on these SNPs' data, the whole-genome synteny analysis indicated that both the *C. maxima* and *C. moschata* genomes underwent a whole-genome duplication (WGD) event and that pairs of *C. maxima* (or *C. moschata*) homoeologous regions are shared between chromosomes corresponding to the two sub-genomes [22]. Montero et al. also identified that the covered regions in most of the *C. pepo* genome had experienced a WGD event [23]. Furthermore, some transcriptomes of *Cucurbita* species have become available, and EST-SSRs were developed from them [24–28]. To date, the development of SSR markers in *Cucurbita* species is still limited.

The whole-genome sequences of *C. moschata*, *C. maxima*, and *C. pepo* have been completely assembled, which will greatly promote the large-scale development of SSR markers, allowing for the construction of high-resolution maps, gene mapping, and genome-wide association studies (GWAS). In this study, we conducted a genome-wide identification of SSR motifs in three *Cucurbita* species, analyzed the distribution and frequency of different repeat types, identified cross-species transferable SSR markers by in silico PCR analysis, and studied the chromosome synteny of *C. pepo* with other Cucurbitaceae crops. In addition, 66 core SSR markers were identified in *Cucurbita* genomes and used to evaluate the genetic diversity and population structure of 61 *C. pepo* germplasms. Our study will be useful for research on the population structure, genetic diversity, molecular-assisted selection, and map-based cloning in *Cucurbita* species.

2. Methods

2.1. Plant Materials

All of the pumpkin accessions used in this study were introduced from the National Crop Germplasm Resource Platform of China (platform of vegetable germplasm resources) in 2018. Four of the accessions came from Russia, one from America, and 56 accessions were from 17 provinces in China. The number and sources are shown in Table S1.

2.2. Genome SSR Identification and Development in Cucurbita Genomes

The genome information of watermelon, melon, cucumber, and pumpkin was downloaded from http://cucurbitgenomics.org/ (2020). To develop a set of higher polymorphic SSR primers for the future study, the criteria used for microsatellite identification in this study was from 2 to 8 bp, and mononucleotides were not considered due to the difficulty in distinguishing bona fide microsatellites from sequencing or assembly error. The microsatellite identification tool (MISA) was used to identify and analyze SSR markers including perfect and compound microsatellites. The specific screening details were as follows: repeats with a minimum length of 18 bp (for di- and tetra-nucleotides), 20 bp (for penta-nucleotides), 24 bp (for hexa-nucleotides), 21 bp (for hepta-nucleotides), and 24 bp (for octa-nucleotides). The oligonucleotide primers for these SSRs were designed according to the flanking genomic sequence using Primer3 software (v.1.1.4). Primers were designed to generate amplicons of 100–300 bp in length with the following minimum, optimum, and maximum values for Primer3 parameters: primer length (bp): 18–20–24; Tm (°C): 50–55–60. Other parameters used the default program values.

2.3. In Silico PCR and Synteny Analysis of Cross-Species SSR Markers

Using the SSR markers from pumpkin (*C. pepo* MU-CU-16) genome as a reference, we comparatively analyzed the genome SSR information of cucumber (Gy14), melon (DH92), watermelon (97103), *C. moschata* cv. Rifu, and *C. maxima* cv. Rimu. This was performed with a custom Perl script that used the NCBI BLASTN program as a search engine with an expected value of 10 and filtering. We allowed up to five nucleotide mismatches at the 5'-end of the primer, no mismatches at the 3'-end, and a minimum of 90% overall match homology. To establish the syntenic relationships of chromosomes between *C. pepo* with *C. sativus*, *C. lanatus*, *C. melo*, *C. maxima*, and *C. moschata*, we discarded these SSR markers with multiply physical locations in the same genome, only retaining the SSR markers in the genomes which had a single in silico PCR product. In addition, these shared SSR markers located on the unanchored scaffolds of the chromosome were further filtered. The SSR marker-based syntenic relationships were finally visualized with visualization blocks in Circos software v.0.55 [29].

2.4. Genomic DNA Extraction, PCR Amplification, and Electrophoresis Detection

Genomic DNA of all the materials was extracted using 1 g of young leaf sample with the cetyl trimethyl ammonium bromide (CTAB) method [30]. The extracted DNA was dissolved in 1× Tris-EDTA (TE) buffer (Solarbio, Cat: T1121). The concentration and purity were detected by the Nanodrop-2000 nucleic acid analyzer. The extracted DNA was diluted to 30 ng/μL as working solution and kept at 4 °C.

Each PCR reaction contained 1 μL of template DNA, 0.5 μM each of forward and reverse primers, 5 μL mastermix (GenStar, Cat: A012-105), and 3 μL ddH$_2$O. The amplification was carried out as follows: An initial denaturing step at 95 °C for 5 min, 94 °C for 30 s, followed by 6 cycles of 68–58 °C for 45 s. Each cycle was reduced by 2 °C, each annealing time was 1 min, and 72 °C for 1 min; 30 cycles of 94 °C for 30 s, 50 °C for 30 s, and 72 °C for 1 min. In the last cycle, primer extension was performed at 72 °C for 10 min.

The PCR products were analyzed by 9% polyacrylamide gel electrophoresis, and a 100 bp DNA ladder was used as the reference marker. After electrophoresis, silver staining was performed to display the PCR products, and photos were taken for preservation.

2.5. Calculation of Clustering

The heterozygosity (He), observer gene number (Na), effective alleles (Ne), observed heterozygosity (Ho), and the Shannon–Weaver index (I) were calculated using Pop-gen software v.1.32 (Canada, University of Alberta). Polymorphic information content (PIC) of SSR markers was computed using EXCEL (China, WPS of JINSHAN). When the PIC of an SSR marker was below 0.25, it was considered as a low polymorphic marker, and a marker was considered highly polymorphic if its PIC was above 0.5.

These amplification bands of each SSR primer were separated using polyacrylamide gel-electrophoresis. The band patterns were visualized with silver staining, and gel images were taken with a digital camera. In the same location, the presence of a band was marked as "1", the absence of a band was marked as "0", and a missing band was marked as "−1". In this study we used Genalex-6 software [31] to conduct the matrix calculation of SSR marker data which had been assigned a value, then transformed it into a triangle matrix, saved it as a mega-file, finally, imported the mega-file into the Mega-6.0 software (USA, Tamura, K team), and selected the unweighted pair group method with arithmetic (UPGMA) algorithm in the "phylogeny" dropdown menu to draw the cluster diagram [32].

The software Structure v.2.3 (USA, UChicago; Britain, Oxon) was used to analyze the population structure [33,34]. An admixture model and correlated allele frequencies were used to estimate the number of the populations. For each of the K-values (ranging from 1 to 5), ten independent runs were performed with a burn-in period of 100,000 followed by 500,000 Markov chain Monte Carlo runs. The optimal K-values depends on the peak of K = mean (| Ln"P(D) |)/(sdLnP(D)). Based on the structure results, the most probable K-value was analyzed using Structure Harvester (http://taylor0.biology.ucla.edu/struct_harvest/, 2020).

3. Result

3.1. The Frequency and Distribution of Different SSR Types in Cucurbita Genomes

A total of 103,056 microsatellite sequences were identified in the *Cucurbita* genome, including 34,375 SSR loci in the 269.9 Mb draft genome sequence of *C. moschata* cv. Rifu, 30,577 SSR loci in the 271.4 Mb draft genome sequence of *C. maxima* cv. Rimu, and 38,104 SSR loci in the 263 Mb draft genome sequence of *C. pepo* MU-CU-16 (Table S2). *Cucurbita pepo* had the largest number of markers with the smallest reference genome size, indicating the highest average density of markers (145 SSR/Mb). To obtain more information, we used *C. pepo* with a higher marker density as the control for the following comparative genomic analysis.

Here, we analyzed repeat types ranging from di-nucleotide to octa-nucleotide. Among all of these nucleotide motifs, di-nucleotide motifs (41.0%) were the most common type, accounting for 41.78%, 39.90%, and 41.01% of the total SSR loci discovered in *C. moschata*, *C. maxima*, and *C. pepo*, respectively, followed by tri-nucleotide motifs (16.97%, 19.19%, and 17.88%, respectively), whereas octa-nucleotide motifs (3.78%, 3.76%, and 3.38%, respectively) were the least represented repeat type in the three *Cucurbita* genomes (Table S2). In general, the frequency of the total SSR loci decreased with the increase in motif length, except for hepta-nucleotide SSRs.

We further examined the distribution of SSR motifs with regard to their repeat numbers (Figure 1). For all the repeat types, with an increase in the repeat number, the SSR frequency decreased sharply, and this change was more obvious in the longer SSR motifs (Figure 1). Consequently, the mean repeat numbers in the di-nucleotides were the highest of all of the repeat types. The analysis of individual SSR types revealed that some specific motifs were more prevalent than others in each class (Figure S1). For example, the AT motif was the most frequent di-nucleotide type in all three genomes, accounting for 31.61%

(in *C. moschata*), 28.81% (in *C. maxima*), and 30.45% (in *C. pepo*) of the total di-nucleotide loci. Similarly, the AAT, AAAT, AAAAT, AAAAAT, AAAAAAT, and AAAAAAAT motifs (AATAATAT motif in *C. maxima*) were the most frequent types in each class. These results indicated that AnT-rich motifs were the most abundant in all SSR motifs in the *C. moschata*, *C. maxima*, and *C. pepo* genomes.

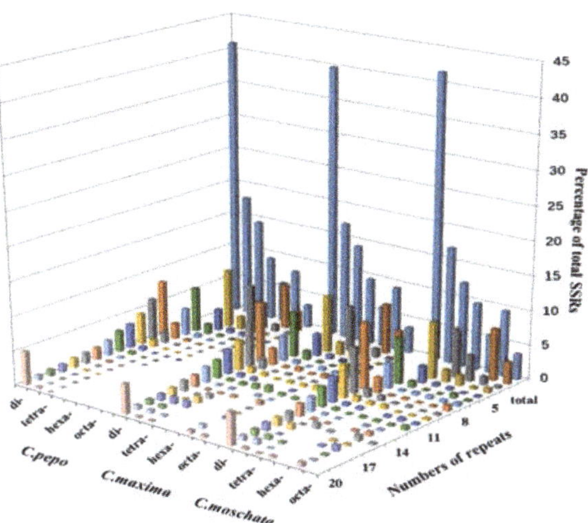

Figure 1. Distribution of SSR motif repeat numbers and relative frequency in *Cucurbita* genome. The vertical axis shows the abundance of microsatellites that have different motif repeat numbers (from 3 to >20) with different colors.

We also investigated the SSR density in each chromosome of the three *Cucurbita* species and found that the density of microsatellite loci was not correlated with the chromosome size (Table S3). For example, in the *C. moschata* genome, the SSR density of the longest chromosome (Chr04) had a medium density of SSRs, while Chr02, which is much shorter than Chr04, had the highest SSR density. A similar trend was also observed in the other two genomes, indicating that the distribution of SSRs was uneven in the *Cucurbita* chromosomes. To better understand the distributions of different SSR motifs, we further checked their frequencies on each chromosome (Figure 2). Our results showed that the distribution of different SSR types on the chromosomes corresponded with their frequencies and SSR density in the *Cucurbita* whole genomes.

The genomic sequences containing these microsatellites were screened for PCR primer design, and 94,272 SSR microsatellite loci were found to contain suitable flanking sites for SSR primer design. While *C. moschata* had the lowest proportion of SSRs suitable for primers design (84.75%), the percentages in *C. maxima* and *C. pepo* reached 94.53% and 95.09%, respectively (Table S2). Though the di-nucleotide repeat types were the most frequent in all three genomes, they did not exhibit good performance in primer design. Interestingly, the hexata-nucleotide repeat types had the highest ratio of SSRs suitable for primer design in all three genomes, followed by penta-nucleotide repeat types, indicating that the longer motifs were more suitable for primer design in *Cucurbita* species. Finally, a total of 91,248 SSR primers (28,194 in *C. moschata*, 28,061 in *C. maxima*, and 34,993 in *C. pepo*) were designed, with some primers including more than one SSR locus as the compound SSR (Tables S4–S6).

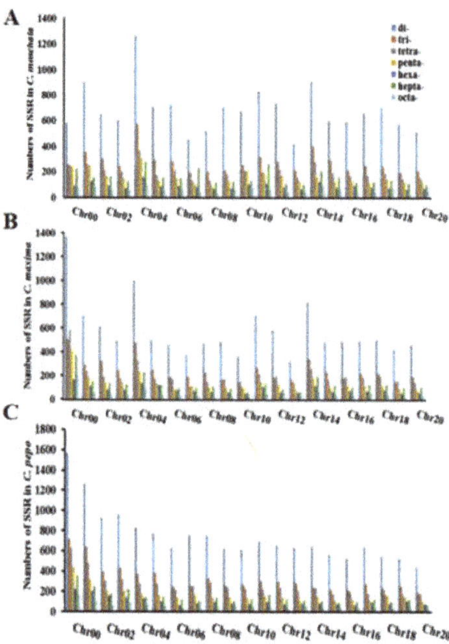

Figure 2. (**A**) The distribution of SSR repeat types on each chromosome in *C. moschata*. (**B**) The distribution of SSR repeat types on each chromosome in *C. maxima*. (**C**) The distribution of SSR repeat types on each chromosome in *C. pepo*. The vertical axis shows the number of microsatellites from di-nucleotide to octo-nucleotide which are discriminated by different colors. The horizontal axis shows different chromosomes of *C. ssp*, and LG00 means all the chromosome unanchored scaffolds.

3.2. Chromosome Synteny Relationships of C. pepo with Other Cucurbitaceae Species

In order to understand the universality and correlation of SSR markers among Cucurbitaceae crops, we compared and analyzed the cross-species SSR markers between *C. pepo* and other Cucurbitaceae species by in silico PCR. We identified 391 cross-species SSR markers between *C. pepo* and *C. sativus*, 425 cross-species SSR markers between *C. pepo* and *C. melo*, 717 cross-species SSR markers between *C. pepo* and *C. lanatus*, 11,732 cross-species SSR markers between *C. pepo* and *C. maxima*, and 15,274 cross-species SSR markers between *C. pepo* and *C. moschata* (Tables S7–S11). The collinear blocks to inversion blocks ratio was 26:26 between the *C. pepo* and *C. sativus* genomes, 25:36 between the *C. pepo* and *C. melo* genomes, 51:38 between the *C. pepo* and *C. lanatus* genomes, 154:158 between the *C. pepo* and *C. maxima* genomes, and 153:152 between the *C. pepo* and *C. moschata* genomes. Interestingly, the ratio of collinear blocks to inversion blocks was nearly 1:1 among the three *Cucurbita* species. Each *C. pepo* chromosome shared 3–36 SSR markers with *C. sativus*, *C. lanatus*, or *C. melo*. However, most of the *C. pepo* chromosome shared a larger number of SSR markers (3-1,436) with *C. maxima* or *C. moschata*. The *C. pepo* syntenic block, CpeCma7, had the largest number of shared SSR markers (i.e., 296) between *C. pepo* chromosome Cpe1 and *C. maxima* chromosome Cma4.

The physical positions of those common shared markers were compared. The main syntenic relationships between *C. pepo* and other Cucurbitaceae species are listed in Table 1, and the syntenic relationships visualized for *C. pepo* with *C. lanatus*, *C. melo*, and *C. sativus* are shown in Figure 3. The main syntenic relationships among the chromosomes revealed complex mosaic patterns. In Figure 3, each *C. pepo* chromosome was syntenic to more than two chromosomes in other Cucurbitaceae species. The *C. pepo* chromosomes Cpe9 and Cpe16 had the simplest syntenic pattern with watermelon, and each of them was mainly syntenic to one watermelon chromosome (Table 1). Cpe9 was syntenic to watermelon chromosome W5, and 14 commonly shared SSR markers were found between Cpe9 and W5. From the markers CpeSSR15544 to CpeSSR16107, there were three blocks belonging to watermelon chromosome W5, and each block contained at least four SSR markers. According to the continuous physical positions of these markers on both of the reference genomes, the syntenic blocks CpeWM37 and CpeWM38 showed an inversion pattern, and the syntenic block CpeWM39 showed a collinear pattern between *C. pepo* and *C. lanatus*. Similar comparisons were carried out between *C. pepo* and *C. sativus* or *C. pepo* and *C. melo* using the cross-species SSR markers. The *C. pepo* chromosomes Cpe7, Cpe8, Cpe11, and Cpe20 had the simplest syntenic pattern with *C. sativus*, and each of them was only syntenic to one cucumber chromosome. Meanwhile, the simplest syntenic patterns between *C. pepo* and *C. melo* were mainly found on chromosomes Cpe15, Cpe18, Cpe19, and Cpe20. The most complicated syntenic pattern was found on *C. pepo* chromosome Cpe1, which corresponded to five chromosomes of *C. moschata*, four chromosomes of *C. maxima*, seven chromosomes of *C. lanatus*, three chromosomes of *C. sativus*, and five chromosomes of *C. melo*.

Table 1. The main syntenic relationships of *C. pepo* with other Cucurbitaceae species.

C. pepo	*C. moschata*	*C. maxima*	*C. lanatus*	*C. sativus*	*C. melo*
Cpe1	Cmo3(4), Cmo4(1,436), Cmo9(5), Cmo10(3), Cmo17(24)	Cma3(5), Cma4(1,103), Cma9(9), Cma17(19)	Cla1(5), Cla5(21), Cla6(4), Cla7(13), Cla8(5), Cla10(3), Cla11(14)	Csa3(6), Csa5(28), Csa6(3)	Cme3(4), Cme6(4), Cme7(3), Cme9(5), Cme10(16)
Cpe2	Cmo1(913), Cmo10(3), Cmo18(12)	Cma1(674), Cma18(8)	Cla5(8), Cla7(8), Cla10(13), Cla11(7)	Csa3(7), Csa4(6), Csa7(3)	Cme1(3), Cme4(10), Cme7(9)
Cpe3	Cmo4(3), Cmo14(1,080)	Cma4(4), Cma14(859)	Cla5(24), Cla7(8), Cla10(36)	Csa3(34), Csa4(4)	Cme4(25), Cme6(17), Cme7(6)
Cpe4	Cmo10(3), Cmo11(822)	Cma11(640)	Cla2(6), Cla3(5), Cla6(12), Cla10(3)	Csa1(15), Csa3(3)	Cme2(11)
Cpe5	Cmo2(904), Cmo10(8)	Cma2(692), Cma10(8)	Cla1(8), Cla2(20), Cla9(5)	Csa3(5), Csa5(8), Csa6(8)	Cme4(3), Cme5(4), Cme9(6), Cme11(10)
Cpe6	Cmo9(551)	Cma9(396)	Cla5(12), Cla8(4), Cla9(5), Cla11(9)	Csa3(6), Csa4(3)	Cme4(6), Cme7(9)
Cpe7	Cmo5(3), Cmo12(587), Cmo14(20)	Cma5(4), Cma12(452)	Cla2(3), Cla8(6)	Csa2(5)	Cme3(5), Cme5(3)
Cpe8	Cmo6(785)	Cma6(431)	Cla5(13), Cla10(16)	Csa3(6)	Cme4(10), Cme6(6)
Cpe9	Cmo18(544), Cmo19(6)	Cma2(3), Cma18(440)	Cla5(14)	Csa1(4), Csa3(5), Csa5(5)	Cme6(4), Cme10(3), Cme12(4)
Cpe10	Cmo3(659), Cmo18(5)	Cma3(547), Cma18(6)	Cla1(23), Cla4(13)	Csa4(3), Csa6(19)	Cme8(20)
Cpe11	Cmo5(707), Cmo10(3)	Cma5(574)	Cla2(4), Cla8(17), Cla11(14)	Csa2(14)	Cme3(11), Cme5(9)
Cpe12	Cmo17(665)	Cma17(516)	Cla6(12), Cla9(15)	Csa6(9), Csa7(14)	Cme1(17), Cme11(6)
Cpe13	Cmo8(9), Cmo15(649)	Cma4(3), Cma8(7), Cma15(468)	Cla1(21), Cla8(11), Cla11(8)	Csa2(3), Csa5(13), Csa6(6)	Cme3(8), Cme9(12)
Cpe14	Cmo16(565)	Cma16(409)	Cla5(4), Cla7(17), Cla10(14)	Csa3(15), Csa4(8)	Cme6(13), Cme7(5)
Cpe15	Cmo19(493)	Cma19(356)	Cla2(16), Cla7(4), Cla9(11)	Csa3(4), Csa7(11)	Cme1(12)
Cpe16	Cmo20(526)	Cma20(394)	Cla2(23)	Csa2(4), Csa6(4)	Cme5(5), Cme11(10)
Cpe17	Cmo4(3), Cmo8(634), Cmo9(7), Cmo14(3), Cmo17(3)	Cma8(472), Cma14(3), Cma17(4)	Cla6(11), Cla9(17)	Csa6(3), Csa7(3)	Cme1(8), Cme11(4)
Cpe18	Cmo10(500), Cmo14(3)	Cma10(354), Cma14(3)	Cla3(8), Cla6(22)	Csa1(16)	Cme2(17)
Cpe19	Cmo7(658)	Cma7(462)	Cla1(20), Cla4(7)	Csa4(4), Csa6(16)	Cme8(23)
Cpe20	Cmo13(468)	Cma13(330)	Cla1(4), Cla3(12), Cla4(4)	Csa1(11)	Cme12(8)

The syntenic relationships among different *Cucurbita* species were simple and clear. For instance, each of the 20 chromosomes in *C. pepo* was mainly syntenic with one chromosome in *C. moschata* or *C. maxima* (Figure 4), implying that the chromosomes in the *Cucurbita* genomes were highly conserved during evolution. Our results also showed that there were three main relationship patterns among the *C. pepo*, *C. maxima*, or *C. moschata* genomes, including (1) the eleven linear relationship chromosomes between *C. pepo* and *C. maxima* or *C. moschata* such as Cpe2–Cmo1–Cma1. Most of the cross-markers in the corresponding chromosomes showed collinear patterns. (2) There were eight inverted relationship chromosomes between *C. pepo* and *C. maxima* or *C. moschata*. For example, the chromosome Cpe1 of *C. pepo* was inverted to the chromosome Cmo4 of *C. moschata* and Cma4 of *C. maxima*. (3) There was a mosaic pattern between *C. pepo* and *C. maxima* or *C. moschata*, for example, Cpe4–Cmo11–Cma11.

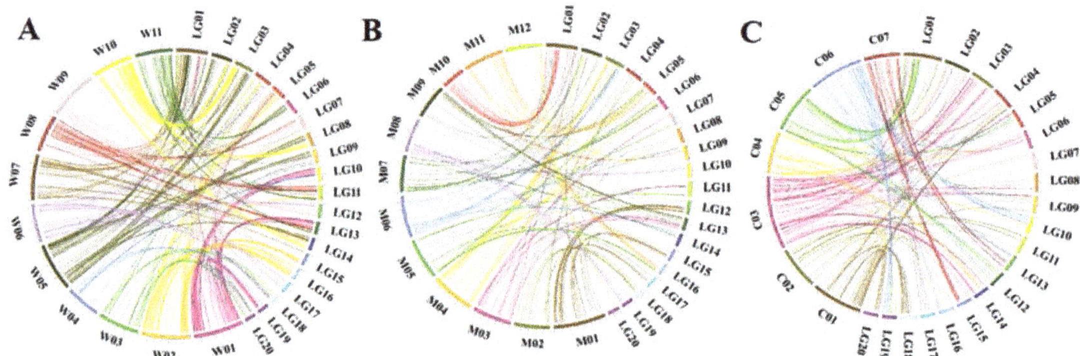

Figure 3. Syntenic relationships of *C. pepo* with (**A**) *C. lanatus*, (**B**) *C. melo*, and (**C**) *C. sativus*. Chromosome synteny between *C. pepo* and *C. sativus* was based on 391 cross-species markers; synteny between *C. pepo* and *C. melo* was based on 425 cross-species markers; synteny between *C. pepo* and *C. lanatus* was based on 717 cross-species markers. W1–W11 represent *C. lanatus*' eleven chromosomes, M01–M12 represent *C. melo*'s twelve chromosomes, C01–C07 represent *C. sativus*'s seven chromosomes, and LG01–LG20 represent *C. pepo*'s twenty chromosomes. Syntenic blocks are connected by the same color lines from *C. pepo* chromosomes.

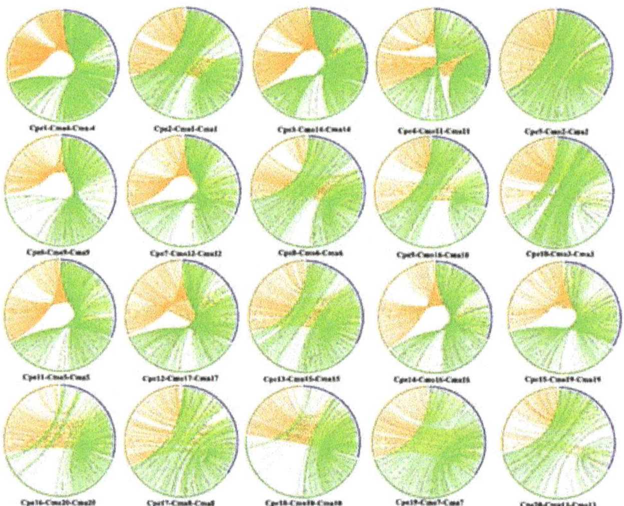

Figure 4. Chromosome synteny of *C. pepo* (blue) with *C. moschata* (green) and *C. maxima* (yellow). The physical positions of chromosomes of each crop in the figure are arranged clockwise. Chromosome synteny between *C. pepo* and *C. moschata* was based on 14,276 cross-species markers; synteny between *C. pepo* and *C. maxima* was based on 10,655 cross-species markers. Cpe1–Cpe20 represent *C. pepo*'s twenty chromosomes, Cmo1–Cmo20 represent *C. moschata*'s chromosomes, and Cma1–Cma20 represent *C. maxima* chromosomes. The syntenic relationship between *C. pepo* and *C. moschata* are connected with the green color lines, and the syntenic relationship between *C. pepo* and *C. maxima* are connected with the yellow color lines.

3.3. The Genetic Diversity and Population Structure Analysis of the C. pepo Germplasm

In our preliminary study, approximately 400 SSR markers were screened using 61 accessions of *C. pepo* germplasm. Finally, a total of 66 core SSR markers were selected based on the allelic number, the genomic coverage, and the efficiency of PCR amplification (Table S12). These markers exhibited clear band spectrums and were evenly distributed on the chromosomes. In this study, 276 alleles were detected by the 66 SSR markers in the 61 *C. pepo* accessions with an average of 4.18 loci per SSR marker. The number of Na ranged from two to nine. The highest number of Na was nine, which was detected by SSR010246, SSR026560, SSR026918, SSR027656, and SSR026980, followed by SSR011546, SSR003315, and SSR026797 with eight alleles. The number of Ne varied from 1.03 to 6.07 with an average of 2.31. The SI ranged from 0.083 to 1.96 with an average of 0.83. The PIC value ranged from 0.03 to 0.83 with an average of 0.43 (Table S13).

We further used a model-based approach for population structure analysis of the 61 *C. pepo* accessions. According to the results of the structural operation, when K = 2, ΔK showed a significant peak value, indicating that the 61 accessions used in this study could be obviously divided into two groups (Figure S2), named group I and group II. The five *C. pepo* subsp. *ovifer* accessions (2, 29, 30, 31, and 45) were clustered into group I (8.20%), and all of them were wild materials. Most of the *C. pepo* subsp. *pepo* accessions were clustered into group II (91.80%), which were all cultivated materials (Figure 5A). This indicated that the SSR markers we used could clearly distinguish the cultivated materials from the wild materials. The backgrounds of the cultivated accessions were narrow, except for accession 45 in group I, which should have a complex genetic background, similar to accession 14 and 16 in group II. The UPGMA analysis revealed that the 61 *C. pepo* accessions were divided into two clusters (Figure 5B), which was consistent with their population structure. The five *C. pepo* subsp. *ovifer* accessions were clustered together at the base of the phylogenetic tree, which further supported our population structure analysis.

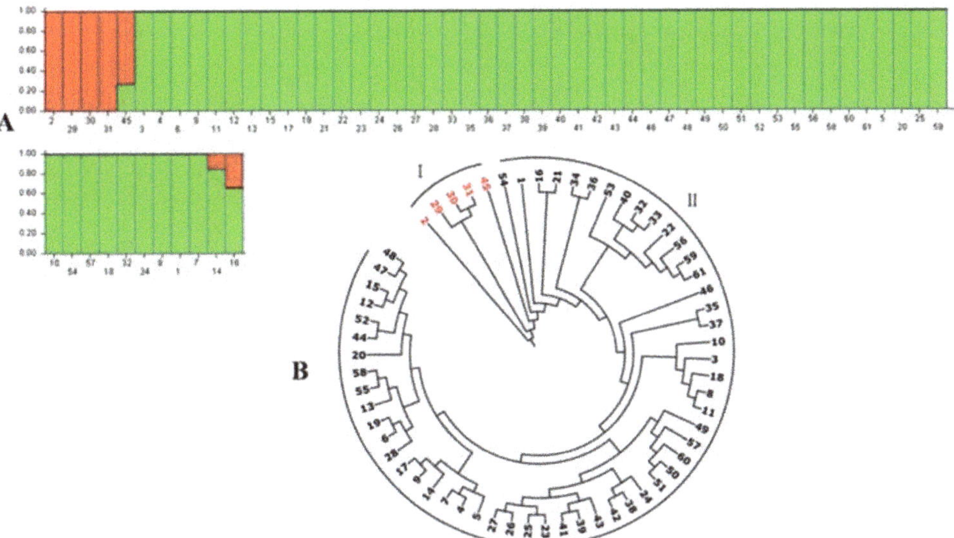

Figure 5. The genetic diversity of the 61 accessions based on SSR markers. (**A**) Population structure of 61 accessions in *C. pepo* by the model-based analysis. The scale of the *y*-axis represents the percentage of genetic components, and the *x*-axis represents the different materials. (**B**) Phylogenetic tree of 61 accessions in the *C. pepo* by UPGMA analysis; Group I (red makers) and Group II correspond to the structure analysis.

4. Discussion

4.1. Frequency, Distribution, and Characterization of Microsatellites in Three Cucurbita Genomes

With the development of sequencing technology, the discovery and mining of genomic SSR loci has successfully been applied in many plant species, such as cotton [35,36], foxtail millet [37], cucumber [11], watermelon [13], tobacco [38], and melon [12]. *Cucurbita moschata*, *C. maxima*, and *C. pepo* are important species that are cultivated worldwide, and their graft genomes were released several years ago. However, there remains little information on the development of genome-wide SSR markers in *Cucurbita* species, which has strongly limited their genetic research. In the present study, genome-wide microsatellites were identified and characterized in the three *Cucurbita* species. A total of 34,375, 30,577, and 38,104 SSR loci were detected in the *C. moschata*, *C. maxima*, and *C. pepo* genomes, respectively. The smallest genome size and maximum number of microsatellites were detected in *C. pepo*, indicating that there was no direct correlation between genome size and the number of microsatellites. The density of the SSR markers in the three *Cucurbita* species was approximately 113–145 SSR/Mb, which is lower than that in cucumber (552 SSR/Mb) but comparable to that in melon (109 SSR/Mb) and watermelon (111 SSR/Mb) [11–13]. In addition to the natural differences among different genomes, many other factors could affect the deviations in SSR density such as the software and parameters used for microsatellite detection. We suspect that the main reason for the difference in SSR density between *Cucurbita* species and cucumber was the different selection criteria for the SSR loci, e.g., the repeat types (di- to octa-nucleotides versus mono- to penta-nucleotides) and the minimum lengths (18 bp versus 12 bp).

We further analyzed the distribution and frequency of microsatellites in the three *Cucurbita* species (Figures 1 and 2). In most cases, a negative correlation was observed between the microsatellite frequency and the number of repeat units. Consistent with previous studies in watermelon and melon, the di-nucleotide repeats were the most abundant SSRs, followed by tri-, tetra-, penta-, hepta-, hexa-, and octo-nucleotide repeats [12,13]. This is something that varies in different species. For example, the density of tetra-nucleotide repeats was highest in *C. sativus* (164.2 SSR/Mb), *Populus trichocarpa* (144.9 SSR/Mb), *Medicago truncatula* (102.8 SSR/Mb), and *Vitis vinifera* (171.3 SSR/Mb), whereas the density of tri-nucleotide repeats was the highest in Arabidopsis thaliana (146.6 SSR/Mb), Glycine max (103.1 SSR/Mb), and Oryza sativa (220.1 SSR/Mb) [11]. Some studies have revealed that the di-nucleotide motifs with high repeat numbers are more abundant and polymorphic compared to those with short repeat units [39]. The reason is that di-nucleotide repeats are much less frequent in coding regions than in non-coding regions [40,41]. It is also reported that the exon region contains more triplet SSRs than other repeats, and triplet SSR motifs may be related to high frequencies of certain amino acids [42,43]. These SSRs in the coding sequence may have the potential to affect all aspects of genetic functions including gene regulation, development, and evolution. However, the function of genes that contain SSRs and the role of these SSR motifs in plant genes are less studied and poorly understood [44]. It is interesting to note that many bacterial SSRs in the intergenic regions have regulatory functions [45], and whether these SSR motifs in the intergenic regions of *Cucurbita* species play a role in specialization or gene regulation should be further studied.

The low number of repeat motifs was predominant, and the AT-rich motifs in particular contributed a large proportion of all types of di-nucleotide repeats in the *Cucurbita* species (Figure S1). The AT or AAT type is more common in dicots [13], which is consistent with our results. Recently, the characterization of SSR markers in bitter gourd showed that the tri-nucleotide repeat units were the main type, with an overrepresentation of A/T, AT/AT, AAT/ATT, and AAAT/ATTT motifs in all kinds of repeat types [46]. This has also been found in other genomes [11,47,48]. On the contrary, the frequency of the GC or CCG type was much lower at the genomic level [49,50], and the GC, TC, or GA types have relatively stable structures. Most of the AT types are distributed in non-genic regions, while the TC/GA types are primarily distributed in coding sequences [38].

4.2. Chromosome Synteny Analysis between C. pepo and Other Cucurbitaceae Species

Chromosome synteny analysis has been conducted in many species, such as cucumber, watermelon, and melon, but few studies have been conducted on the chromosome synteny among different *Cucurbita* species or between *Cucurbita* species and other Cucurbitaceae crops. In this study, the genome-wide SSR development from the three *Cucurbita* genomes provided the possibility to identify their syntenic relationships at a high-resolution level via in silico PCR analysis. Though the sizes of the pumpkin genomes are similar to that of other sequenced Cucurbitaceae species, the number of cross-species SSR markers in the *Cucurbita* genus is much higher. Compared to hundreds of shared markers in previous studies [14], we identified many more cross-species transferable SSR markers in the *Cucurbita* genus that were used for chromosome synteny analysis. The WGD event in *Cucurbita*, which has not been observed in other sequenced Cucurbitaceae species, such as cucumber [8], melon [10], and watermelon [9], may be a possible reason leading to the high abundance of SSR markers.

According to the cross-species transferable SSR markers, 52, 61, and 89 syntenic blocks distributed on all chromosomes were identified between *C. pepo* with cucumber, melon, and watermelon, respectively (Figure 3). Similar homoeologous blocks were detected by whole-genome comparison [22], suggesting that the cross-species transferable SSR markers are useful and reliable in genome comparisons and chromosome synteny analyses. In most cases, there were multiple synteny blocks detected between *C. pepo* and other Cucurbitaceae species due to the fact of chromosome fission. The most complicated syntenic pattern existed on chromosome Cpe1 of *C. pepo*, which was syntenic to seven watermelon chromosomes, indicating that complicated structural changes occurred after their divergence from a common ancestor. The ratio of collinear blocks to inversion blocks was nearly 1:1 in *Cucurbita*, and the reason for this may be that genome duplication and inter-chromosomal exchanges occurred randomly during chromosome evolution.

Based on the cross-species transferable SSR markers, we identified more highly conserved syntenic blocks among *Cucurbita* species than melon, cucumber, or watermelon. We found that each block among three *Cucurbita* species of the same genus contained many more shared common SSR markers, and these homoeologous chromosomes were much conserved, which further confirmed their close evolutionary relationships in the Cucurbitaceae family. For example, the *C. pepo* syntenic block contained more markers than that in melon [12]. Due to the WGD during chromosome evolution and speciation, the number of the chromosomes and cross-markers increased. However, those blocks were highly conserved during chromosome evolution among different Cucurbitaceae species. The chromosomal pair analysis by cross-species SSR markers showed that there were eight large-scale inversions on different chromosomes between *C. pepo* and *C. moschata* or between *C. pepo* and *C. maxima*, indicating that *C. pepo* experienced more complex evolutionary processes (Figure 4). Interestingly, Chr4 contained a mosaic region among *Cucurbita* species. The reason might be due to the fact of genome duplication, large-scale inter-chromosomal exchanges, or long-term evolutionary forces. Whether the partial inversion of chromosome 4 in *C. pepo* will affect the mapping, cloning, and study of some traits is worth exploring in the future.

4.3. The Genetic Diversity and Population Structure of C. pepo Germplasm

Previously, because of the scarcity of genomic sequences, there were limited molecular markers available to study the genetic diversity and population structure of *Cucurbita* species. Though the genetic diversity of *Cucurbita* species has been evaluated using sequence-related amplified polymorphism (SRAP), AFLP, SSR, RAPD, and inter-simple sequence repeats (ISSRs), most of the markers used have high randomness, lack precise location information, and have low genomic coverage and poor polymorphism, which greatly limit their application [18,51,52]. With the draft genome available for three cucurbit crops, we developed 91,248 SSR markers with precise physical locations on chromosomes and evaluated the genetic diversity of 61 pumpkin accessions using 66 core SSR markers.

The population structure of 61 accessions revealed that the background of some materials was mixed between group I and group II, suggesting that these accessions may have undergone gene exchange between two subspecies. The materials were collected from different provinces in China, and they were obviously classified into two subspecies, subsp. *ovifer* (or subsp. *texana*) and subsp. *pepo*, which is consistent with previous studies [21,51]. However, the three subspecies of *C. pepo* classified by Decker are *C. pepo* subsp. *fraterna* (Bailey) Andres, *C. pepo* subsp. *texana* (Scheele) Filov, and *C. pepo* subsp. *pepo* [53]. The putative ancestor for *C. pepo*, namely, subsp. *fraterna* from northeastern Mexico, has been considered a wild gourd [54]. The population structure and UPGMA results indicated that these accessions of *C. pepo* in China come from the common ancestor. Thus, there have great prospects for germplasm improvement.

The *Cucurbita* genus contains several economically important crops, but its breeding has lagged behind the other Cucurbitaceous crops. Limited high-quality cultivars cannot meet the production requirements. Thus, different breeding programs can be facilitated using marker assisted selection. The whole-genome SSR markers detected in this study will promote the development and utilization in basic and applied research of *Cucurbita* species.

Supplementary Materials: The following are available online at https://www.mdpi.com/article/10.3390/horticulturae7060143/s1, Figure S1: The top five types of each SSR repeat motif and their frequencies in *C. moschata*, *C. maxima*, and *C. pepo*, Figure S2: The optimal K-values analysis by using Structure Harvester, Table S1: The list of the *C. pepo* introduction accessions, Table S2: The distribution of different nucleotide repeats in the genome of three *Cucurbita* species, Table S3: The distribution of SSR loci on different chromosomes in *C. moschata*, *C. maxima*, and *C. pepo*, Table S4: The identified SSR markers in *C. moschata*, Table S5: The identified SSR markers in *C. maxima*, Table S6: The identified SSR markers in *C. pepo*, Table S7: List of cross-species SSR markers between *C. pepo* and *C. sativus* identified by in silico PCR, Table S8: List of cross-species SSR markers between *C. pepo* and melon identified by in silico PCR, Table S9: List of cross-species SSR markers between *C. pepo* and watermelon identified by in silico PCR, Table S10: List of cross-species SSR markers between *C. pepo* and *C. maxima* identified by in silico PCR, Table S11: List of cross-species SSR markers between *C. pepo* and *C. moschata* identified by in silico PCR, Table S12: The total SSR markers in *C. pepo* genetic diversity and population structure analysis, Table S13: Polymorphism and allelic diversity of SSR markers in *C. pepo* materials.

Author Contributions: L.Z. and H.Z. performed the data analysis and wrote the manuscript. Y.L., Y.W. (Yong Wang), and X.W. conducted validation of SSR polymorphism, genetic diversity, and population structure analysis. J.L., Z.Z., and Y.W. (Yanjiao Wang). participated in the SSR genetic diversity analysis. J.H., S.Y., and Y.L. conducted the germplasm collection. S.S. and L.Y. designed the experiments and revised the manuscript. All authors have read and agreed to the published version of the manuscript.

Funding: This research was supported by the Major Public Welfare Project of Henan Province (202102110202) and the Science and Technology Project of China National Tobacco Corporation (Henan Tobacco Company) (2018410000270095).

Data Availability Statement: All generated or analyzed data during this study are reflected in the present article.

Acknowledgments: The authors are highly grateful to Yufeng Wu of Nanjing Agricultural University with the Circos analysis. The *C. pepo* accessions used in this study were introduced from the National Crop Germplasm Resource Platform of China. We are also indebted to the editor and reviewers for critically evaluating the manuscript and providing constructive comments for improvement.

Conflicts of Interest: The authors declare no conflict of interest.

References

1. George, E.B.; Ronald, J.T. Toxic plants of North America. *Choice Curr. Rev. Acad. Libraries* **2013**, *12*, 2202–2203.
2. Loy, J.B. Morpho-physiological aspects of productivity and quality in squash and pumpkins (*Culcurbita* spp.). *Crit. Rev. Plant Sci.* **2004**, *23*, 337–363. [CrossRef]

3.   Savage, J.A.; Haines, D.F.; Holbrook, N.M. The making of giant pumpkins: How selective breeding changed the phloem of *Cucurbita maxima* from source to sink. *Plant Cell Environ.* **2015**, *38*, 1543–1554. [CrossRef] [PubMed]
4.   Davis, A.R.; Perkins-Veazie, P.; Sakata, Y.; Lopez-Galarza, S.; Maroto, J.V.; Lee, S.G.; Huh, Y.C.; Sun, Z.Y.; Miguel, A.; King, S.R.; et al. Cucurbit grafting. *Crit. Rev. Plant Sci.* **2008**, *27*, 50–74. [CrossRef]
5.   Lee, J.M.; Kubota, C.; Tsao, S.J.; Bie, Z.; Echevarria, P.H.; Morra, L.; Oda, M. Current status of vegetable grafting: Diffusion, grafting techniques, automation. *Sci. Hortic.* **2010**, *127*, 93–105. [CrossRef]
6.   Lv, J.; Qi, J.J.; Shi, Q.X.; Shen, D.; Zhang, S.P.; Shao, G.J.; Li, H.; Sun, Z.Y.; Weng, Y.Q.; Shang, Y.; et al. Genetic Diversity and Population Structure of Cucumber (*Cucumis sativus* L.). *PLoS ONE* **2012**, *7*, e46919. [CrossRef] [PubMed]
7.   Li, Y.; Wen, C.; Weng, Y. Fine mapping of the pleiotropic locus B for black spine and orange mature fruit color in cucumber identifies a 50 kb region containing a R2R3-MYB transcription factor. *Theor. Appl. Genet.* **2013**, *126*, 2187–2196. [CrossRef] [PubMed]
8.   Huang, S.W.; Li, R.Q.; Zhang, Z.H.; Li, L.; Gu, X.F.; Fan, W.; Lucas, W.J.; Wang, X.W.; Xie, B.Y.; Ni, P.X.; et al. The genome of the cucumber, *Cucumis sativus* L. *Nat. Genet.* **2009**, *41*, 1275–1281. [CrossRef] [PubMed]
9.   Guo, S.G.; Zhang, J.G.; Sun, H.H.; Salse, J.; Lucas, W.J.; Zhang, H.Y.; Zheng, Y.; Mao, L.Y.; Ren, Y.; Wang, Z.W.; et al. The draft genome of watermelon (*Citrullus lanatus*) and resequencing of 20 diverse accessions. *Nat. Genet.* **2013**, *45*, 51–58. [CrossRef]
10.   Garcia-Mas, J.; Benjak, A.; Sanseverino, W.; Bourgeois, M.; Mir, G.; González, V.M.; Hénaff, E.; Câmara, F.; Cozzuto, L.; Lowy, E.; et al. The genome of melon (*Cucumis melo* L.). *Proc. Natl. Acad. Sci. USA* **2012**, *109*, 11872–11877. [CrossRef]
11.   Cavagnaro, P.F.; Senalik, D.A.; Yang, L.; Simon, P.W.; Harkins, T.T.; Kodira, C.D.; Huang, S.; Weng, Y. Genome-wide characterization of simple sequence repeats in cucumber (*Cucumis sativus* L.). *BMC Genom.* **2010**, *11*, 569. [CrossRef]
12.   Zhu, H.; Guo, L.; Song, P.; Luan, F.; Hu, J.; Sun, X.; Yang, L. Development of genome-wide SSR markers in melon with their cross-species transferability analysis and utilization in genetic diversity study. *Mol. Breed.* **2016**, *36*, 153. [CrossRef]
13.   Zhu, H.; Song, P.; Koo, D.H.; Guo, L.; Li, Y.; Sun, S.; Weng, Y.; Yang, L. Genome wide characterization of simple sequence repeats in watermelon genome and their application in comparative mapping and genetic diversity analysis. *BMC Genom.* **2016**, *17*, 557. [CrossRef] [PubMed]
14.   Li, D.W.; Cuevas, H.E.; Yang, L.M.; Li, Y.H.; Garcia-Mas, J.; Zalapa, J.; Staub, J.E.; Luan, F.S.; Reddy, U.; He, X.M.; et al. Syntenic relationships between cucumber (*Cucumis sativus* L.) and melon (*C. melo* L.) chromosomes as revealed by comparative genetic mapping. *BMC Genom.* **2011**, *12*, 396. [CrossRef] [PubMed]
15.   Yang, L.M.; Koo, D.H.; Li, D.W.; Zhang, T.; Jiang, J.M.; Luan, F.S.; Renner, S.S.; Henaff, E.; Sanseverino, W.; Garcia-Mas, J.; et al. Next-generation sequencing, FISH mapping and synteny-based modeling reveal mechanisms of decreasing dysploidy in Cucumis. *Plant J.* **2014**, *77*, 16–30. [CrossRef] [PubMed]
16.   Hi, L.Y.; Jeong, J.H.; Hong, K.H.; Dong, K.B. Use of Random Amplified Polymorphic DNAs for Linkage Group Analysis in Interspecific Hybrid F_2 Generation of *Cucurbita*. *Hortic. Environ. Biotechnol.* **1995**, *36*, 323–330.
17.   Brown, R.N.; Myers, J.R. A genetic map of squash (*Cucurbita* sp.) with randomly amplified polymorphic DNA markers and morphological markers. *J. Am. Soc. Hortic.* **2002**, *127*, 568–575. [CrossRef]
18.   Paris, H.S.; Yonash, N.; Portnoy, V.; Mozes-Daube, N.; Tzuri, G.; Katzir, N. Assessment of genetic relationships in *Cucurbita pepo* (Cucurbitaceae) using DNA markers. *Theor. Appl. Genet.* **2003**, *106*, 971–978. [CrossRef]
19.   Zraidi, A.; Stift, G.; Pachner, M.; Shojaeiyan, A.; Gong, L.; Lelley, T. A consensus map for *Cucurbita pepo*. *Mol. Breed.* **2007**, *20*, 375–388. [CrossRef]
20.   Gong, L.; Stift, G.; Kofler, R.; Pachner, M.; Lelley, T. Microsatellites for the genus *Cucurbita* and an SSR-based genetic linkage map of *Cucurbita pepo* L. *Theor. Appl. Genet.* **2008**, *117*, 37–48. [CrossRef]
21.   Esteras, C.; Gomez, P.; Monforte, A.J.; Blanca, J.; Vicente-Dolera, N.; Roig, C.; Nuez, F.; Pico, B. High-throughput SNP genotyping in *Cucurbita pepo* for map construction and quantitative trait loci mapping. *BMC Genom.* **2012**, *13*, 80. [CrossRef] [PubMed]
22.   Sun, H.H.; Wu, S.; Zhang, G.Y.; Jiao, C.; Guo, S.G.; Ren, Y.; Zhang, J.; Zhang, H.Y.; Gong, G.Y.; Jia, Z.C.; et al. Karyotype Stability and Unbiased Fractionation in the Paleo-Allotetraploid *Cucurbita* Genomes. *Mol. Plant* **2017**, *10*, 1293–1306. [CrossRef] [PubMed]
23.   Montero-Pau, J.; Blanca, J.; Bombarely, A.; Ziarsolo, P.; Esteras, C.; Marti-Gomez, C.; Ferriol, M.; Gomez, P.; Jamilena, M.; Mueller, L.; et al. De novo assembly of the zucchini genome reveals a whole-genome duplication associated with the origin of the *Cucurbita* genus. *Plant Biotechnol. J.* **2018**, *16*, 1161–1171. [CrossRef]
24.   Blanca, J.; Canizares, J.; Roig, C.; Ziarsolo, P.; Nuez, F.; Pico, B. Transcriptome characterization and high throughput SSRs and SNPs discovery in *Cucurbita pepo* (Cucurbitaceae). *BMC Genom.* **2011**, *12*, 104. [CrossRef]
25.   Wyatt, L.E.; Strickler, S.R.; Mueller, L.A.; Mazourek, M. An acorn squash (*Cucurbita pepo* ssp. ovifera) fruit and seed transcriptome as a resource for the study of fruit traits in Cucurbita. *Hortic. Res.* **2015**, *2*, 14070. [CrossRef]
26.   Xanthopoulou, A.; Psomopoulos, F.; Ganopoulos, I.; Manioudaki, M.; Tsaftaris, A.; Nianiou-Obeidat, I.; Madesis, P. De novo transcriptome assembly of two contrasting pumpkin cultivars. *Genom. Data* **2016**, *7*, 200–201. [CrossRef] [PubMed]
27.   Aliki, X.; Ganopoulos, I.; Psomopoulos, F.; Manioudaki, M.; Moysiadis, T.; Kapazoglou, A.; Osathanunkul, M.; Michailidou, S.; Kalivas, A.; Tsaftaris, A.; et al. De novo comparative transcriptome analysis of genes involved in fruit morphology of pumpkin cultivars with extreme size difference and development of EST-SSR markers. *Gene* **2017**, *622*, 50–66. [CrossRef]
28.   Vitiello, A.; Scarano, D.; D'Agostino, N.; Digilio, M.C.; Pennacchio, F.; Corrado, G.; Rao, R. Unraveling zucchini transcriptome response to aphids. *Peer J. PrePrints* **2016**, *4*. [CrossRef]

29. Krzywinski, M.; Schein, J.; Birol, I.; Connors, J.; Gascoyne, R.; Horsman, D.; Jones, S.J.; Marra, M.A. Circos: An information aesthetic for comparative genomics. *Genome Res.* **2009**, *19*, 1639–1645. [CrossRef] [PubMed]
30. Murray, M.G.; Thompson, W.F. Rapid isolation of high molecular weight plant DNA. *Nucleic Acids Res.* **1980**, *8*, 4321–4325. [CrossRef]
31. Peakall, R.; Smouse, P.E. GenAlEx 6.5: Genetic analysis in Excel. Population genetic software for teaching and research—An update. *Bioinformatics* **2012**, *28*, 2537–2539. [CrossRef]
32. Tamura, K.; Stecher, G.; Peterson, D.; Filipski, A.; Kumar, S. MEGA6: Molecular Evolutionary Genetics Analysis Version 6.0. *Mol. Biol. Evol.* **2013**, *30*, 2725–2729. [CrossRef] [PubMed]
33. Pritchard, J.K.; Stephens, M.; Donnelly, P. Inference of population structure using multilocus genotype data. *Genetics* **2000**, *155*, 945–959. [CrossRef]
34. Evanno, G.; Regnaut, S.; Goudet, J. Detecting the number of clusters of individuals using the software STRUCTURE: A simulation study. *Mol. Ecol.* **2005**, *14*, 2611–2620. [CrossRef]
35. Lu, C.R.; Zou, C.S.; Zhang, Y.P.; Yu, D.Q.; Cheng, H.L.; Jiang, P.F.; Yang, W.C.; Wang, Q.L.; Feng, X.X.; Prosper, M.A.; et al. Development of chromosome-specific markers with high polymorphism for allotetraploid cotton based on genome-wide characterization of simple sequence repeats in diploid cottons (*Gossypium arboreum* L. and *Gossypium raimondii* Ulbrich). *BMC Genom.* **2015**, *16*, 55. [CrossRef] [PubMed]
36. Wang, Q.; Fang, L.; Chen, J.D.; Hu, Y.; Si, Z.F.; Wang, S.; Chang, L.J.; Guo, W.Z.; Zhang, T.Z. Genome-Wide Mining, Characterization, and Development of Microsatellite Markers in *Gossypium* Species. *Sci. Rep.* **2015**, *5*, 10638. [CrossRef] [PubMed]
37. Zhang, S.; Tang, C.J.; Zhao, Q.; Li, J.; Yang, L.F.; Qie, L.F.; Fan, X.K.; Li, L.; Zhang, N.; Zhao, M.C.; et al. Development of highly polymorphic simple sequence repeat markers using genome-wide microsatellite variant analysis in Foxtail millet [*Setaria italica* (L.) P. Beauv.]. *BMC Genom.* **2014**, *15*, 78. [CrossRef] [PubMed]
38. Wang, X.W.; Yang, S.; Chen, Y.D.; Zhang, S.M.; Zhao, Q.S.; Li, M.; Gao, Y.L.; Yang, L.; Bennetzen, J.L. Comparative genome-wide characterization leading to simple sequence repeat marker development for Nicotiana. *BMC Genom.* **2018**, *19*, 500. [CrossRef]
39. Weber, J.L. Informativeness of human (dC-dA)n.(dG-dT)n polymorphisms. *Genomics* **1990**, *7*, 524–530. [CrossRef]
40. Li, Y.C.; Korol, A.B.; Fahima, T.; Beiles, A.; Nevo, E. Microsatellites: Genomic distribution, putative functions and mutational mechanisms: A review. *Mol. Ecol.* **2002**, *11*, 2453–2465. [CrossRef] [PubMed]
41. Wang, Z.; Weber, J.L.; Zhong, G.; Tanksley, S.D. Survey of plant short tandem DNA repeats. *Theor. Appl. Genet.* **1994**, *88*, 1–6. [CrossRef] [PubMed]
42. Morgante, M.; Hanafey, M.; Powell, W. Microsatellites are preferentially associated with nonrepetitive DNA in plant genomes. *Nat. Genet.* **2002**, *30*, 194–200. [CrossRef] [PubMed]
43. Li, Y.C.; Korol, A.B.; Fahima, T.; Nevo, E. Microsatellites within genes: Structure, function, and evolution. *Mol. Biol. Evol.* **2004**, *21*, 991–1007. [CrossRef] [PubMed]
44. Varshney, R.K.; Graner, A.; Sorrells, M.E. Genic microsatellite markers in plants: Features and applications. *Trends Biotechnol.* **2005**, *23*, 48–55. [CrossRef] [PubMed]
45. Zhao, Z.X.; Guo, C.; Sutharzan, S.; Li, P.; Echt, C.S.; Zhang, J.; Liang, C. Genome-Wide Analysis of Tandem Repeats in Plants and Green Algae. *G3 Genes Genomes Genet.* **2014**, *4*, 67–78. [CrossRef]
46. Cui, J.; Cheng, J.; Nong, D.; Peng, J.; Hu, Y.; He, W.; Zhou, Q.; Dhillon, N.P.S.; Hu, K. Genome-Wide Analysis of Simple Sequence Repeats in Bitter Gourd (*Momordica charantia*). *Front. Plant Sci.* **2017**, *8*, 1103. [CrossRef] [PubMed]
47. Kim, T.S.; Booth, J.G.; Gauch, H.G., Jr.; Sun, Q.; Park, J.; Lee, Y.H.; Lee, K. Simple sequence repeats in Neurospora crassa: Distribution, polymorphism and evolutionary inference. *BMC Genom.* **2008**, *9*, 31. [CrossRef]
48. Cheng, J.; Zhao, Z.; Li, B.; Qin, C.; Wu, Z.; Trejo-Saavedra, D.L.; Luo, X.; Cui, J.; Rivera-Bustamante, R.F.; Li, S.; et al. A comprehensive characterization of simple sequence repeats in pepper genomes provides valuable resources for marker development in Capsicum. *Sci. Rep.* **2016**, *6*, 18919. [CrossRef] [PubMed]
49. Tangphatsornruang, S.; Somta, P.; Uthaipaisanwong, P.; Chanprasert, J.; Sangsrakru, D.; Seehalak, W.; Sommanas, W.; Tragoonrung, S.; Srinives, P. Characterization of microsatellites and gene contents from genome shotgun sequences of mungbean (*Vigna radiata* (L.) Wilczek). *BMC Plant Biol.* **2009**, *9*, 137. [CrossRef] [PubMed]
50. Portis, E.; Lanteri, S.; Barchi, L.; Portis, F.; Valente, L.; Toppino, L.; Rotino, G.L.; Acquadro, A. Comprehensive Characterization of Simple Sequence Repeats in Eggplant (*Solanum melongena* L.) Genome and Construction of a Web Resource. *Front. Plant Sci.* **2018**, *9*, 401. [CrossRef] [PubMed]
51. Ferriol, M.; Pico, B.; Nuez, F. Genetic diversity of a germplasm collection of Cucurbita pepo using SRAP and AFLP markers. *Theor. Appl. Genet.* **2003**, *107*, 271–282. [CrossRef] [PubMed]
52. Ntuli, N.R.; Tongoona, P.B.; Zobolo, A.M. Genetic diversity in *Cucurbita pepo* landraces revealed by RAPD and SSR markers. *Sci. Hortic.* **2015**, *189*, 192–200. [CrossRef]
53. Decker, D.S. Origin (s), Evolution, and Systematics of *Cucurbita pepo* (Cucurbitaceae). *Econ. Bot.* **1988**, *42*, 4–15. [CrossRef]
54. Paris, H.S. Germplasm enhancement of *Cucurbita pepo* (pumpkin, squash, gourd: Cucurbitaceae): Progress and challenges. *Euphytica* **2015**, *208*, 415–438. [CrossRef]

Article

Genome-Wide Identification and Characterization of *Hsf* and *Hsp* Gene Families and Gene Expression Analysis under Heat Stress in Eggplant (*Solanum melongema* L.)

Chao Gong [1,†], Qiangqiang Pang [1,2,3,†], Zhiliang Li [1], Zhenxing Li [1], Riyuan Chen [2], Guangwen Sun [2,*] and Baojuan Sun [1,*]

1 Guangdong Key Laboratory for New Technology Research of Vegetables, Vegetable Research Institute, Guangdong Academy of Agricultural Sciences, Guangzhou 510640, China; gongchao@gdaas.cn (C.G.); qiangq113@163.com (Q.P.); vri_li@163.com (Z.L.); lizxgdaas@163.com (Z.L.)
2 College of Horticulture, South China Agricultural University, Guangzhou 510640, China; rychen@scau.edu.cn
3 Vegetable Research Institute, Hainan Academy of Agricultural Sciences, Haikou 571100, China
* Correspondence: sungw1968@scau.edu.cn (G.S.); baojuansun@gdaas.cn (B.S.); Tel.: +86-020-3846-9456 (B.S.)
† These authors contributed equally to this work.

Abstract: Under high temperature stress, a large number of proteins in plant cells will be denatured and inactivated. Meanwhile Hsfs and Hsps will be quickly induced to remove denatured proteins, so as to avoid programmed cell death, thus enhancing the thermotolerance of plants. Here, a comprehensive identification and analysis of the *Hsf* and *Hsp* gene families in eggplant under heat stress was performed. A total of 24 *Hsf*-like genes and 117 *Hsp*-like genes were identified from the eggplant genome using the interolog from Arabidopsis. The gene structure and motif composition of *Hsf* and *Hsp* genes were relatively conserved in each subfamily in eggplant. RNA-seq data and qRT-PCR analysis showed that the expressions of most eggplant *Hsf* and *Hsp* genes were increased upon exposure to heat stress, especially in thermotolerant line. The comprehensive analysis indicated that different sets of *SmHsps* genes were involved downstream of particular *SmHsfs* genes. These results provided a basis for revealing the roles of *SmHsps* and *SmHsp* for thermotolerance in eggplant, which may potentially be useful for understanding the thermotolerance mechanism involving *SmHsps* and *SmHsp* in eggplant.

Keywords: eggplant; heat shock factor (Hsf); heat shock protein (Hsp); heat stress; thermotolerance

Citation: Gong, C.; Pang, Q.; Li, Z.; Li, Z.; Chen, R.; Sun, G.; Sun, B. Genome-Wide Identification and Characterization of *Hsf* and *Hsp* Gene Families and Gene Expression Analysis under Heat Stress in Eggplant (*Solanum melongema* L.). *Horticulturae* **2021**, *7*, 149. https://doi.org/10.3390/horticulturae7060149

Academic Editor: Yuyang Zhang

Received: 16 May 2021
Accepted: 3 June 2021
Published: 10 June 2021

Publisher's Note: MDPI stays neutral with regard to jurisdictional claims in published maps and institutional affiliations.

Copyright: © 2021 by the authors. Licensee MDPI, Basel, Switzerland. This article is an open access article distributed under the terms and conditions of the Creative Commons Attribution (CC BY) license (https://creativecommons.org/licenses/by/4.0/).

1. Introduction

Plants live in complex environments where multiple abiotic stresses, such as salt, drought and extreme temperature, may seriously restrict their growth and development [1]. As sessile organisms, plants cannot move to avoid these stresses and, thus, they have developed mechanisms, such as enhanced expression of tolerance-related genes, in response to heat stress [2,3]. To survive and acclimatize under adverse environment conditions, plants have established self-defense mechanisms during the course of long-term evolution. Previous studies have shown that under heat stress (HS), plant cells respond rapidly to high temperatures by inducing the expression of genes encoding heat shock proteins (Hsps), which are involved in preventing heat-related damage and confer plant thermotolerance in strawberry, walnut, barley and grapevines [4–7]. Many Hsps function as molecular chaperones in preventing protein misfolding and aggregation, consequently maintaining protein homeostasis in cells and inducing acquired thermotolerance in plants [8]. The expression of Hsps is controlled and regulated by specific types of transcription factors called heat shock factors (Hsfs), which normally exist as inactive proteins [9].

Currently, many plant *Hsf* and *Hsp* genes from various species have been isolated and comprehensively studied. Based on their approximate molecular weights and sequence homologies, Hsps are classified into five families, namely, the small *Hsp* (*sHsp*), *Hsp60s*,

Hsp70s, *Hsp90s* and *Hsp100s* [10]. The expression of *sHsp* is positively correlated with thermostability [11]. As chaperones, *Hsp60* proteins participate in the folding and aggregation of many proteins transported to organelles, such as chloroplasts and mitochondria [12]. *Hsp70* chaperones, together with their co-chaperones, make up a set of prominent cellular machines that assist with a wide range of protein folding processes in almost all cellular compartments [13]. In Arabidopsis TU8 mutants, the downregulation of *Hsp90* expression leads to mutants that are more sensitive to heat. In *Arabidopsis thaliana* seedlings, fungi producing *Hsp90* inhibitors increase the expression of the *Hsp101* and *Hsp70* genes, resulting in the enhancement of plant heat resistance [14]. Arabidopsis has at least 21 *Hsf*s [15]. *HsfA1a*, *HsfA1b* and *HsfA1d* act as the main positive regulators of the heat shock response [16] and HsfA2 can enhance the thermotolerance of plants [17]. Above all, *Hsfs* and *Hsps* play crucial roles in plant thermotolerance. The *Hsf* and *Hsp* gene families have been extensively studied in the model plant *Arabidopsis thaliana* and in non-model plants, such as rice (*Oryza sativa*) [18], poplar (*Populus trichocarpa*) [19], maize (*Zea mays*) [20] and Chinese cabbage (*Brassica rapa* L. ssp. *pekinensis*) [21].

Eggplant (*Solanum melongema* L.) is an important economic solanaceous crop, ranking third, after potato and tomato. Eggplant is primarily cultivated in East Asia, South Asia, the Middle East and northern Africa. The optimal temperature for eggplant growth and development ranges from 22 °C to 30 °C. With global warming, the temperature in subtropical and tropical regions is often above 35 °C, resulting in serious heat injury in eggplant, including limited plant growth, reduced productivity and damaged quality [2]. Thermotolerance is an important agronomic trait for eggplants, but the molecular mechanisms of heat tolerance remain elusive. Hsfs and Hsps play core roles in the signal transduction pathways involved in plant response to heat stress. Due to the vital regulatory functions of *Hsf* and *Hsp* genes in plant responses to heat stress, *Hsf* and *Hsp* genes in eggplant under heat stress were studied. The eggplant genome was sequenced and assembled [22], enabling the characterization of the eggplant *Hsf* and *Hsp* families and their responses to heat stresses at the molecular level. Therefore, genome-wide identification of *Hsf* and *Hsp* genes in eggplant was conducted to infer their expansion and evolutionary history. RNA-seq data and quantitative real-time RT-PCR analyses were used to explore their expression difference in the thermotolerant line 05-4 and the thermosensitive line 05-1 as elicited by naturally increased temperature. The results provide a relatively complete profile of the *Hsf* and *Hsp* gene families in eggplant and elucidate their relationship with thermotolerance, which provides a foundation for further functional research on these genes in eggplant. Furthermore, these findings could potentially be useful for understanding the mechanism of thermotolerance mediated by Hsfs and Hsps in eggplant.

2. Materials and Methods

2.1. Identification and Classification of Hsf and Hsp Family Members in Eggplant

Published Arabidopsis *Hsf* and *Hsp* sequences [23] were retrieved and used as queries in BLAST searches against the eggplant genome database (http://eggplant.kazusa.or.jp/, accessed on 6 June 2021) to identify potential eggplant *Hsfs* and *Hsps*. All output genes identified according to Arabidopsis *Hsf* and *Hsp* sequences were collected and confirmed using Pfam (http://pfam.xfam.org/search, accessed on 6 June 2021) and SMART (http://smart.embl-heidelberg.de/, accessed on 6 June 2021). The isoelectric points and molecular weights were predicted using the Compute pI/Mw tool from ExPASy (http://web.expasy.org/compute_pi, accessed on 6 June 2021).

2.2. Phylogenetic Analysis

Alignments of the full eggplant Hsf and Hsp proteins were performed using clustal X2.1 [24]. Phylogenetic trees were constructed using the neighbor-joining (NJ) method in MEGA (version 5.0) [25] with bootstrap values from 1000 replicates indicated at each node. To identify signature domains, the Hsf and Hsp protein sequences were compared with Arabidopsis and tomato. *SmHsfs* and *SmHsps* (*sHsp*, *Hsp60s*, *Hsp70s*, *Hsp90s* and *Hsp100s*) were named based on the subfamily classification and their phylogenetic relationships with the corresponding *AtHsfs* and *AtHsps* and gene names of eggplant *sHsps* were revised according to their molecular weights in the eggplant genome database based on Hirakawa et al. [22].

2.3. Gene Structures, Conserved Motifs and Protein Functional Network Analysis

The exon and intron structures were illustrated using the Gene Structure Display Server (GSDS, http://gsds.cbi.pku.edu.cn accessed on 6 June 2021) [26] by aligning the predicted cDNA sequences with their corresponding genomic DNA sequences. The conserved motifs in the encoded proteins were analyzed using the MEME online program (http://meme.sdsc.edu, v4.9.0, accessed on 6 June 2021) [27]. MEME was run locally with the following parameters: number of repetitions = any, maximum number of motifs = 20 and optimum motif width = 6–100 residues for *Hsf*, *sHsp*, *Hsp60*, *Hsp70* and *Hsp100*. The STRING protein interaction database (http://string-db.org/, accessed on 6 June 2021) was used to analyze the interaction networks of Hsf and Hsp proteins in the highly specific protein and parameter selection model plant species *Arabidopsis thaliana*.

2.4. Plant Materials, Growth Conditions and Stress Treatments

In the present study, two inbred eggplant lines (selected by the Vegetable Research Institute, Guangdong Academy of Agricultural Sciences, Guangzhou, China), the thermotolerant line 05-4 and the thermosensitive line 05-1, were used. Eggplant seedlings were cultivated under 25/20 °C day/night conditions and a 16/8 h day/night photoperiod in a growth chamber until the four true leaves period for treatments. For the HS treatment, the seedlings of 05-1 and 05-4 with four leaves were directly placed in the 42 °C light incubator (RXZ-1000B3, Jiangnan Instrument Factory, Ningbo, China). For the heat treatment used for RNA-seq, the 3rd mature leaves of treated seedlings were collected at 0 and 6 h after HS treatment and 10 plants were used for each treatment. For the heat treatment used for the qRT-PCR, the 3rd mature leaves from two different lines were harvested at 0 and 6 h. The samples were harvested, immediately frozen in liquid nitrogen and stored at −80 °C for RNA extraction. Three biological replicates were performed and each replicate had 10 plants.

2.5. RNA Extraction and Quantitative Real-Time PCR Analysis

Total RNA was extracted using a TransZol Plant kit (TransGen Biotech/TransBionovo, Beijing, China) and the cDNA was synthesized according to the manufacturer's instructions (Takara, Dalian, China). Primers with amplicon lengths of 80–150 bp were designed using Primer5 software. All primer sequences are listed in Table S14. Real-time qRT-PCR was conducted on a Real-Time PCR Detection System (Bio-Rad, Hercules, CA, USA) using the SYBR Premix Ex Taq kit (TaKaRa, Dalian, China) according to the manufacturer's instructions. The 10 μL reaction system contained 5 μL of SYBR Green Supermix (2×), 4 μL of cDNA template (30 ng/μL) and 0.5 μL of each primer (10 μM). The qRT-PCR reaction was performed using the following parameters: pre-denaturation at 95 °C for 30 s, followed by 39 cycles of denaturation at 95 °C for 5 s, annealing at 60 °C for 15 s and extension at 72 °C for 15 s. The fluorescent signal was measured at the end of each cycle and the melting curve analysis was performed by heating the PCR product from 65 °C to 90 °C to verify the specificity of the primers. Three independent biological replicates were performed and the qPCR of each replicate was performed in triplicate. The relative expression levels of eggplant *Hsf* and *Hsp* genes were calculated using the $2^{-\triangle Ct}$ method [28]. The *SmEF1a* genes were used as internal controls.

3. Results
3.1. Genome-Wide Identification and Analysis of Hsf and Hsp Gene Family Members in Eggplant

To search for *Hsf* and *Hsp* genes in eggplant, we used the conserved Hsf and Hsp domain consensus sequences of several proteins as BLASTP queries against the eggplant genome database (http://eggplant.kazusa.or.jp/, accessed on 6 June 2021). In addition, homology searches using identified protein sequences of *Arabidopsis thaliana* were performed. After automated database searching and a manual review, 24 and 117 genes were identified as members of the *Hsf* and *Hsp* families in eggplant, respectively, whose classification and naming were based on the rules of the *Hsp* gene families from Arabidopsis and tomato, including *sHsp*, *Hsp60*, *Hsp70*, *Hsp90* and *Hsp100*. The *Hsf* and *Hsp* gene families in eggplant were relatively large compared with those in Arabidopsis and those in tomato and rice, respectively. The numbers of identified genes in the *Hsf*, *sHsp*, *Hsp60*, *Hsp70*, *Hsp90* and *Hsp100* families of eggplant were 24, 39, 21, 30, 17 and 10, respectively (Table 1).

Table 1. Numbers of *Hsf* and *Hsp* genes in Arabidopsis, eggplant, tomato and rice.

Family	Arabidopsis	Eggplant	Tomato	Rice
Hsf	22	24	23	25
Hsp20	27	39	23	39
Hsp60	18	21	16	20
Hsp70	19	30	22	24
Hsp90	7	17	8	9
Hsp100	8	10	13	10

As shown in Supplementary Materials Table S1, the amino acid lengths for Hsfs ranged from 111 (*SmHsfA1c*) to 496 (*SmHsfA1b*), with deduced molecular weights from 12.2 kDa to 54.8 kDa and the predicted isoelectric points of Hsfs were divergent, ranging from 4.60 (*SmHsfA3*) to 9.64 (*SmHsfA1d*). The length of sHsp proteins ranged from 87 (*Sm10.2-sHsp*) to244 amino acids (*Sm27.2-sHsp*) and the predicted molecular weights were between 10.2 kDa (*Sm10.2-sHsp*) and 27.2 kDa (*Sm10.2-sHsp*). In addition, the predicted pI-values of sHsp proteins ranged from 4.56 (*Sm10.2-sHsp*) to 10.49 (*Sm12.7-sHsp*). The amino acids lengths were consistent with the molecular weights of Hsp60s. The amino acid number and molecular weight for Smcpn60-4 was the highest, while that for SmCpn60-7.3 was the lowest and the predicted pI-values ranged from 5.26 (*SmCpn60-a1*) to 10.29 (*SmCpn60-7.3*). The deduced length of the Hsp70 proteins ranged from 85 (*SmmtHsc70-3*) to 914 (*SmHsp70-18*) amino acids and the highest- and lowest-molecular-weight SmHsp70s were *SmHsp70-18* (103.1 kDa) and *SmHsp70-5* (11.5 kDa), respectively, while the pI values ranged from 4.52 (*SmmtHsc70-3*) to 9.35 (*SmHsp70-19*). The length of Hsp90 proteins ranged from 137 (*SmHsp90-4.4*) to 782 (*SmHsp90-6*) amino acids, the predicted molecular weights of Hsp90s were between 16.2 kDa (*SmHsp90-4.4*) and 89.6 kDa (*SmHsp90-7.1*) and the predicted isoelectric points ranged from 4.78 (*SmHsp90-5*) to 9.55 (*SmHsp90-2.1*). The longest amino acids lengths and highest molecular weights in Hsp100s were *SmHsp100-ClpB1*, with 979 amino acids and 110.2 kDa, respectively. In contrast, the smallest was *SmHsp100-ClpC3*; the predicted isoelectric points ranged from 5.38 (*SmHsp100-ClpB3*) to 9.07 (*SmHsp100-ClpC1*) and these proteins were distributed from the alkaline to acidic.

3.2. Phylogenetic and Sequence Structure Analysis of Hsf and Hsp Proteins in Eggplant

To evaluate the evolutionary relationship of the eggplant Hsf and Hsp proteins, a phylogenetic analysis of each family was performed based on the full-length amino acid sequences from Arabidopsis, eggplant and tomato and each family could be classified into different subfamilies. The *SmHsf* family contained three subfamilies: type A (18 genes), type B (5 genes) and type C (1 gene). Based on the phylogenetic tree, class HsfA had the maximum number of subclasses among the three classes and was closer to tomato Hsf proteins, which coincided with the botanical classification (Table S2). A total of 39 *sHsp* genes could be grouped into 12 distinct subfamilies, containing 6 groups of cytosolic *sHsp* genes, C-I, C-II, C-III, C-IV, C-V and C-VI and 2 groups of mitochondrial *sHsp* genes, MT I and MT II. Notably, the C-I *sHsp* group in the eggplant genome was large, containing 24 genes, compared with 6 in Arabidopsis (Table S3). The *Hsp60* family was divided into 4 subfamilies, including cytosol-localized Cpn60 (12 genes), mitochondrion-localized Hsp60

(4 genes) and chloroplast-localized Cpn60-a (2 genes) and Cpn60-b (3 genes) (Table S4). The *Hsp70* family contains genes encoding 19 cytosolic *Hsp70s*, 4 binding proteins (BIPs, Hsp70 homologs in the ER), 3 mitochondrial Hsp70s (mtHsc70s) and 2 chloroplastid Hsp70s (cpHsc70s) (Table S5). Seventeen *Hsp90* family genes could be divided into cytoplasm (Cyt), mitochondrial (MT), endoplasmic reticulum (ER) and chloroplast, containing 8, 3, 2 and 1 proteins, respectively (Table S6). The Hsp100 family can be classified into ClpB, C, D and X classes as follows: 3 ClpB proteins (designated as B1, B2 and B3), 4 ClpC proteins (C1, C2, C3 and C4), 1 ClpD protein (D1) and 2 ClpX proteins (X1 and X2) (Table S7).

3.3. Structure of Hsf and Hsp Genes and Conserved Motifs of Hsf and Hsp Proteins in Eggplant

To obtained further insights into the structural diversity of *Hsf* and *Hsp* genes in eggplant, we used the Multiple Expectation maximization for Motif Elicitation (MEME) [27] to predict the conserved motifs shared among the related proteins within these families. In each family, 20 putative motifs were identified. The details of these motifs are listed in Tables S8–S13. Most of the closely related members in the phylogenetic tree shared common motif compositions.

The exon/intron structures of eggplant *Hsf* and *Hsp* members were analyzed based on their coding sequences and the corresponding genome sequences. The eggplant *Hsfs* shared highly conserved exon/intron structures with 0–3 intron phases (Figure 1A). The intron phases were remarkably well conserved among family members. Most of the eggplant *sHsps* did not contain introns and only a few had 1–3 introns (Figure 1B). Interestingly, in the *Hsp60* family, two members, *SmCpn60-8* and *SmHsp60-3*, had no introns in their coding regions, while the other eggplant Hsp60s contained several introns (1–22) (Figure 1C). In the *Hsp70* family, cytosolic *Hsp70s* had 0–13 introns, ER-localized BIPs had 4–7 introns, mitochondrion-localized *mtHsc70s* had 0–4 introns and chloroplast-localized *cpHsc70s* had 6 introns, while truncated *Hsp70ts* had no introns (Figure 2A). With the exception of *SmHsp90-1.1*, each *Hsp90s* member contained 1–17 introns (Figure 2B). The number of exons and introns of *Hsp100* family members differed greatly. For example, *SmHsp100-ClpX1* contained up to 16 introns, but *SmHsp100-ClpB3* and *SmHsp100-ClpC3* only had 4 introns (Figure 2C).

Figure 1. Phylogenetic relationships, gene structures and motif compositions of *Hsf*, *sHsp* and *Hsp60* family members in eggplant. Multiple alignment of the Hsf (**A**), sHsp (**B**) and Hsp60 (**C**) proteins from eggplant (Sm) was performed with MEGA 5.0 using the neighbor-joining (NJ) method with 1000 bootstrap replicates (left panel). A schematic representation of conserved motifs (obtained using MEME) in the Hsf and sHsp proteins is displayed in the middle panel. Different motifs are represented by differently colored boxes. Details of the individual motifs are in Tables S8–S10. Exon/intron structures of the *Hsf* and *sHsp* genes are shown in the right panel. Green boxes represent exons and black lines represent introns.

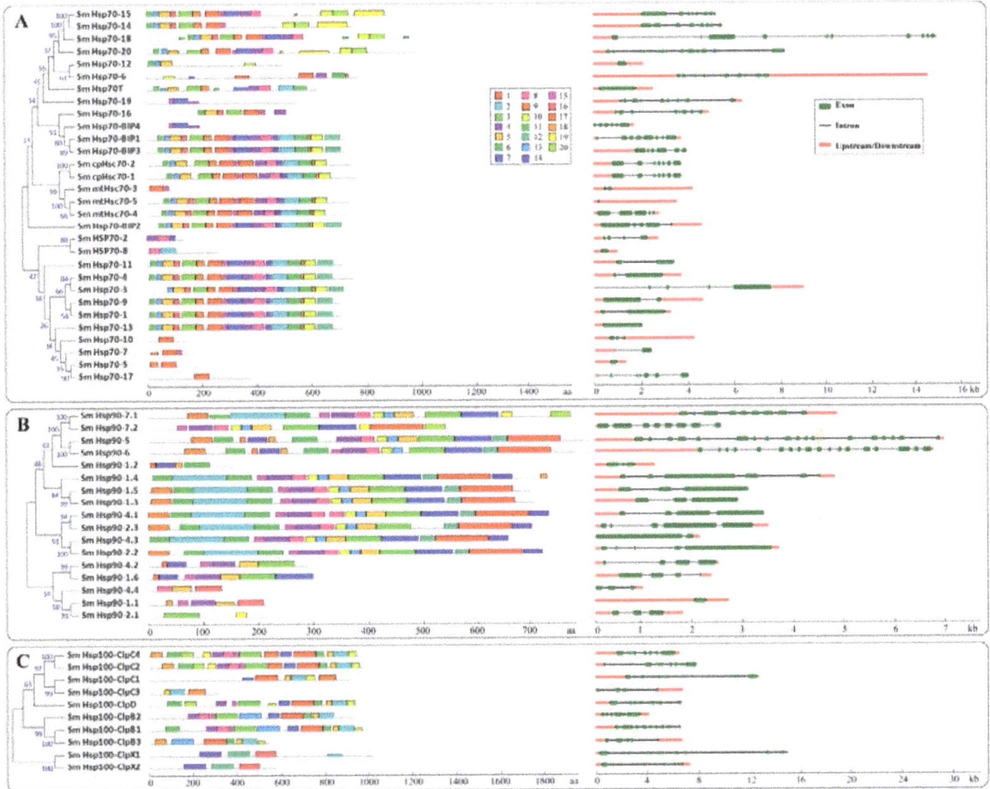

Figure 2. Phylogenetic relationships, gene structures and motif compositions of the Hsp70, Hsp90 and Hsp100 family members in *S. melongena* (Sm). Multiple alignment of the Hsp70 (**A**), Hsp90 (**B**) and Hsp100 (**C**) proteins from *S. melongena* (Sm) was performed with MEGA 5.0 using the neighbor-joining (NJ) method with 1000 bootstrap replicates (left panel). A schematic representation of conserved motifs (obtained using MEME) in the Hsp70 (**A**), Hsp90 (**B**) and Hsp100 (**C**) proteins is displayed in the right panel. Different motifs are represented by differently colored boxes. Details of the individual motifs are in Tables S11–S13. The exon/intron structures of the Hsp70 (**A**), Hsp90 (**B**) and Hsp100 (**C**) genes are shown in the middle panel. Green boxes represent exons and black lines represent introns.

3.4. Expression Patterns of Eggplant Hsf and Hsp Genes

To examine the heat response for *Hsfs* and *Hsps* in eggplant, an RNA sequencing profile (data not shown) in leaves of thermosensitive line 05-1 and thermotolerant line 05-4, at 0 and 6 h after HS treatment, was used. *Hsf* and *Hsp* genes were selected according to annotations and their expression profiles were analyzed. We analyzed the transcription levels of 18 *Hsf*, 25 *sHsp*, 6 *Hsp60*, 18 *Hsp70*, 11 *Hsp90* and 6 *Hsp100* genes in the leaves. As shown in Figure 3, for the thermosensitive line 05-1, 16 genes (89%) of the *Hsf* family were upregulated and two members, *SmHsfA4e* and *SmHsfB3*, were downregulated under HS conditions, which were more than in the thermotolerant line 05-4, in which 17 *Hsf* genes (94%) were upregulated by HS and only *SmHsfA8* was downregulated. In contrast to line 05-1, *SmHsfA4e* and *SmHsfB3* were strongly induced in treated 05-4 leaves. In the leaves of the thermotolerant line 05-4, among the upregulated members, the expression levels of most A (A1a, A1b, A3, A4a, A4b, A4d, A4e, A5, A6a and A6b), B1 and B2a were higher than those of other members under HS.

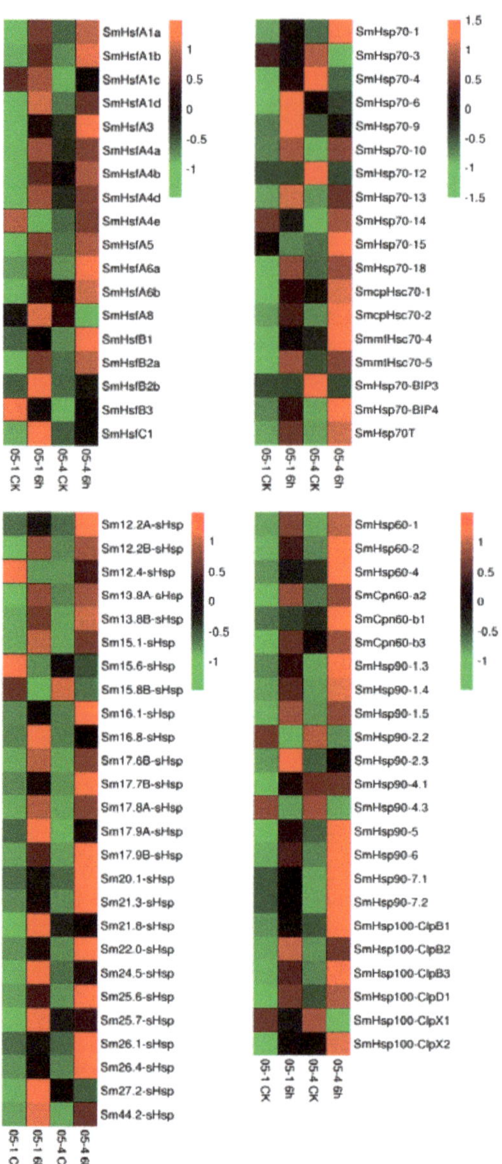

Figure 3. Expression analysis of eggplant *Hsf* and *Hsp* genes. Raw data were from RNA-seq data, in response to HS treatment in 05-1 and 05-4 leaves. HS treatment: 42 °C for 6 h; 05-1: eggplant thermosensitive line; 05-4: eggplant thermotolerant line.

A strong response to HS in all of the 25 *sHsp* genes from both lines (05-1 and 05-4) was observed, in which a majority of these genes were upregulated and only *Sm15.6-sHsp* and *Sm15.8-sHsp* were downregulated. After high-temperature treatment, the expression of *SmHsp60-1*, *SmHsp60-2*, *SmCpn60-a2* and *SmCpn60-b3* was increased in the two inbred lines, while *SmHsp60-4* and *SmCpn60-b3* was increased in 05-4 and no significant difference could be observed in 05-1. Among the 18 *Hsp70* genes, the expression was remarkably changed in response to heat treatment in the thermosensitive 05-1 and thermotolerant

05-4 leaves and these genes were upregulated in both plants. Among these upregulated genes, the expression quantity of *SmHsp70-1*, *SmcpHsp70-2* and *SmHsp70-BIP4* was higher in 05-4, compared with 05-1. However, *SmHsp70-14* and *SmHsp70-15* were increased in the thermotolerant line, but decreased in the thermosensitive line. Considering the Hsp90 genes, the expression levels of most genes (*SmHsp90-1.3*, *SmHsp90-1.4*, *SmHsp90-1.5*, *SmHsp90-2.3*, *SmHsp90-5*, *SmHsp90-6*, *SmHsp9-7.1* and *SmHsp90-7.2*) were increased and only *SmHsp90-2.2* and *SmHsp90-4.3* were downregulated in the two lines. Among the upregulated *Hsp90* genes, gene expression levels of six genes in 05-4 were obviously higher than in 05-1. After heat treatment, *SmHsp100-ClpB1*, *SmHsp100-ClpB2*, *SmHsp100-ClpB3*, *SmHsp100-ClpD1* and *SmHsp100-ClpX2* expressions in the two lines were significantly increased. Among these genes, *SmHsp100-ClpB1*, *SmHsp100-ClpB3* and *SmHsp100-ClpX2* showed higher expression in 05-4 than in 05-1, but *SmHsp100-ClpB2* was more abundant in the thermosensitive line.

3.5. Validation of Hsf and Hsp Gene Expression Levels by qRT-PCR

To verify the accuracy of the transcriptome sequencing, the expressions of 12 randomly selected genes were validated using quantitative real-time RT-PCR (qRT-PCR). The results showed that the expression pattern of each tested gene was similar to that of the transcriptome sequencing and the increase rate of all these *Hsf* and *Hsp* genes in the thermotolerant line 05-4 were significantly higher than those in thermosensitive line 05-1 (Figure 4).

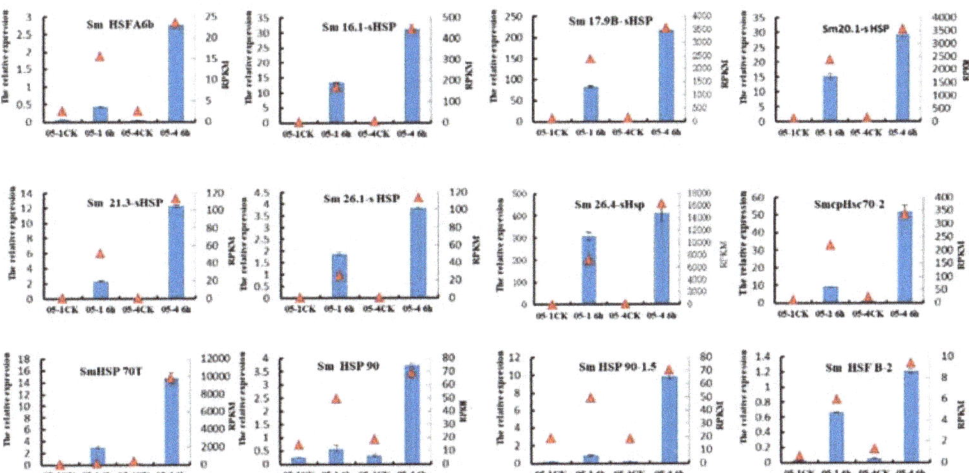

Figure 4. Comparison of transcripts expression results from RNA-seq and qRT-PCR analysis. Abscissa: Sample number; the ordinate (left): the relative expression of gene validated using qRT-PCR, represented by bar chart; coordinates (right): RPKM value obtained from the transcriptome sequencing, represented by triangle scatter diagram.

4. Discussion

Many studies have suggested that *Hsfs* and *Hsps* play central roles in plant developmental and defense processes [29,30]. Benefiting from genome availability, the functions of the *Hsf* and *Hsp* family genes have been characterized in many plants. Although *Hsfs* and *Hsps* exist in all living organisms, their numbers vary in different plants. There are 22 *Hsfs* in Arabidopsis, 25 *Hsfs* in rice [18], 30 *Hsfs* in maize [20], 25 *Hsfs* in pepper [31] and 52 *Hsfs* in soybean [32]. Compared to the 27 *sHsp* genes in Arabidopsis [33], there are 35, 51 and 27 *sHsp* genes in pepper [31], soybean [34] and Chinese cabbage [35], respectively. Previous studies have identified 18 *Hsp70* genes in Arabidopsis and 32 genes in rice [36]. The grapevine genome contains at least seven genes encoding members of the

Hsp90 super family [37]. Zhang et al. (2015) reported 28 *Hsf*, 37 *sHsp*, 28 *Hsp60*, 20 *Hsp70* and 5 *Hsp100* genes in the poplar genome [19]. However, with the limited investigations into the molecular mechanism of heat tolerance, little is known about the *Hsf* family in eggplant.

In the present study, we identified 24 *Hsf* genes, 39 *sHsp* genes, 21 *Hsp60* genes, 30 *Hsp70* genes, 17 *Hsp90* genes and 10 *Hsp100* genes based on the eggplant genome (Table 1). Although the total number of *Hsf* and *Hsp* genes was similar to that of Arabidopsis [18,38–40], rice [18,41] and tomato [42], the members of some specific *Hsf* and *Hsp* subclasses in eggplant were different from the other three species. Two members were identified that belonged to subclass HsfC2 in rice, while no eggplant *Hsf* members were classified into subclass HsfC2 and the same events were also observed in *Arabidopsis thaliana* [18] and pepper [31]. Rice is the model plant use for the monocot lineage and we inferred that the gene duplications led to the unique HsfC2 subclass in monocot species [17,42], which was the most marked difference between monocots and eudicots. In contrast, similar to tomato and *Arabidopsis thaliana* [18,43], eggplant also has members that were partitioned into the HsfA6 subclass, but no rice *Hsf* members were classified into subclass HsfA6 [44]. This finding suggested that *Hsf* genes were doubled and gained new functions during the evolution of the eggplant genome. Another interesting observation was that the subclass *HsfA9* had 1 member in eggplant, compared with 4 members in pepper [31] and *Eucalyptos grandis* (Myrtaceae) contained at least 17 closely related *HsfA9*-encoding genes [17], suggesting a gene loss event during the evolutionary process of eggplant. However, there were two *HsfA4* subclass genes in eggplant, more than in pepper *CaHsfA4*, which showed that some *Hsfs* might have the similar functions, as in maize [20]. The reasons for the increase in the *HsfA9* genes need further investigation.

The phylogenetic analysis revealed that eggplant *Hsf* and *Hsp* members were more closely related to those from tomato than to those from Arabidopsis, which was consistent with the fact that both eggplant and tomato are members of the Solanaceae family [45]. Based on the previous analysis of the evolution of *Hsfs* and *Hsps* in Chinese cabbage [21,35], rice [46] and soybean [47], *Hsf* and *Hsp* genes essentially cover all the subfamilies and are relatively stable and conserved in the evolutionary process of eggplant and most of the *Hsf* and *Hsp* gene families were closely related to the evolutionary species.

Divergences in coding regions, particularly those that change the function of the gene, reflect amino acid altering substitutions and/or alterations in exon–intron structure [19]. The differences in intron and exon structure play important roles in the evolution of family genes. Structural analyses showed that the eggplant *Hsf* genes contained 0–7 introns and there were significant differences in the intron length; similar results were also obtained in cucumber [48], rice [49] and chickpea [50], but this result was different from that of pepper [31], for which all members have one intron. The number of introns of the *Hsp* gene family members in eggplant also showed differences, similar to the results of previous studies on poplar *sHsp*, *Hsp60*, *Hsp70* and *Hsp100* [19]. Qiao (2015), researching the pear *Hsf* and Guo (2015), researching the pepper *Hsp20*, showed a lack of conserved motifs among all the family genes and none of these genes contained the whole sequence, consistent with the eggplant *Hsfs* and *Hsps* in the present study [31,51]. We speculated that the deletion of introns and domains leads to structural changes during evolution, leading to functional diversity in *Hsf* and *Hsp* genes in eggplant; however, this theory needs experimental confirmation.

Hsfs, as transcriptional activators of *Hsps*, cooperate with *Hsps* to form a network responding to various stresses. These factors play a broad role in the tolerance to multiple environmental stress treatments apart from heat stress [52,53]. The comprehensive analysis of the expression for individual *Hsf* and *Hsp* members under HS was necessary for further functional analyses in plant thermotolerance [23,54]. The present study showed that most members of the eggplant *sHsp*, *Hsp60*, *Hsp70*, *Hsp90* and *Hsp100* families were induced by HS treatment in lines 05-1 and 05-4 and only a few members were significantly downregulated. Several studies have indicated that the expression and accumulation of heat shock proteins and heat shock transcription factors can enhance the thermostability of tomato [55], wheat [56] and rice [57]. *Hsfs* are activated under HS conditions and subsequently bind the HSE elements of the promoters of the *Hsp* genes to regulate the expression of downstream *Hsp* genes [17]. The accumulation of the *Hsps* effectively reduces the damage from HS and enhances thermotolerance by binding denaturing proteins and preventing them from irreversible aggregation [58]. Thus far, only *sHsp* has been shown to play a major role in improving plant thermotolerance in the form of molecular chaperones and cell membrane stabilizing factors [59]. However, the specific mechanisms of other *Hsp* genes are less well established. Previous studies have shown that the response of plants to high temperature was a quantitative trait controlled by multiple genes; some normal genes were closed and some stress tolerance-related genes were induced under high-temperature stress, thus altering plant morphogenesis, physiological functions and biochemical and molecular structures, which in turn influenced the growth of plants [60]. In addition, heat shock proteins are different from other stress proteins and have their own unique characteristics. In the present study, *Hsps* (*sHsp*, *Hsp60*, *Hsp70*, *Hsp90* and *Hsp100*) showed species diversity, universal distribution and instantaneous response and structural conservation. For example, the synthesis of heat shock protein was fast, beginning between the first few minutes and tens of minutes and the expression lasted for up to several hours, occasionally continuing for 12 or more hours (Figure 4). Similar results were also observed in poplar [19] and grape [61].

In Arabidopsis, there are four members of the HsfA1 family, A1a, A1b, A1c and A1d [62]. Studies have shown that HsfA1a can directly sense heat stress and become activated and the same treatments also induced the binding to *Hsp18.2* and *Hsp70* promoters, as examined by chromatin immunoprecipitation [63]. Overexpressing *HsfA1a* enhances diverse stress tolerance by promoting stress-induced *Hsp18.2* and *Hsp70* gene expression [64]. In addition, *AtHsfA1* was also related to drought stress [65] and programmed cell death [66]. Thus, in eggplant, *HsfA1* may also play a similar function to *AtHsfA1* and simultaneously communicate with *Hsps*. Increasing evidence suggests that *Hsp* is one of the most important heat stress proteins regulated by *Hsf* and is the material basis of the response of plant cells to high temperature damage [67–69]. Once exposed to high temperature, most of *Hsf* and *Hsp* genes in eggplant were induced to express rapidly and the expression level of these genes in the thermotolerant line was much higher than that in the thermosensitive line. Therefore, the Hsf–Hsp involved protein degradation pathway is also the main pathway of eggplant response to high temperature stress and may play an important role in the production of heat-tolerance in eggplant. The results provide a foundation for further functional research of these genes in eggplant, which could potentially be useful for elucidating the mechanism of thermotolerance in eggplant, even in other solanaceous plants.

5. Conclusions

In the present study, 24 *Hsf* genes and 117 *Hsp* genes (including *sHsp*, *Hsp60*, *Hsp70* and *Hsp100*) were identified from the eggplant genome. The phylogeny, gene structure, expression profiles and heat stress responses of these genes were analyzed. The total number of *Hsf* and *Hsp* genes of eggplant was similar to that of Arabidopsis, tomato and rice, covering all the subfamilies, and the gene structure and motif composition were relatively stable and conserved in the evolutionary process. *SmHsf* genes, as key transcriptional activators of *Hsp* genes, regulated different subfamilies of *Hsps* in eggplant. Most of *Hsf* and *Hsp* genes are highly induced by HS in eggplant leaves, which indicated these genes participate in the response to heat stress. The expression levels of these genes in the thermotolerant line were enhanced significantly higher than those of in the thermosensitive line under HS in eggplant, which may be the main reason for strong thermotolerance in thermotolerant eggplant. According to the above results, it is expected to evaluate the thermotolerance of different eggplant resources by analyzing the expression change of specific *Hsf* and *Hsp* genes under HS and even the thermotolerance of other solanaceae species resources. The present study was undertaken to establish a solid foundation for functional research on the eggplant *Hsf* and *Hsp* gene families and broaden our understanding of the mechanism of thermotolerance mediated by *Hsf* and *Hsp* genes in solanaceous plants.

Supplementary Materials: The following are available online at https://www.mdpi.com/article/10.3390/horticulturae7060149/s1, Table S1: *Hsf* and *Hsp* gene families in eggplant. Table S2: Phylogenetic analysis of Hsf proteins in eggplant, Arabidopsis, and tomato. Table S3: Phylogenetic analysis of sHsp proteins in eggplant, Arabidopsis, and tomato. Table S4: Phylogenetic analysis of Hsp60 proteins in eggplant, Arabidopsis, and tomato. Table S5: Phylogenetic analysis of Hsp70 proteins in eggplant, Arabidopsis, and tomato. Table S6: Phylogenetic analysis of Hsp90 proteins in eggplant, Arabidopsis, and tomato. Table S7: Phylogenetic analysis of Hsp100 proteins in eggplant, Arabidopsis, and tomato. Table S8: Sequence logos for the conserved motifs of Hsf proteins in Arabidopsis and eggplant. Table S9: Sequence logos for the conserved motifs of sHsp proteins in Arabidopsis and eggplant. Table S10: Sequence logos for the conserved motifs of Hsp60 proteins in Arabidopsis and eggplant. Table S11: Sequence logos for the conserved motifs of Hsp70 proteins in Arabidopsis and eggplant. Table S12: Sequence logos for the conserved motifs of Hsp90 proteins in Arabidopsis and eggplant. Table S13: Sequence logos for the conserved motifs of Hsp100 proteins in Arabidopsis and eggplant. Table S14: Primers used in qRT-PCR analysis.

Author Contributions: Conceptualization, B.S.; methodology, G.S.; software, R.C.; validation, C.G., G.S. and B.S.; formal analysis, B.S.; investigation, Q.P.; resources, Z.L. (Zhiliang Li) and Z.L. (Zhenxing Li); data curation, Q.P.; writing—original draft preparation, C.G. and Q.P.; writing—review and editing, C.G., B.S. and G.S.; visualization, G.S.; supervision, B.S.; project administration, Z.L.; funding acquisition, C.G. and B.S. All authors have read and agreed to the published version of the manuscript.

Funding: This research was funded by the National Natural Science Foundation of China, grant number 31801875, the Natural Science Foundation of Guangdong Province, grant number 2020A1515011168, the Department of agriculture and rural areas of Guangdong province of China, grant number 2018KCZX06, 2020KJ106 and 2020KJ110 and the Special fund for scientific innovation strategy-construction of high level Academy of Agriculture Science, grant number R2019PY-QF009, R2018QD-040.

Data Availability Statement: The data supporting the findings of this study are available from the corresponding author upon reasonable request.

Conflicts of Interest: The authors declare no conflict of interest.

References

1. Cramer, G.R.; Urano, K.; Delrot, S.; Pezzotti, M.; Shinozaki, K. Effects of abiotic stress on plants: A systems biology perspective. *BMC Plant Biol.* **2011**, *11*, 163. [CrossRef]
2. Li, Y.; Li, Z.; Luo, S.; Sun, B. Effects of heat stress on gene expression in eggplant (*Solanum melongema* L.) seedlings. *Afr. J. Biotechnol.* **2011**, *10*, 18078–18084. [CrossRef]
3. Zhang, A.; Zhu, Z.; Shang, J.; Zhang, S.; Shen, H.; Wu, X.; Zha, D. Transcriptome profiling and gene expression analyses of eggplant (*Solanum melongena* L.) under heat stress. *PLoS ONE* **2020**, *15*. [CrossRef] [PubMed]
4. Kesici, M.; Ipek, A.; Ersoy, F.; Ergin, S.; Gülen, H. Genotype-dependent gene expression in strawberry (*Fragaria x ananassa*) plants under high temperature stress. *Biochem. Genet.* **2020**, *58*, 848–866. [CrossRef] [PubMed]
5. Yang, G.; Gao, X.; Ma, K.; Li, D.; Jia, C.; Zhai, M.; Xu, Z. The walnut transcription factor *JrGRAS2* contributes to high temperature stress tolerance involving in Dof transcriptional regulation and HSP protein expression. *BMC Plant Biol.* **2018**, *18*, 1. [CrossRef]
6. Sadura, I.; Libik-Konieczny, M.; Jurczyk, B.; Gruszka, D.; Janeczko, A. HSP transcript and protein accumulation in brassinosteroid barley mutants acclimated to Low and high temperatures. *IJMS* **2020**, *21*, 1889. [CrossRef]
7. Zha, Q.; Xi, X.; Jiang, A.; Wang, S.; Tian, Y. Changes in the protective mechanism of photosystem II and molecular regulation in response to high temperature stress in grapevines. *Plant Physiol. Bioch.* **2016**, *101*, 43–53. [CrossRef] [PubMed]
8. Sung, D.-Y.; Kaplan, F.; Lee, K.-J.; Guy, C.L. Acquired tolerance to temperature extremes. *Trends Plant Sci.* **2003**, *8*, 179–187. [CrossRef]
9. Wang, F.; Dong, Q.; Jiang, H.; Zhu, S.; Chen, B.; Xiang, Y. Genome-wide analysis of the heat shock transcription factors in *Populus trichocarpa* and *Medicago truncatula*. *Mol. Biol. Rep.* **2012**, *39*, 1877–1886. [CrossRef] [PubMed]
10. Lindquist, S.; Craig, E.A. The heat-shock protein. *Annu. Rev. Genet.* **1988**, *22*, 631–677. [CrossRef] [PubMed]
11. Lopez-Matas, M.A.; Nuñez, P.; Soto, A.; Allona, I.; Casado, R.; Collada, C.; Guevara, M.A.; Aragoncillo, C.; Gomez, L. Protein cryoprotective activity of a cytosolic small heat shock protein that accumulates constitutively in chestnut stems and is up-regulated by low and high temperatures. *Plant Physiol.* **2004**, *134*, 1708–1717. [CrossRef] [PubMed]

12. Suzuki, K.; Nakanishi, H.; Bower, J.; Yoder, D.W.; Osteryoung, K.W.; Miyagishima, S.Y. Plastid chaperonin proteins Cpn60α and Cpn60β are required for plastid division in Arabidopsis thaliana. *BMC Plant Biol.* **2009**, *9*, 38. [CrossRef] [PubMed]

13. Usman, M.G.; Rafii, M.Y.; Ismail, M.R.; Malek, M.A.; Latif, M.A. Expression of target gene *Hsp70* and membrane stability determine heat tolerance in chili pepper. *J. Am. Soc. Hortic. Sci.* **2015**, *140*, 144–150. [CrossRef]

14. McLellan, C.A.; Turbyville, T.J.; Wijeratne, E.M.; Kerschen, A.; Vierling, E.; Queitsch, C.; Whitesell, L.; Gunatilaka, A.A.L. A rhizosphere fungus enhances Arabidopsis thermotolerance through production of an HSP90 inhibitor. *Plant Physiol.* **2007**, *145*, 174–182. [CrossRef] [PubMed]

15. Nover, L.; Bharti, K.; Döring, P.; Mishra, S.K.; Ganguli, A.; Scharf, K.D. Arabidopsis and the heat stress transcription factor world: How many heat stress transcription factors do we need? *Cell Stress Chaperon* **2001**, *6*, 177–189. [CrossRef]

16. Yoshida, T.; Ohama, N.; Nakajima, J.; Kidokoro, S.; Mizoi, J.; Nakashima, K.; Maruyama, K.; Kim, J.M.; Seki, M.; Todaka, D.; et al. Arabidopsis *HsfAl* transcription factors function as the main positive regulators in heat shock-responsive gene expression. *Mol. Genet. Genom.* **2011**, *286*, 321–332. [CrossRef]

17. Scharf, K.-D.; Berberich, T.; Ebersberger, I.; Nover, L. The plant heat stress transcription factor (Hsf) family: Structure function and evolution. *Biochim. Biophys. Acta.* **2012**, *1819*, 104–119. [CrossRef]

18. Guo, J.; Wu, J.; Ji, Q.; Wang, C.; Luo, L.; Yuan, Y.; Wang, Y.H.; Wang, J. Genome-wide analysis of heat shock transcription factor families in rice and Arabidopsis. *J. Genet. Genom.* **2008**, *35*, 105–118. [CrossRef]

19. Zhang, J.; Liu, B.; Li, J.; Zhang, L.; Wang, Y.; Zheng, H.; Lu, M.; Chen, J. Hsf and Hsp gene families in Populus: Genome-wide identification organization and correlated expression during development and in stress responses. *BMC Genom.* **2015**, *16*, 181–200. [CrossRef]

20. Lin, Y.; Jiang, H.; Chu, Z.; Tang, X.; Zhu, S.; Cheng, B. Genome-wide identification classification and analysis of heat shock transcription factor family in maize. *BMC Genom.* **2011**, *12*, 76. [CrossRef]

21. Song, X.; Liu, G.; Duan, W.; Liu, T.; Huang, Z.; Ren, J.; Li, Y.; Hou, X. Genome-wide identification classification and expression analysis of the heat shock transcription factor family in Chinese cabbage. *Mol. Genet. Genom.* **2014**, *289*, 541–551. [CrossRef]

22. Hirakawa, H.; Shirasawa, K.; Miyatake, K.; Nunome, T.; Negoro, S.; Ohyama, A.; Yamaguchi, H.; Sato, S.; Isobe, S.; Tabata, S.; et al. Draft Genome Sequence of eggplant (*Solanum melongena* L.): The representative solanum species indigenous to the old world. *DNA Res.* **2014**, *21*, 649–660. [CrossRef] [PubMed]

23. Swindell, W.R.; Huebner, M.; Weber, A.P. Transcriptional profiling of Arabidopsis heat shock proteins and transcription factors reveals extensive overlap between heat and non-heat stress response pathways. *BMC Genom.* **2007**, *8*, 125. [CrossRef] [PubMed]

24. Larkin, M.A.; Blackshields, G.; Brown, N.P.; Chenna, R.; McGettigan, P.A.; McWilliam, H.; Valentin, F.; Wallace, I.M.; Wilm, A.; Lopez, R.; et al. Clustal W and Clustal X version2.0. *Bioinformatics* **2007**, *23*, 2947–2948. [CrossRef]

25. Tamura, K.; Peterson, D.; Peterson, N.; Stecher, G.; Nei, M.; Kumar, S. MEGA5: Molecular evolutionary genetics analysis using maximum likelihood evolutionary distance and maximum parsimony methods. *Mol. Biol. Evol.* **2011**, *28*, 2731–2739. [CrossRef] [PubMed]

26. Guo, A.Y.; Zhu, Q.H.; Chen, X.; Luo, J.C. GSDS: A gene structure display server. *Yi Chuan* **2007**, *29*, 1023–1026. [CrossRef]

27. Bailey, T.L.; Williams, N.; Misleh, C.; Li, W.W. MEME: Discovering and analyzing DNA and protein sequence motifs. *Nucl. Acids Res.* **2006**, *34*, W369–W373. [CrossRef]

28. Schmittgen, T.D.; Livak, K.J. Analyzing real-time PCR data by the comparative C T method. *Nat Protoc.* **2008**, *3*, 1101–1108. [CrossRef]

29. Giorno, F.; Wolters-Arts, M.; Grillo, S.; Scharf, K.D.; Vriezen, W.H.; Mariani, C. Developmental and heat stress-regulated expression of HsfA2 and small heat shock proteins in tomato anthers. *J. Exp. Bot.* **2010**, *61*, 453–462. [CrossRef]

30. Pérez-Salamó, I.; Papdi, C.; Rigó, G.; Zsigmond, L.; Vilela, B.; Lumbreras, V.; Nagy, I.; Horváth, B.; Domoki, M.; Darula, Z.; et al. The heat shock factor A4A confers salt tolerance and is regulated by oxidative stress and the mitogen-activated protein kinases MPK3 and MPK6. *Plant Physiol.* **2014**, *165*, 319–334. [CrossRef]

31. Guo, M.; Lu, J.; Zhai, Y.; Chai, W.; Gong, Z.; Lu, M. Genome-wide analysis expression profile of heat shock factor gene family (*CaHsfs*) and characterization of CaHsfA2 in pepper (*Capsicum annuum* L.). *BMC Plant Biol.* **2015**, *15*, 151. [CrossRef] [PubMed]

32. Chung, E.; Kim, K.M.; Lee, J.H. Genome-wide analysis and molecular characterization of heat shock transcription factor family in Glycine max. *J. Genet. Genom.* **2013**, *3*, 127–135. [CrossRef]

33. Siddique, M.; Gernhard, S.; von Koskull-Doring, P.; Vierling, E.; Scharf, K.D. The plant sHSP superfamily: Five new members in Arabidopsis thaliana with unexpected properties. *Cell Stress Chaperon* **2008**, *13*, 183–197. [CrossRef] [PubMed]

34. Lopescaitar, V.S.; Carvalho, M.C.D.; Darben, L.M.; Kuwahara, M.K.; Nepomuceno, A.L.; Dias, W.P.; Abdelnoor, R.V.; Marcelino-Guimarães, F.C. Genome-wide analysis of the *Hsp20* gene family in soybean: Comprehensive sequence genomic organization and expression profile analysis under abiotic and biotic stresses. *BMC Genom.* **2013**, *14*, 577. [CrossRef]

35. Tao, P.; Guo, W.; Li, B.; Wang, W.; Yue, Z.; Lei, J.; Zhong, X. Genome-wide identification classification and expression analysis of sHSP genes in Chinese cabbage (*Brassica rapa* ssp. pekinensis). *Genet. Mol. Res.* **2015**, *14*, 11975–11993. [CrossRef] [PubMed]

36. Sarkar, N.K.; Kundnani, P.; Grover, A. Functional analysis of Hsp70 superfamily proteins of rice (Oryza sativa). *Cell Stress Chaperon* **2013**, *18*, 427–437. [CrossRef]

37. Banilas, G.; Korkas, E.; Englezos, V.; Nisiotou, A.A.; Hatzopoulos, P. Genome-wide analysis of the heat shock protein 90 gene family in grapevine (Vitis vinifera L.). *Aust. J. Grape Wine Res.* **2012**, *18*, 29–38. [CrossRef]

38. Krishna, P.; Gloor, G. The *Hsp90* family of proteins in Arabidopsis thaliana. *Cell Stress Chaperon* **2001**, *6*, 238–246. [CrossRef]

39. Lee, U.; Rioflorido, I.; Hong, S.; Larkindale, J.; Waters, E.R.; Vierling, E. The Arabidopsis ClpB/Hsp100 family of proteins: Chaperones for stress and chloroplast development. *Plant J.* **2007**, *49*, 115–127. [CrossRef]

40. Scharf, K.D.; Siddique, M.; Vierling, E. The expanding family of Arabidopsis thaliana small heat stress proteins and a new family of proteins containing alpha-crystallin domains (Acd proteins). *Cell Stress Chaperon* **2001**, *6*, 225–237. [CrossRef]

41. Singh, A.; Singh, U.; Mittal, D.; Grover, A. Genome-wide analysis of rice *ClpB/HSP100 ClpC* and *ClpD* genes. *BMC Genom.* **2010**, *11*, 95. [CrossRef]

42. Wang, G.D.; Kong, F.Y. Research advancement of heat shock factors in tomato. *Plant Physiol. J.* **2013**, *49*, 217–224.

43. Chauhan, H.; Khurana, N.; Agarwal, P.; Khurana, P. Heat shock factors in rice (*Oryza sativa* L.): Genome-wide expression analysis during reproductive development and abiotic stress. *Mol. Genet. Genom.* **2011**, *286*, 171–187. [CrossRef]

44. Mueller, L.A.; Tanksley, S.D. The SOL Genomics Network: A comparative resource for Solanaceae biology and beyond. *Plant Physiol.* **2005**, *138*, 1310–1317. [CrossRef]

45. Xu, G.; Guo, C.; Shan, H.; Kong, H. Divergence of duplicate genes in exon-intron structure. *Proc. Natl. Acad. Sci. USA* **2012**, *109*, 1187–1192. [CrossRef]

46. Wang, Y.; Lin, S.; Song, Q.; Tao, H.; Huang, J.; Chen, X.; Que, S.; He, H. Genome-wide identification of heat shock proteins (Hsps) and Hsp interactors in rice: Hsp70s as a case study. *BMC Genom.* **2014**, *15*, 344. [CrossRef]

47. Zhang, L.; Zhao, H.K.; Dong, Q.L.; Zhang, Y.Y.; Wang, Y.M.; Li, H.Y.; Xing, G.J.; Li, Q.Y.; Dong, Y.S. Genome-wide analysis and expression profiling under heat and drought treatments of *HSP70* gene family in soybean (*Glycine max* L.). *Front Plant Sci.* **2015**, *6*, 773. [CrossRef]

48. Zhou, S.; Zhang, P.; Jing, Z.; Shi, J. Genome-wide identification and analysis of heat shock transcription factor family in cucumber (*Cucumis sativus* L.). *Plant Omics.* **2013**, *6*, 449–455. [CrossRef]

49. Wan, B.L.; Zha, Z.P.; Du, X.S. Expression profile analysis of rice heat shock transcription factor (*HSF*) genes in response to plant growth regulators and abiotic stresses. *China Biotechnol.* **2010**, *30*, 22–32. [CrossRef]

50. Zafar, S.A.; Hussain, M.; Raza, M.; Ahmed, M.D.; Rana, I.A.; Sadia, B. Genome wide analysis of heat shock transcription factor (*HSF*) family in chickpea and its comparison with Arabidopsis. *Plant Omics* **2016**, *9*, 136–141. [CrossRef]

51. Qiao, X.; Li, M.; Li, L.; Yin, H.; Wu, J.; Zhang, S. Genome-wide identification and comparative analysis of the heat shock transcription factor family in Chinese white pear (*Pyrus bretschneideri* L.) and five other Rosaceae species. *BMC Plant Biol.* **2015**, *15*, 12. [CrossRef]

52. Wahid, A.; Gelani, S.; Ashraf, M.; Foolad, M.R. Heat tolerance in plants: An overview. *Environ. Exp. Bot.* **2007**, *61*, 199–223. [CrossRef]

53. Hahn, A.; Bublak, D.; Schleiff, E.; Scharf, K.D. Crosstalk between *Hsp90* and *Hsp70* chaperones and heat stress transcription factors in tomato. *Plant Cell* **2011**, *23*, 741–755. [CrossRef]

54. Hu, W.; Hu, G.; Han, B. Genome-wide survey and expression profiling of heat shock proteins and heat shock factors revealed overlapped and stress specific response under abiotic stresses in rice. *Plant Sci.* **2009**, *176*, 583–590. [CrossRef]

55. Mishra, S.K.; Tripp, J.; Winkelhaus, S.; Tschiersch, B.; Theres, K.; Nover, L.; Scharf, K.D. In the complex family of heat stress transcription factors *HsfA1* has a unique role as master regulator of thermotolerance in tomato. *Genes Dev.* **2002**, *16*, 1555–1567. [CrossRef]

56. Chauhan, H.; Khurana, N.; Nijhavan, A.; Khurana, J.P.; Khurana, P. The wheat chloroplastic small heat shock protein (sHsp26) is involved in seed maturation and germination and imparts tolerance to heat stress. *Plant Cell Environ.* **2012**, *35*, 1912–1931. [CrossRef]

57. Kim, S.R. and An, G. Rice chloroplast-localized heat shock protein 70 OsHsp70CP1 is essential for chloroplast development under high-temperature conditions. *J. Plant Physiol.* **2013**, *170*, 854–863. [CrossRef]

58. Park, H.S.; Jeong, W.J.; Kim, E.C.; Jung, Y.; Lim, J.M.; Hwang, M.S.; Park, E.J.; Ha, D.S.; Choi, D.W. Heat shock protein gene family of the porphyra seriata and enhancement of heat stress tolerance by *PsHsp70* in Chlamydomonas. *Mar. Biotechnol.* **2012**, *14*, 332–342. [CrossRef]

59. Li, D.; Yang, F.; Lu, B.; Chen, D.; Yang, W. Thermotolerance and molecular chaperone function of the small heat shock protein HSP20 from hyperthermophilic archaeon *Sulfolobus solfataricus* P2. *Cell Stress Chaperon* **2012**, *17*, 103–108. [CrossRef]

60. Bondino, H.G.; Valle, E.M.; Ten, H.A. Evolution and functional diversification of the small heat shock protein/alpha-crystallin family in higher plants. *Planta* **2012**, *235*, 1299–1313. [CrossRef]

61. Liu, G.; Wang, J.; Cramer, G.; Dai, Z.; Duan, W.; Xu, H.; Wu, B.; Fan, P.; Wang, L.; Li, S. Transcriptomic analysis of grape (*Vitis vinifera* L.) leaves during and after recovery from heat stress. *BMC Plant Biol.* **2012**, *12*, 174. [CrossRef]

62. Liu, H.; Liao, H.; Charng, Y. The role of class A1 heat shock factors (*HSFA1s*) in response to heat and other stresses in Arabidopsis. *Plant Cell Environ.* **2011**, *34*, 738–751. [CrossRef]

63. Liu, Y.; Zhang, C.; Chen, J.; Guo, L.; Li, X.; Li, W.; Yu, Z.; Deng, J.; Zhang, P.; Zhang, K.; et al. Arabidopsis heat shock factor *HsfA1a* directly senses heat stress pH changes and hydrogen peroxide via the engagement of redox state. *Plant Physiol. Bioch.* **2013**, *64*, 92–98. [CrossRef]

64. Qian, J.; Chen, J.; Liu, Y.; Yang, L.; Li, W.; Zhang, L. Overexpression of Arabidopsis *HsfA1a* enhances diverse stress tolerance by promoting stress-induced *Hsp* expression. *Genet Mol. Res.* **2014**, *13*, 1233–1243. [CrossRef]

65. Bechtold, U.; Albihlal, W.S.; Lawson, T.; Fryer, M.J.; Sparrow, P.A.C.; Richard, F. Arabidopsis heat shock transcription factor *A1b* overexpression enhances water productivity resistance to drought and infection. *J. Exp. Bot.* **2013**, *64*, 3467–3481. [CrossRef]

66. Li, N.; Zeng, J.; Xu, Y.; Guo, L. Effects of Heat Shock Factor *AtHsfA1a* on programmed cell death in Arabidopsis thaliana under cold stress. *Agric. Biotechnol.* **2016**, *5*, 57–59.
67. Wang, W.; Vinocur, B.; Shoseyov, O.; Altman, A. Role of plant heat-shock proteins and molecular chaperones in the abiotic stress response. *Trends Plant Sci.* **2004**, *9*, 244–252. [CrossRef]
68. Ostling, P.; Björk, J.K.; Roos-Mattjus, P.; Mezger, V.; Sistonen, L. Heat shock factor 2 (HSF2) contributes to inducible expression of *hsp* genes through interplay with HSF1. *J. Biol. Chem.* **2007**, *282*, 7077–7086. [CrossRef]
69. Singh, A.; Mittal, D.; Lavania, D.; Agarwal, M.; Mishra, R.C.; Grover, A. *OsHsfA2c* and *OsHsfB4b* are involved in the transcriptional regulation of cytoplasmic *OsClpB* (*Hsp100*) gene in rice (*Oryza sativa* L.). *Cell Stress Chaperones* **2012**, *17*, 243–254. [CrossRef]

Review

Research Advances in Allelopathy of Volatile Organic Compounds (VOCs) of Plants

Yiqi Xie [1,2], **Libo Tian** [2], **Xu Han** [1,*] and **Yan Yang** [1,*]

[1] Tropical Crops Genetic Resources Institute, Chinese Academy of Tropical Agricultural Sciences, Haikou 571101, China; xieyiqi277@163.com
[2] College of Horticulture, Hainan University, Haikou 570228, China; faiy7play@163.com
* Correspondence: hanxu@catas.cn (X.H.); yangyan@catas.cn (Y.Y.);
Tel.: +86-186-8976-3830 (X.H.); +86-135-1803-1806 (Y.Y.)

Abstract: Allelopathy is an ecological phenomenon in which organisms interfere with each other. As a management strategy in agricultural systems, allelopathy can be mainly used to control weeds, resist pests, and disease and improve the interaction of soil nutrition and microorganisms. Volatile organic compounds (VOCs) are allelochemicals volatilized from plants and have been widely demonstrated to have different ecological functions. This review provides the recent advance in the allelopathic effects of VOCs on plants, such as growth, competition, dormancy, resistance of diseases and insect pests, content of reactive oxygen species (ROS), enzyme activity, respiration, and photosynthesis. VOCs also participate in plant-to-plant communication as a signaling substance. The main methods of collection and identification of VOCs are briefly summarized in this article. It also points out the disadvantages of VOCs and suggests potential directions to enhance research and solve mysteries in this emerging area. It is necessary to study the allelopathic mechanisms of plant VOCs so as to provide a theoretical basis for VOC applications. In conclusion, allelopathy of VOCs released by plants is a more economical, environmentally friendly, and effective measure to develop substantial agricultural industry by using the allelopathic effects of plant natural products.

Keywords: allelochemicals; VOCs properties; VOCs action; VOCs detection; green agriculture

Citation: Xie, Y.; Tian, L.; Han, X.; Yang, Y. Research Advances in Allelopathy of Volatile Organic Compounds (VOCs) of Plants. *Horticulturae* **2021**, *7*, 278. https://doi.org/10.3390/horticulturae7090278

Academic Editors: Yuyang Zhang and Juan Capel

Received: 4 June 2021
Accepted: 30 August 2021
Published: 2 September 2021

Publisher's Note: MDPI stays neutral with regard to jurisdictional claims in published maps and institutional affiliations.

Copyright: © 2021 by the authors. Licensee MDPI, Basel, Switzerland. This article is an open access article distributed under the terms and conditions of the Creative Commons Attribution (CC BY) license (https://creativecommons.org/licenses/by/4.0/).

1. Introduction

The concept of "allelopathy" was first proposed by Austrian scientist Hans Molisch in 1937 and mainly referred to the chemical relationship of plant interaction. Allelopathy is an ecological phenomenon and plays an important role in the ecological adaptation of plants [1,2]. The allelopathic effects have both positive and negative effects. Various studies have reported the advantages of allelopathic effects in agricultural systems, such as weed control [3–6], inhibition of pests [7–10], disease [11,12], improvement of soil nutrition [13,14], and microbial interactions [15,16]. Ultimately, allelopathy of most plants has effect on plant growth [10,17,18]. Plants can synthesize various secondary metabolites during growth and development. Plant VOCs vary by species, and they are related to the abundance of neighboring plant species and plant species composition [19,20]. These secondary metabolites can be beneficial or harmful to other organisms when stored or released into the environment, such as secondary metabolites stored in plants that can prevent animal feeding and microbial infestation, while volatiles released into the air can attract insect pollinators [21]. Plants communicate with organisms in the environment through VOCs, thereby achieving a wide range of ecological functions, such as affecting their growth, development, defense, reproduction, and life cycle [22]. In 1984, allelopathy was defined as "any direct or indirect harmful effect by one plant (including microorganisms) on another through production of chemical compounds released into the environment" by Rice [23]. These products of secondary metabolism, called allelochemicals, can be found in any organ of the plant (leaves, stems, flowers, seeds, fruits, and/or roots) and can be released from the producing plant by different routes: volatilization, foliar leaching, root exudations, and decomposition of plant residue (Figure 1).

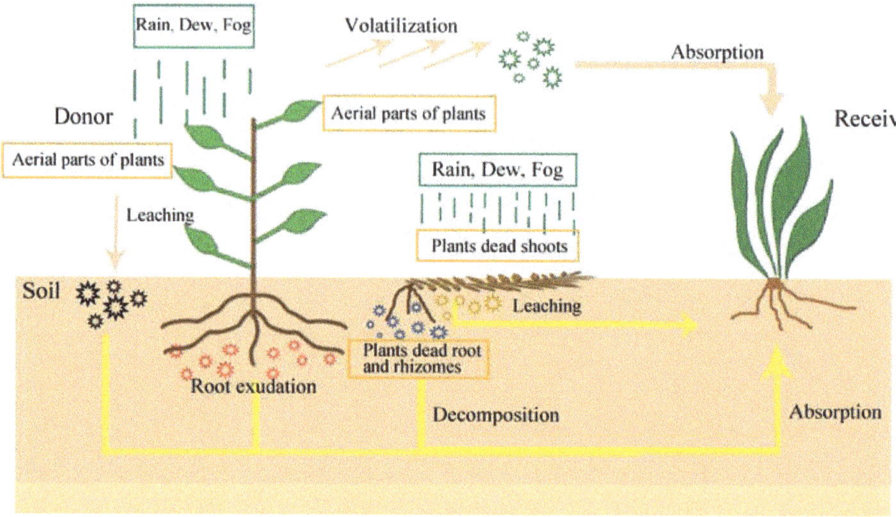

Figure 1. The allelopathy pathways of plants.

VOCs are secondary metabolites volatilized by plants and ubiquitous allelochemicals of plants [24]. Shikimate/phenylalanine, the mevalonic acid (MVA), the methylerythritol phosphate (MEP), and lipoxygenase (LOX) pathways are the four main synthesis pathways of VOCs, and plants can synthesize and release various VOCs including terpenoids, phenylpropanoids/benzenoids, and fatty acid derivatives [25]. The VOCs released by these plants often have different ecological functions, such as chemical communication, kin recognition, attracting or repelling insects, and many other effects [21,26–29]. Although the researches of plants VOCs are mainly aboveground some chemical signal, more and more studies show that VOCs also play an integral part in belowground plant–plant interactions [30]. In fact, the phenomenon that plants release allelochemicals through the volatile pathway has been noticed for a long time. One of the first empirical studies of allelopathy involving VOCs was researched by Molisch, who found that VOCs released by apples and pears could inhibit potato germination [31]. VOCs have been widely demonstrated to defend primarily against herbivorous insects [32,33], microbes, and pathogens [34–36], thereby reducing extreme environmental stress [37,38] and promoting nutritional acquisition [11]. Muller et al. [39] researched the volatiles of annual grassland species in *Salvia leucophylla* Greene and *Artemisia californica* communities, and this revealed that volatile allelochemicals had the interspecific allelopathic effects on the woody herbaceous plants, which would negatively affected the recipient plant species [40,41] and changed soil microorganisms [42,43]. Besides, in addition to VOCs released from plant shoots, root volatiles may also have allelopathic effects on neighboring plants; for example, VOCs from big sagebrush (*Artemisia tidentata* Nutt.) root inhibited seed germination of wild tobacco (*Nicotiana attenuata*) [44]. Most allelochemicals produced by plant roots are considered as "root exudates" [45], but the few allelochemicals released by volatilization of roots are called VOCs, which play an important ecological role in the soil ecosystem and have not been studied thoroughly [46,47].

Allelopathy has received high attention and become one of the central scientific problems in ecology [48]. Allelopathy is forming an independent scientific system, and we are conducting in-depth and extensive research from both theoretical and practical aspects. VOCs released by plants are one of the main ways to achieve allelopathic effects. It is a more economical, environmentally friendly, and effective measure to use the allelopathic effect of plant natural products to develop the agricultural production [49]. Several excellent reviews have summarized the relevant research on potential applications of VOCs [20,22,49], but the studies on allelopathy of plants VOCs have not been systematically reviewed and reported. VOCs are a kind of natural and environmentally friendly chemical substances that volatilize from plants and are used as natural herbicides and fungicides to protect neighboring plants from stress and increase crop yields [49]. We think VOCs have a much broader range of the potential applications. So, in this context, study on the allelopathy of VOCs is particularly important to the future development of green agriculture. The review mainly focuses on the recent studies of allelopathy of plants VOCs, regarding resisting diseases and preventing pests of plant, impacting on competition (inhibiting weed hazards), breaking dormancy, regulating plant growth, affecting reactive oxygen species (ROS) content and enzyme activity, modulating plant respiration and photosynthesis, and their role as a signal conducting substance. We present the evidence from the references to illustrate these roles to deepen the understanding of allelopathy of plants VOCs.

2. Allelopathy of VOCs of Plants

With the increasing attention of experts on the allelopathy of volatile, the potential role of VOCs in agriculture has been gradually discovered. The allelopathy of VOCs on plants are summarized based on existing research, which involved the main allelopathy of plant VOCs. (Figure 2).

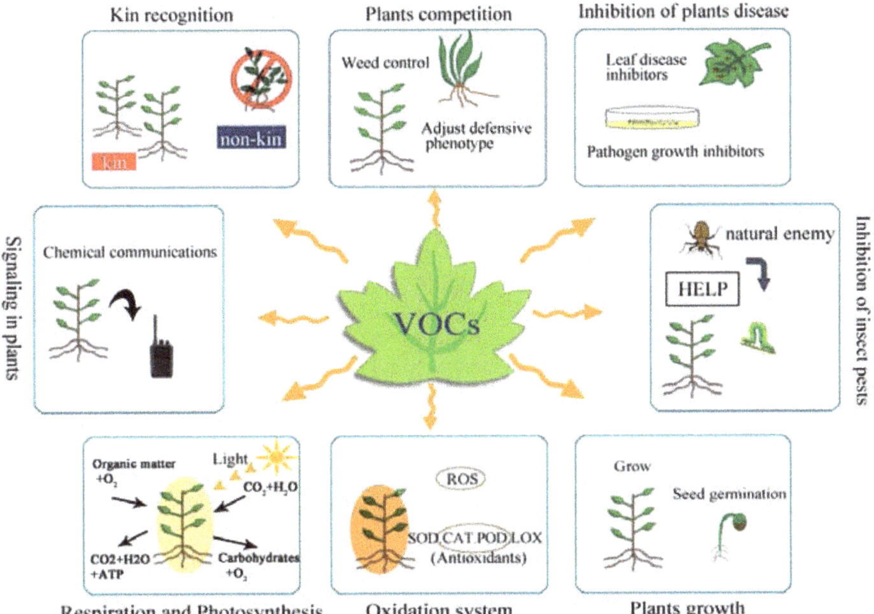

Figure 2. The main allelopathy of plants VOCs. (Note: This figure is widely inspired and adapted from Brilli et al. (2019) [49]).

2.1. VOCs and Plants Growth

Numerous results showed that VOCs have an effect on plant cell growth and differentiation [50–53], such as diallyl disulfide (DADS) of garlic VOCs, which can affect mitotic activity and cell length of tomato roots by impacting on cell division, endogenous plant hormone levels, expansin gene expression, and sulfate assimilation and glutathione (GSH) metabolism [53,54].

Similarly, many studies confirm that plant volatiles can inhibit seed germination and the growth of root and seedling [50,55–60]. β-terpineol, linalool, eugenol, and tetradecanoic acid are the VOCs released from tomato (*Solanum lycopersicon* Mill.) foliage, and they could inhibit seed germination of the tropical plant *Amaranthus mangostanus* L. [55]. The brassica species exude allelochemicals, which are glucosinolates [61] that could break down into several biological action compounds, such as isothiocyanates, which have biologically active and inhibit germination and growth of exposed plant species [57,62]. VOCs released from pine needles and the roots of *Pinus halepensis* L. mainly inhibited the seed germination and root growth of two herbaceous target species *Lactuca sativa* L. and *Linum strictum* L. [63]. VOCs affect plant growth, mainly to change plant morphology and reduce plant biomass. Therefore, seed germination, seedling root length, and seedling height are often used as intuitive indicators to evaluate allelopathy. Low concentrations of DADS promoted the growth of cucumber roots and induce elongation of the main roots by up-regulating the expression of CsCDKA and CsCDKB genes and regulating the hormone balance of the roots [64]. The VOCs released by *Atriplex cana* Ledeb. (Amaranthaceae) significantly inhibited seedling growth of *Amaranthus retroflexus* L. and *Poa annua* L., and 5 μg/mL essential oil completely inhibited the seed germination of *A. retroflexus*, *Medicago sativa* L., *P. annua*, and *Echinochloa crusgalli* L. [65]. Besides, Effah et al. found that plant VOCs mediate multiple ecological networks, and they may mediate the allelopathic effects of the germination or growth of competitors seeds [66]. Monoterpenoids are considered as effective inhibitors of seed germination and seedling growth [67,68].

VOCs affect plant growth, mainly to inhibit plant growth, but some studies have found that VOCs have dual effects on germination and plant growth, both promoting and inhibiting; for example, Arroyo et al. found that volatile chemicals from *Artemisia herba-alba* Asso. inhibited the germination of *Pinus halepensis* Mill. seeds, promoted the growth of *P. halepensis* seedlings, and reduced the root biomass of *Salsola vermiculata* L. seedlings [69]. VOCs also have an effect on plants growth direction. Runyon et al. found that *Cuscuta pentagona* seedlings were favored by not only the growth of host tomato plants, but also the direction of tomato VOCs [70]. The extracted VOCs of tomato and wheat were placed on both sides of *C. pentagona* seedlings, and the *C. pentagona* seedlings continued to grow in the direction of tomato VOCs. It was also found that β-phellandrene and β-myrcene of tomato VOCs could significantly attract the growth of *C. pentagona* seedlings.

2.2. VOCs and Weed Control

Allelopathy gain extensive attention in biological weed control [3–6,71]. Boydston et al. [3] found that mustard seed meal has the potential for weed control in organic production systems. The release of volatile organic compounds from the leaves can cause allelopathic effects and damage the growth of other competitive plants [72]. The allelopathy of the volatile allelochemicals may perform a vital part in inhibiting the competitive ability of weed species, be one of the alternatives to control weed infestation, and it has excited the greatest interest [67,73–75]. Therefore, the best way to control weeds is to use the crops' own competition and allelopathy. Wei et al. reported that the volatile oil released by *Atriplex cana* Ledeb. significantly inhibited the growth of seedlings of four weed species, including *Amaranthus retroflexus* L. and *Poa annua* L., and it had a high value for further use as a biological herbicide [65].

Brassicas produce the allelochemicals glucosinolates throughout their plant parts and released them into the environment by volatilization [76]. In the natural environment, glucosinolates are broken down into several compounds, and the most important compound of them is isothiocyanate [77], which can inhibit the growth and development of plant or weed [78]. *Digitaria sanguinalis* is a common non-irrigated weed that severely affects crop yields. Pardo-Muras et al. showed that the oxygenated monoterpenes in VOCs produced and released by both *Ulex europaeus* and *Cytisus scoparius* inhibited the germination and early growth of two weeds, *A. retroflexus* and *D. sanguinalis* [79]. Many studies have reported that in addition to independent effects of VOCs, VOCs also have synergistic or antagonistic effects. These synergistic and antagonistic effects lead to the final allelopathy [80,81].

Besides, VOCs from invasive species can also reduce interspecific performance and improve the performance of invasive species [82]. VOCs might also be perceived by neighboring plants to adjust their defensive phenotype according to the present risk of attack. Allelochemicals as natural herbicide have an attractive prospect, and some plants may be expected to develop into a new generation of herbicides or fungicides.

2.3. VOCs and Plants Dormancy

Dormancy is a physiological state in which plants respond to stress [83]. DADS is considered as main allelochemical of VOCs in garlic [53]. Hosoki et al. reported that the sulfur-containing compounds from garlic VOCs could break the bud dormancy in some corms, tubers, and ornamental trees [84,85]. Kubota et al. found that exposure to volatile diallyl disulfides and trisulfides was the most effective treatment to promote the bud break of single-bud cuttings of "Kyoho" (*Vitis vinifera* × *labruscana Bailey*) [86]. So, the allelochemicals in garlic that break the dormancy of vine buds are sulfur compounds, particularly DADS. In addition to DADS, more and more studies can prove that VOCs can affect plant dormancy. In a recent study, Shukla et al. studied the breaking dormancy of potato tubers effects on 20 essential oils from medicinal and aromatic plants [87]. The essential oils could induce or inhibit the sprouting process of potato tubers by altering the accumulation of reducing sugars, ethylene production, and expression of genes, thereby affecting the dormancy of plants [87]. Besides, eugenol from clove essential oil and carvone from caraway and dill essential oils have been reported to inhibit potato tubers sprouting [88–90]. The results showed that the essential oils of lemon grass and clove were the most effective VOCs for breaking dormancy and inducing germination of potato tubers. The oils of palmarosa and ajwain inhibited the sprout of potato tubers [87]. Owolabi et al. found the essential oils of *Lippia multiflora*, *Cymbopogon citratus*, and *Zingiber officinale* could control potato tubers dormancy [91]. They are suitable for application as sprout suppressants. At present, there are few studies on the effects of VOCs on plant dormancy, and the specific mechanism of action needs to be studied in the future.

2.4. VOCs and the Inhibition of Plants Diseases and Insect Pests

2.4.1. Inhibition of Plants Disease

VOCs not only have an inhibitory effect on plants, but also on pathogenic bacteria. A number of experimental trials showed that leaves' VOCs inhibited germination and the growth of plant pathogens and had stronger activity than commercial fungicides [92–94]. The essential oil of oregano triggers the expression of hundreds of genes involved in the grapevine immune system, so it can prevent *Plasmopara viticola* infection in grapevine (*Vitis Vinifera*) and primes plant immunity mechanisms [95]. The inhibition rate of volatile allelochemicals from leaves of *Ocimum adscendens* to 29 different kinds of mycelium was over 98%, and the inhibition rate of mycelium growth of Aspergillus reached 100%; especially, the activity of *A. lavus* was 10~100 times higher than another five commercial fungicides [96]. Chaturvedi et al. found that volatiles released by *Adenocalymma allicea* could effectively control leaf spot in rice and kill the pathogenic fungus *Drechslera oryzae*. When volatile oil obtained from *A.alicea* plants compared with the activity of the synthetic fungicides blitox-50 and m-45, the activity of volatile oil to *D. oryzae* was up to four times higher than those synthetic fungicides [97]. In particular, this volatile oil had no harmful effect on the growth and development of rice seedlings. The volatile oil released from the leaves of the same genus *Ocimum basilicum* also inhibited the growth of other fungal mycelia by more than 85%, and the dosage was only 1/4 of the commercial fungicide [98]. Therefore, using VOCs to replace commercial fungicide is no longer a dream. Phenols are a kind of allelochemicals that have been concerned and studied. Eugenol is an important class of phenolic allelochemicals. Cloves have a long history of use, and the main volatile component eugenol has a strong inhibitory effect on fungi and bacteria [99,100].

In a recent study, Quintana-Rodriguez et al. performed a screening on the efficacy of 22 VOCs, which were known to be volatilized from infected plants leaves, against the fungal pathogens *Colletotrichum lindemuthianum*, *Fusarium oxysporum*, and *Botrytis cinerea*. The work results showed that nonanal, (+)-carvone, citral, trans-2-decenal, L-linalool, nerolidol, and eugenol significantly inhibited the growth of the three fungal species, and eugenol had the most active among them. Therefore, the VOCs of plants have the disease resistance function [101].

In addition to eugenol, there are many VOCs that can also inhibit pathogens. DADS is a volatile organosulfur compound derived from garlic (*Allium sativum* L.) bulbs, and it is known as an allelochemical because of the potential allelopathy of garlic. A large number of study results show that it has a strong inhibitory effect on a variety of pathogenic bacteria [102]. In a recent study, Yang et al. demonstrated that the VOCs (DADS) from green garlic (*Allium sativum* L.) increased the accumulation of H_2O_2 and the disease resistance of cucumber [103]. Sekine et al. reported that other VOCs such as cuminaldehyde and *p*-cymene also have been demonstrated to possession antifungal activity against *B. cinerea*, *F. oxysporum*, *Verticillium dahliae*, and *Alternaria mali* [104]. According to the work of Mandal and Mandal, linalool, a substance with antifungal and antioxidant potential, was found in the volatile oil of coriander (*Coriandrum sativum* L.) [105].

2.4.2. Inhibition of Plants Insect Pests

Numerous studies showed that plants not only produce toxins and hormones directly in response to insect feeding, but also release VOCs to attract predators [106,107]. A class of VOCs produced as a response to herbivore attacking are mainly terpenoids, predominantly monoterpenoids, and sesquiterpenoids [107], and they benefited the host plant by interacting with herbivores. For example, when the larva of *Spodoptera exigua* Hübner. were feeding on corn, the corn released volatile terpenoids to attract the parasitic wasp, which was the natural enemy of *S. exigua*. If it was not mechanically damaged, the corn did not release VOCs to attract the natural enemy. Further studies found that the oral saliva of the beet moth contained volicitin, when the beet moth bit corn, its messenger jasmonic acid derivatives were activated by volicitin to release terpenoids to lure *Cotesia marginiventris* Cresson [108]. Kessler and Baldwin found that wild tobacco could release VOCs to attract mealoptera, the natural predators of caterpillars, when caterpillars ate them. Additionally, the VOCs released by wild tobacco also could prevent caterpillars from laying eggs on the leaves [21].

Under natural conditions, plants can also achieve effective control of related pests by releasing volatile substances to attract natural enemies. It is confirmed that the indirect chemical defense of plants by releasing VOCs to attract natural enemies is a chemical relationship in nature. *Ageratum conyzoides* L. released volatile terpenoid into the air by stalks, leaves, and flowers to attract predatory mites (*Amblyseius* SPP.) and maintain their population density [26]. Predatory mites are the most effective natural enemies of red mites (*Panonychus citri*). So, the population density of red mites was reduced [26]. Degenhardt et al. found that VOCs from corn root (E)-b-caryophyllene attracted insect-killing nematodes to control a major pest [109]. Therefore, the sustainable control of pests and diseases can be achieved through the natural chemical mechanisms that exist in the field to regulate plant–organism interactions. As a natural fungicide, VOCs have no harmful effect on plant growth and development, and using VOCs is a more environmentally friendly and economical way to kill bacteria.

2.5. VOCs and Plants Respiration and Photosynthesis

Previous studies showed that allelochemical can affect plant respiration by interfering in various stages of respiration, including the generation of carbon dioxide (CO_2) by electron transport, oxidative phosphorylation, and the activity of ATPase, and it has the potential to inhibit plant growth and development [41,110,111]. VOCs released from the leaves of *A. tridentata* Nutt. *var. vaseyana* and *Sasa cernua* Makino. inhibited the respiration of germinating seeds [110,112]. In fact, studies of allelopathy processes in shrubs in the 1960s found that volatile terpenes could reduce respiration in the young leaves of some plants and increase respiration in mature leaves [113]. Similarly, terpenes in eucalyptus volatile oil could affect target plants by inhibiting cellular respiration [114]. The influence mechanism of these allelochemicals on plant respiration deserves more in-depth study.

In addition, photosynthesis plays an important role in realizing the energy conversion in nature and maintains the carbon–oxygen balance in the atmosphere. The high concentration of allelochemicals involved in multiple metabolic steps may lead to the inhibition of plant photosynthesis, or even block the mechanism of photosynthesis by inhibiting electron receptors, energy coupling, or destroying photosynthetic pigments and enzymes [115]. Isoprene volatilized from the foliage of many woody species was thought to increase the rate of photosynthesis by stabilizing thylakoid membranes, so that adjacent plants could tolerate high temperatures [116]. Kaur et al. found the volatile oils α-Pinene and 1,8-cineole from *Eucalyptus tereticornis* Sm. could significantly reduce the respiration and photosynthetic pigment content of *Amaranthus viridis* Linn. Seedlings; thereby, the negative effects of the oil on photosynthetic machinery was explained [117]. These studies can confirm that olefin compounds can affect the photosynthesis of plants. In addition, Tsubo et al. found that exposure to low concentrations of the volatile chemicals released by *A. adamsii* Besser stimulated the photosynthetic rates of *Stipa krilovii* Roshev [118]. The volatile oil of *Artemisia ordosica* Krasch. inhibited the growth and photosynthesis of *Palmellococcus miniatus* through oxidative damage [119]. Zhao et al. studied the effects of eucalyptol and limonene, the main terpenoids in cyanobacteria VOCs on the photosynthetic capacity of *Chlorella vulgaris* [120]. The results showed that the compounds could induce the degradation of photosynthetic pigments and reduce the photosynthetic abilities of other algae. These studies can confirm that VOCs have the ability to affect plant photosynthesis.

2.6. VOCs and Plants ROS Content and Enzymatic Activity

Reactive oxygen species (ROS) play a vital role in the plant defense against stresses. The balance between ROS generation and scavenging is considered as paramount in cellular homeostasis. In recent years, more and more attention has been paid to the effectiveness and feasibility of monomer organic sulfide extracted from garlic as an anti-tumor drug, and the research and development of monomer organic sulfide has become an important research topic [121]. Similarly, DADS had an effect on the ROS content of plants. Yang et al. used cucumber and garlic as test materials to study the allelopathy of VOCs from green garlic on the scavenging of cucumber ROS [103]. The results showed that DADS, a volatile substance in garlic, reduced superoxide anions and increased hydrogen peroxide accumulation in cucumber seedlings. The effects of VOCs on antioxidant enzymatic activities were species dependent. They can regulate the activity of antioxidant enzymes (SOD, CAT, and POD) of cucumber seedlings in response to oxidative stress. VOCs released from *Acacia dealbata* Link. leaves increased the activity of superoxide dismutase (SOD) and peroxidase (POD) in *L. multiflorum* flowers, but decreased SOD activity in *T. subterraneum* [60]. The volatile allelochemical myrcene rapidly induced ROS production and significantly increased the activity of lipoxygenase (LOX) in rice roots [122]. In other cases, Mutlu et al. reported that the aerial parts of *Nepeta meyeri* Benth. contained two volatile oils, Germacrene-d and Caryophyllene oxide, and they could reduce the SOD activity of six weed species [123]. Jin et al. reported that the essential oil such as carvacrol, cinnamaldehyde, perillaldehyde, and linalool enhanced the SOD and POD activities of Chinese bayberries, and carvacrol had the best effect [124].

2.7. VOCs and Plants Signal Transduction
2.7.1. Chemical Communications

Plants, similar to animals, do not exist in isolation. Plant individuals and populations maintain population relationships and resist external stress through chemical communication. VOCs volatilized from some plants are involved in plant–plant communication.

Through more than 20 years of research, it has been found that when plants were stressed by insect feeding, microbial infection, and mechanical damage, they could use volatile organic substances to carry out inter-chemical and intra-chemical chemical communication [125,126]. Plants sent VOCs signals under attack or stress, then neighboring plants received these VOCs signals directly or indirectly to turn on the chemical defense mechanisms and produce phenolic alkaloids chemical defense compounds including directly terpenoids and other defense substances. Indirect chemical defense the use of VOCs to attract the natural enemies of pests, through the methods of predation and parasitism to eliminate pests. For example, the VOCs signals released by injured *A. tidentata* could induce direct chemical defense by protease inhibitors in *Nicotiana attenuate*, and the VOC signals released by maize leaves when insects fed could induce plants to rapidly release monoterpenoids that attracted insect predators for indirect defense [127,128]. Wild lima bean quickly synthesized and secreted excess nectar to attract natural enemies after receiving VOC signals from plants that have been harmed by foraging [129]. In recent years, more and more research has gone deep into the mechanism of action of these chemical communication signaling molecules. Baldwin et al. [125] found when in response to the attack, infested leaves released (E)-β-ocimene (typical signaling chemicals of volatile terpenoids) to increase the resistance of un-infested leaves and induce the expression of defense-related genes in neighboring un-infected leaves [130]. In future research, based on the clarification of the biochemical and metabolic processes of the phytochemical signal substances that have been discovered, the response mechanism of the recipient plants to these signal substances should be further explored. Therefore, although the research of phytochemical communication faces huge challenges, the clarification of phytochemical identification and communication relationships will broaden the horizons of the interspecific and intraspecific relationships of plants.

2.7.2. Plant Kin Recognition

Kin recognition is simply the ability of an individual to distinguish the relationship between genetically close related kin and non-kin. The kin recognition of plants has very important ecological and evolutionary significance. Increasing evidence shows that plants' recognition of neighboring allogenous and heterogeneous plants is mostly mediated by chemicals [131,132]. When plants are attacked by herbivores, they will emit volatile signals to surrounding plants. Some plants, such as *A. tidentata*, suffered less damage than other plants that receive volatile signals from non-self-wounds [133,134]. This shows that VOCs play a role in plant self- and non-self-recognition. Because VOCs are the simplest and fastest chemical signal that can send to neighboring plants, plant recognition of volatiles helps plants to establish corresponding response strategies in the early stages of competition. This can avoid wasting resources in the competition between self and relatives as much as possible [130]. After a series of studies recently, although the kin recognition of plants is ubiquitous, most studies show that some plants do not have kin recognition behaviors in order to avoid meaningless competition [27,135–137]. Plants' kin recognition has always been a hot topic in the study of behavioral ecology and evolution. From the perspective of genetic recognition, it is of great theoretical and practical significance to re-examine the ecological interaction of individual crops and groups and the environment.

3. Method of VOCs Collection and Identification

The allelopathy of VOCs has attracted widespread attention in recent years and a lot of research work has been done. With the advancement of organic separation and identification technology and the participation of more and more experts, the collection and identification technology of plants VOCs is no longer a difficult issue in research.

3.1. Collection of Plants VOCs

The volatiles mostly are organic compounds with a molecular weight between 100 and 200, such as hydrocarbons, alcohols, ketones, organic acids, nitrogen compounds, and organic sulfur [138]. Most of them have high chemical activity. Different collection methods may directly affect the type and proportion of VOCs, so it is particularly important to choose the appropriate method.

Traditional distillation collection techniques include steam distillation (SD), simultaneous distillation and solvent extraction (SDE), microwave-assisted hydrodistillation extraction (MWHD), ultrasound-assisted extraction (USE), and solid-phase trapping solvent extraction (SPTE). They have certain disadvantages in the isolation and purification of chemical constituents from plants tissues, such as long extraction time, high volumes solvent, and low efficiency [139]. In addition, many natural products are thermally unstable and may degrade during thermal extraction or distillation.

The most mainstream approach is headspace solid-phase microextraction (HS-SPME). It has some advantages over SD, SDE, and SPTE, such as rapid solvent-free extraction, no apparent thermal degradation, less laborious manipulation and sample requirement, and so on [140]. Moreover, due to the relatively low temperature and short headspace solid-phase extraction time, the risk of thermal artifacts is extremely low compared to other techniques [141]. Additionally, it is easy to standardize and fully integrate into the analysis system [142]. Thus, HS-SPME is an ideal technology of plants' VOC collection.

3.2. Identification of Plants VOCs

Identification of allelochemicals involves both quantitative and qualitative measurement. Qualitative identification is the identification of the type and structure of the allelochemicals. Qualitative identification involves methods such as gas chromatography (GC), mass spectrometry (MS), nuclear magnetic resonance (NMR), Fourier transform infrared (FT-IR) spectroscopy, and many other methods [143–145]. These are the analysis methods of VOCs, but the analytical difficulties and required instruments are completely different. Quantitative identification means the determination of the concentration of allelochemicals on the premise of clarifying the type of them. The method of chromatography is used to detect the concentration of known substances. Different methods should be selected for qualitative and quantitative identification, and the selection criteria are determined according to the characteristics of VOCs.

The existing identification techniques include gas chromatography mass spectrometry (GC-MS), high-performance liquid chromatography (HPLC), proton-transfer-reaction mass spectrometry (PTR-MS), and so on. PTR-MS has the potential to sample VOCs on-line and make quantitative analysis fast without any sample preparation [146,147]. The most widely used of these identification techniques is GC-MS [148]. Although PTR-MS can better achieve quantitative identification, most of the volatiles identified are preliminary [149]. GC-MS has a higher selectivity and sensitivity in the identification of VOCs and efficient separation and identification of the analytes.

4. Conclusions

This review summarizes the allelopathy of VOCs of plants including growth, competition, dormancy, resistance of pests and diseases, respiration, photosynthesis, ROS content, enzyme activity, and signaling. It also summarizes the main methods of collection and identification of VOCs. The study of allelopathy is quite a complicated work, because it involves a variety of disciplines such as chemistry, ecology, biology, microbiology, and so on. Scientists in these fields need to work together to conduct research. The study of allelopathy on plants VOCs is still a new field. Most of the researches still focus on the expression of the allelopathic phenomenon, but the depth and breadth of them are far from enough, such as the lack of research on allelopathy mechanisms of plants, the relationship between chemical recognition and communication mechanisms and allelopathy mechanisms, and so on. In recent years, we have seen more and more reports on VOCs. VOC transmission, emission, and accumulation are also hot topics in research, which deserve more research attention. There are still many issues that need to be further explored. Plant VOCs deserve more research attention.

5. Prospective

VOCs may have a wide range of potential allelopathic effects in agriculture. VOCs can not only manage weeds and pests as natural substance, but also regulate plant growth, competition, dormancy, respiration and photosynthesis, ROS content, enzyme activity, and diseases resistance. The most important thing is that they come from plants and meet our requirements for developing green agriculture. It has more economic value and is a more environmentally friendly and effective measure to use the allelopathic effects of plant natural products to develop agricultural industry. At the same time, there are some limitations in using allelopathy of VOCs of plants; for example, in field trials, VOCs are not easy to control. The volatilization of VOCs of plants is specific to species, cultivars, genotypes, and organs, as well as environments. In fact, the release of VOCs by plants are not single VOCs, but a complex mixture [150]. In agricultural production, the release of plant VOCs is affected by environmental and meteorological conditions, such as wind speed and direction, humidity and rain, and temperature, among others. These factors make the release of VOCs difficult to control. Moreover, the concentration of VOCs in open field experiments is often lower than in laboratory experiments. Therefore, the allelopathy of plant VOCs in agricultural production is more suitable for an easily controlled greenhouse.

With the development of allelochemicals identification technology and the participation of more and more chemists, the identification of allelochemicals is no longer a difficult problem in the study of allelopathy. However, the molecular mechanism of VOCs has not been studied, and it is not yet clear how VOCs are perceived by plants, and little is known about the dynamics of the active release pathway of VOCs of plants. Therefore, the focus of future research is to explore the nature and regularity of plant VOCs allelopathy and dynamic process of VOCs release, mainly to decipher the perception mechanism of VOCs within plant tissues. Over the past ten years, a large number of studies have proved that VOCs are involved in signal transduction among plants, and many allelochemicals involved in chemical communication have been identified. Now the problem we are facing is how the mechanism of chemical recognition and communication and the mechanism of allelopathy interacts. The general and specific chemical identification and the research on the identification and transfer mode of the communicating substance are urgent issues to be solved. Because the VOCs of plants are not easy to control, identifying the natural concentration of VOCs is also a major problem we face with.

At present, with increasing demands for environmental protection and sustainable development, VOCs have a dominant position in agricultural development and will become more competitive in the future, especially in the greenhouse. The theoretical research and practical application of allelopathy of plants VOCs have profound significance for sustainable development of agricultural production, for maintenance and improvement of natural resources, and for the rational arrangement of rotation and intercropping, the construction of efficient planting systems, and the improvement of natural resources' utilization efficiency, for the construction of efficient planting systems and rational arrangement of rotation and intercropping or controlling pests and weeds. Not only that, the use of allelopathy of plant volatiles can also affect plant life activity by regulating plant growth, dormancy, respiration and photosynthesis, ROS content, and enzyme activity, or by the chemical communication between plants. In summary, the allelopathy of VOCs of plants has inestimable potential in agricultural development. Based on this, it is of great significance to develop and practice the application potential of allelopathy of plant VOCs.

Author Contributions: Writing—review and editing, Y.X., X.H. and L.T.; funding acquisition, X.H. and Y.Y. All authors have read and agreed to the published version of the manuscript.

Funding: This study was funded by the Youth Program of National Natural Science Foundation of China (No. 31601747), High-Level Talents Project of Hainan Natural Science Foundation (No. 320RC723), Innovation Team Project of Chinese Academy of Tropical Agricultural Sciences (No. 1630032017029), Departmental Budget Project of Ministry of Agriculture and Rural Affairs of the People's Republic of China, and Major Science and Technology Project in Hainan Province (No. ZDKJ2017001).

Institutional Review Board Statement: Not applicable.

Informed Consent Statement: Not applicable.

Acknowledgments: We acknowledge Ziji Liu for comments on a previous version of the manuscript.

Conflicts of Interest: The authors declare that they have no conflict of interest.

References

1. Farooq, M.; Jabran, K.; Cheema, Z.A.; Wahid, A.; Siddique, K.H.M. The role of allelopathy in agricultural pest management. *Pest Manag. Sci.* **2011**, *67*, 493–506. [CrossRef] [PubMed]
2. Pan, L.; Li, X.-Z.; Yan, Z.-Q.; Guo, H.-R.; Qin, B. Phytotoxicity of umbelliferone and its analogs: Structure-activity relationships and action mechanisms. *Plant Physiol. Biochem.* **2015**, *97*, 272–277. [CrossRef] [PubMed]
3. Boydston, R.A.; Morra, M.J.; Borek, V.; Clayton, L.; Vaughn, S.F. Onion and weed response to mustard (*Sinapis alba*) seed meal. *Weed Sci.* **2011**, *59*, 546–552. [CrossRef]
4. Awan, F.K.; Rasheed, M.; Ashraf, M.; Khurshid, M.Y. Efficacy of brassica sorghum and sunflower aqueous extracts to control wheat weeds under rainfed conditions of pothwar. *Pakistan J. Anim. Plant Sci.* **2012**, *22*, 715–721.
5. Bajwa, A.A.; Mahajan, G.; Chauhan, B.S. Nonconventional weed management strategies for modern agriculture. *Weed Sci.* **2015**, *63*, 723–747. [CrossRef]
6. Tavella, L.B.; Lima e Silva, P.S.; Monteiro, A.L.; de Oliveira, V.R.; de Oliveira Fernandes, P.L. *Gliricidia sepium* intercropping for weed management in immature corn ear production. *Rev. Cienc. Agron.* **2017**, *48*, 650–656. [CrossRef]
7. Avato, P.; D'Addabbo, T.; Leonetti, P.; Argentieri, M.P. Nematicidal potential of brassicaceae. *Phytochem. Rev.* **2013**, *12*, 791–802. [CrossRef]
8. Liu, T.; Cheng, Z.; Meng, H.; Ahmad, I.; Zhao, H. Growth, yield and quality of spring tomato and physicochemical properties of medium in a tomato/garlic intercropping system under plastic tunnel organic medium cultivation. *Sci. Hortic.* **2014**, *170*, 159–168. [CrossRef]
9. Glinwood, R.; Ninkovic, V.; Pettersson, J. Chemical interaction between undamaged plants—Effects on herbivores and natural enemies. *Phytochemistry* **2011**, *72*, 1683–1689. [CrossRef]
10. Singh, A.; Weisser, W.W.; Hanna, R.; Houmgny, R.; Zytynska, S.E. Reduce pests, enhance production: Benefits of intercropping at high densities for okra farmers in Cameroon. *Pest Manag. Sci.* **2017**, *73*, 2017–2027. [CrossRef]
11. Bertin, C.; Yang, X.; Weston, L.A. The role of root exudates and allelochemicals in the rhizosphere. *Plant Soil* **2003**, *256*, 67–83. [CrossRef]
12. Abdel-Monaim, M.F.; Abo-Elyousr, K.A.M. Effect of preceding and intercropping crops on suppression of lentil damping-off and root rot disease in New Valley—Egypt. *Crop Prot.* **2012**, *32*, 41–46. [CrossRef]
13. Ma, Y.-h.; Fu, S.-l.; Zhang, X.-p.; Zhao, K.; Chen, H.Y.H. Intercropping improves soil nutrient availability, soil enzyme activity and tea quantity and quality. *Appl. Soil Ecol.* **2017**, *119*, 171–178. [CrossRef]
14. Ndungu-Magiroi, K.W.; Wortmann, C.S.; Kibunja, C.; Senkoro, C.; Mwangi, T.J.K.; Wamae, D.; Kifuko-Koech, M.; Msakyi, J. Maize-bean intercrop response to nutrient application relative to maize sole crop response. *Nutr. Cycl. Agroecosyst.* **2017**, *109*, 17–27. [CrossRef]
15. Bressan, M.; Roncato, M.-A.; Bellvert, F.; Comte, G.; Haichar, F.e.Z.; Achouak, W.; Berge, O. Exogenous glucosinolate produced by *Arabidopsis thaliana* has an impact on microbes in the rhizosphere and plant roots. *Isme J.* **2009**, *3*, 1243–1257. [CrossRef]
16. Zhao, M.; Jones, C.M.; Meijer, J.; Lundquist, P.-O.; Fransson, P.; Carlsson, G.; Hallin, S. Intercropping affects genetic potential for inorganic nitrogen cycling by root-associated microorganisms in *Medicago sativa* and *Dactylis glomerata*. *Appl. Soil Ecol.* **2017**, *119*, 260–266. [CrossRef]
17. Farooq, M.; Bajwa, A.A.; Cheema, S.A.; Cheema, Z.A. Application of allelopathy in crop production. *Int. J. Agric. Biol.* **2013**, *15*, 1367–1378.
18. Alemayehu, A.; Tamado, T.; Nigussie, D.; Yigzaw, D.; Kinde, T.; Wortmann, C.S. Maize-common bean intercropping to optimize maize-based crop production. *J. Agric. Sci.* **2017**, *155*, 1124–1136. [CrossRef]
19. Vivaldo, G.; Masi, E.; Taiti, C.; Caldarelli, G.; Mancuso, S. The network of plants volatile organic compounds. *Sci. Rep.* **2017**, *7*, 11050. [CrossRef] [PubMed]
20. Kigathi, R.N.; Weisser, W.W.; Reichelt, M.; Gershenzon, J.; Unsicker, S.B. Plant volatile emission depends on the species composition of the neighboring plant community. *BMC Plant Biol.* **2019**, *19*, 58. [CrossRef]
21. Kessler, A.; Baldwin, I.T. Defensive function of herbivore-induced plant volatile emissions in nature. *Science* **2001**, *291*, 2141–2144. [CrossRef] [PubMed]
22. Bouwmeester, H.; Schuurink, R.C.; Bleeker, P.M.; Schiestl, F. The role of volatiles in plant communication. *Plant J.* **2019**, *100*, 892–907. [CrossRef] [PubMed]
23. Rice, E.L. *Allelopathy*, 2nd ed.; Academic Press: Orlando, FL, USA, 1984.

24. Adebesin, F.; Widhalm, J.R.; Boachon, B.; Lefevre, F.; Pierman, B.; Lynch, J.H.; Alam, I.; Junqueira, B.; Benke, R.; Ray, S.; et al. Emission of volatile organic compounds from petunia flowers is facilitated by an ABC transporter. *Science* **2017**, *356*, 1386–1388. [CrossRef]

25. Dudareva, N.; Klempien, A.; Muhlemann, J.K.; Kaplan, I. Biosynthesis, function and metabolic engineering of plant volatile organic compounds. *New Phytol.* **2013**, *198*, 16–32. [CrossRef] [PubMed]

26. Kong, C.; Hu, F.; Xu, X.; Zhang, M.; Liang, W. Volatile allelochemicals in the *Ageratum conyzoides* intercropped citrus orchard and their effects on mites *Amblyseius newsami* and Panonychus citri. *J. Chem. Ecol.* **2005**, *31*, 2193–2203. [CrossRef]

27. McNickle, G.G.; St Clair, C.C.; Cahill, J.F., Jr. Focusing the metaphor: Plant root foraging behaviour. *Trends Ecol. Evol.* **2009**, *24*, 419–426. [CrossRef] [PubMed]

28. Dicke, M.; Baldwin, I.T. The evolutionary context for herbivore-induced plant volatiles: Beyond the "cry for help". *Trends Plant Sci.* **2010**, *15*, 167–175. [CrossRef]

29. Erb, M.; Veyrat, N.; Robert, C.A.M.; Xu, H.; Frey, M.; Ton, J.; Turlings, T.C.J. Indole is an essential herbivore-induced volatile priming signal in maize. *Nat. Commun.* **2015**, *6*. [CrossRef]

30. Gfeller, V.; Huber, M.; Foerster, C.; Huang, W.; Koellner, T.G.; Erb, M. Root volatiles in plant-plant interactions I: High root sesquiterpene release is associated with increased germination and growth of plant neighbours. *Plant Cell Environ.* **2019**, *42*, 1950–1963. [CrossRef]

31. Molish, H. *Der Einfluss Einer Pflanze auf die Andere-Allelopathie*; Gustav Fischer Verlag: Jena, Germany, 1937.

32. Simms, E.L.; Rausher, M.D. Costs and benefits of plant resistance to herbivory. *Am. Nat.* **1987**, *130*, 570–581. [CrossRef]

33. Kim, J.; Felton, G.W. Priming of antiherbivore defensive responses in plants. *Insect Sci.* **2013**, *20*, 273–285. [CrossRef] [PubMed]

34. Langenheim, J.H. Higher plant terpenoids: A phytocentric overview of their ecological roles. *J. Chem. Ecol.* **1994**, *20*, 1223–1280. [CrossRef] [PubMed]

35. Ameye, M.; Audenaert, K.; De Zutter, N.; Steppe, K.; Van Meulebroek, L.; Vanhaecke, L.; De Vleesschauwer, D.; Haesaert, G.; Smagghe, G. Priming of wheat with the green leaf volatile Z-3-hexenyl acetate enhances defense against fusarium graminearum but boosts deoxynivalenol production. *Plant Physiol.* **2015**, *167*, 1671–1684. [CrossRef]

36. Siri-Udom, S.; Suwannarach, N.; Lumyong, S. Applications of volatile compounds acquired from *Muscodor heveae* against white root rot disease in rubber trees (*Hevea brasiliensis* Mull. Arg.) and relevant allelopathy effects. *Fungal Biol.* **2017**, *121*, 573–581. [CrossRef]

37. Lerdau, M.; Gray, D. Ecology and evolution of light-dependent and light-independent phytogenic volatile organic carbon. *New Phytol.* **2003**, *157*, 199–211. [CrossRef]

38. Cofer, T.M.; Engelberth, M.; Engelberth, J. Green leaf volatiles protect maize (*Zea mays*) seedlings against damage from cold stress. *Plant Cell Environ.* **2018**, *41*, 1673–1682. [CrossRef]

39. Muller, C.H.; Muller, W.H.; Haines, B.L. Volatile growth Inhibitors produced by aromatic shrubs. *Science* **1964**, *143*, 471–473. [CrossRef] [PubMed]

40. Muller, C.H. Inhibitory terpenes volatilized from salvia shrubs. *Bull. Torrey Bot. Club* **1965**, *92*, 38–45. [CrossRef]

41. Abrahim, D.; Braguini, W.L.; Kelmer-Bracht, A.M.; Ishii-Iwamoto, E.L. Effects of four monoterpenes on germination, primary root growth, and mitochondrial respiration of maize. *J. Chem. Ecol.* **2000**, *26*, 611–624. [CrossRef]

42. Norton, J.M.; Harman, G.E. Responses of soil microorganisms to volatile exudates from germinating pea seeds. *Can. J. Bot.* **1985**, *63*, 1040–1045. [CrossRef]

43. Won Yun, K.; Kil, B.S.; Han, D.M. Phytotoxic and antimicrobial activity of volatile constituents of *Artemisia princeps* var. orientalis. *J. Chem. Ecol.* **1993**, *19*, 2757–2766. [CrossRef] [PubMed]

44. Jassbi, A.R.; Zamanizadehnajari, S.; Baldwin, I.T. Phytotoxic volatiles in the roots and shoots of *Artemisia tridentata* as detected by headspace solid-phase microextraction and gas chromatographic-mass spectrometry analysis. *J. Chem. Ecol.* **2010**, *36*, 1398–1407. [CrossRef]

45. Hütsch, B.W.; Augustin, J.; Merbach, W. Plant rhizodeposition—An important source for carbon turnover in soils. *J. Plant Nutr. Soil Sci.* **2002**, *165*, 397–407. [CrossRef]

46. Lin, C.; Owen, S.M.; Peñuelas, J. Volatile organic compounds in the roots and rhizosphere of *Pinus* spp. *Soil Biol. Biochem.* **2007**, *39*, 951–960. [CrossRef]

47. Delory, B.M.; Delaplace, P.; Fauconnier, M.-L.; du Jardin, P. Root-emitted volatile organic compounds: Can they mediate belowground plant-plant interactions? *Plant Soil* **2016**, *402*, 1–26. [CrossRef]

48. Fitter, A. Making allelopathy respectable. *Science* **2003**, *301*, 1337–1338. [CrossRef] [PubMed]

49. Brilli, F.; Loreto, F.; Baccelli, I. Exploiting plant volatile organic compounds (VOCs) in agriculture to Improve sustainable defense strategies and productivity of crops. *Front. Plant Sci.* **2019**, *10*. [CrossRef]

50. Oleszek, W. Allelopathic effects of volatiles from some Cruciferae species on lettuce, barnyard grass and wheat growth. *Plant Soil* **1987**, *102*, 271–273. [CrossRef]

51. Romagni, J.G.; Allen, S.N.; Dayan, F.E. Allelopathic effects of volatile cineoles on two weedy plant species. *J. Chem. Ecol.* **2000**, *26*, 303–313. [CrossRef]

52. Schmidt-Silva, V.; Pawlowski, Â.; Kaltchuk-Santos, E.; Zini, C.; Soares, G. Cytotoxicity of essential oils from two species of Heterothalamus (Asteraceae). *Aust. J. Bot.* **2011**, *59*, 682–691. [CrossRef]

53. Cheng, F.; Cheng, Z.; Meng, H.; Tang, X. The garlic allelochemical diallyl disulfide affects tomato root growth by influencing cell division, phytohormone balance and expansin gene expression. *Front. Plant Sci.* **2016**, *7*, 1199. [CrossRef]

54. Cheng, F.; Cheng, Z.-H.; Meng, H.-W. Transcriptomic insights into the allelopathic effects of the garlic allelochemical diallyl disulfide on tomato roots. *Sci. Rep.* **2016**, *6*, 38902. [CrossRef]

55. Kim, Y.S.; Kil, B.-S. Allelopathic effects of some volatile substances from the tomato plant. *J. Crop Prod.* **2001**, *4*, 313–321. [CrossRef]

56. Alves, M.d.C.S.; Medeiros Filho, S.; Innecco, R.; Torres, S.B. Alelopatia de extratos voláteis na germinação de sementes e no comprimento da raiz de alface. *Pesqui. Agropecuária Bras.* **2004**, *39*, 1083–1086. [CrossRef]

57. Norsworthy, J.K.; Meehan, J.T. Use of isothiocyanates for suppression of Palmer amaranth (*Amaranthus palmeri*), pitted morning-glory (*Ipomoea lacunosa*), and yellow nutsedge (*Cyperus esculentus*). *Weed Sci.* **2017**, *53*, 884–890. [CrossRef]

58. Horiuchi, J.-i.; Badri, D.V.; Kimball, B.A.; Negre, F.; Dudareva, N.; Paschke, M.W.; Vivanco, J.M. The floral volatile, methyl benzoate, from snapdragon (*Antirrhinum majus*) triggers phytotoxic effects in *Arabidopsis thaliana*. *Planta* **2007**, *226*, 1–10. [CrossRef] [PubMed]

59. Silva, E.R.; Overbeck, G.E.; Soares, G.L.G. Phytotoxicity of volatiles from fresh and dry leaves of two Asteraceae shrubs: Evaluation of seasonal effects. *S. Afr. J. Bot.* **2014**, *93*, 14–18. [CrossRef]

60. Souza-Alonso, P.; Novoa, A.; González, L. Soil biochemical alterations and microbial community responses under *Acacia dealbata* Link invasion. *Soil Biol. Biochem.* **2014**, *79*, 100–108. [CrossRef]

61. Haramoto, E.R.; Gallandt, E.R. Brassica cover cropping for weed management: A review. *Renew. Agric. Food Syst.* **2007**, *19*, 187–198. [CrossRef]

62. Morra, M.J.; Kirkegaard, J.A. Isothiocyanate release from soil-incorporated *Brassica tissues*. *Soil Biol. Biochem.* **2002**, *34*, 1683–1690. [CrossRef]

63. Santonja, M.; Bousquet-Mélou, A.; Greff, S.; Ormeño, E.; Fernandez, C. Allelopathic effects of volatile organic compounds released from *Pinus halepensis* needles and roots. *Ecol. Evol.* **2019**, *9*, 8201–8213. [CrossRef]

64. Ren, K.; Hayat, S.; Qi, X.; Liu, T.; Cheng, Z. The garlic allelochemical DADS influences cucumber root growth involved in regulating hormone levels and modulating cell cycling. *J. Plant Physiol.* **2018**, *230*, 51–60. [CrossRef]

65. Wei, C.; Zhou, S.; Li, W.; Jiang, C.; Yang, W.; Han, C.; Zhang, C.; Shao, H. Chemical composition and allelopathic, phytotoxic and pesticidal activities of *Atriplex cana* Ledeb. (Amaranthaceae) essential oil. *Chem. Biodivers.* **2019**, *16*, e1800595. [CrossRef]

66. Effah, E.; Holopainen, J.K.; McCormick, A.C. Potential roles of volatile organic compounds in plant competition. *Perspect. Plant Ecol. Evol. Syst.* **2019**, *38*, 58–63. [CrossRef]

67. Macias, F.A.; Molinillo, J.M.; Varela, R.M.; Galindo, J.C. Allelopathy—A natural alternative for weed control. *Pest Manag Sci* **2007**, *63*, 327–348. [CrossRef] [PubMed]

68. Luiza Ishii-Iwamoto, E.; Marusa Pergo Coelho, E.; Reis, B.; Sebastiao Moscheta, I.; Moacir Bonato, C. Effects of monoterpenes on physiological processes during seed germination and seedling growth. *Curr. Bioact. Compd.* **2012**, *8*, 50–64. [CrossRef]

69. Arroyo, A.I.; Pueyo, Y.; Pellissier, F.; Ramos, J.; Espinosa-Ruiz, A.; Millery, A.; Alados, C.L. Phytotoxic effects of volatile and water soluble chemicals of *Artemisia herba-alba*. *J. Arid Environ.* **2018**, *151*, 1–8. [CrossRef]

70. Runyon, J.B.; Mescher, M.C.; De Moraes, C.M. Volatile chemical cues guide host location and host selection by parasitic plants. *Science* **2006**, *313*, 1964–1967. [CrossRef] [PubMed]

71. Hunt, N.D.; Hill, J.D.; Liebman, M. Reducing freshwater toxicity while maintaining weed control, profits, and productivity: Effects of increased crop rotation diversity and reduced herbicide usage. *Environ. Sci. Technol.* **2017**, *51*, 1707–1717. [CrossRef]

72. Arimura, G.-i.; Shiojiri, K.; Karban, R. Acquired immunity to herbivory and allelopathy caused by airborne plant emissions. *Phytochemistry* **2010**, *71*, 1642–1649. [CrossRef]

73. Verdeguer, M.; Blázquez, M.; Boira, H. Phytotoxic effects of *Lantana camara*, *Eucalyptus camaldulensis* and *Eriocephalus africanus* essential oils in weeds of Mediterranean summer crops. *Biochem. Syst. Ecol.* **2009**, *37*, 362–369. [CrossRef]

74. Benvenuti, S.; Cioni, P.L.; Flamini, G.; Pardossi, A. Weeds for weed control: Asteraceae essential oils as natural herbicides. *Weed Res.* **2017**, *57*, 342–353. [CrossRef]

75. Mushtaq, W.; Ain, Q.; Siddiqui, M.B.; Alharby, H.; Hakeem, K.R. Allelochemicals change macromolecular content of some selected weeds. *S. Afr. J. Bot.* **2020**, *130*, 177–184. [CrossRef]

76. Fahey, J.W.; Zalcmann, A.T.; Talalay, P. The chemical diversity and distribution of glucosinolates and isothiocyanates among plants. *Phytochemistry* **2001**, *56*, 5–51. [CrossRef]

77. Halkier, B.A.; Gershenzon, J. Biology and biochemistry of glucosinolates. *Annu. Rev. Plant Biol.* **2006**, *57*, 303–333. [CrossRef] [PubMed]

78. Petersen, J.; Belz, R.; Walker, F.; Hurle, K. Weed suppression by release of Isothiocyanates from turnip-rape mulch. *Agron. J.* **2001**, *93*, 37–43. [CrossRef]

79. Pardo-Muras, M.; Puig, C.G.; López-Nogueira, A.; Cavaleiro, C.; Pedrol, N. On the bioherbicide potential of *Ulex europaeus* and *Cytisus scoparius*: Profiles of volatile organic compounds and their phytotoxic effects. *PLoS ONE* **2018**, *13*, e0205997. [CrossRef]

80. Pardo-Muras, M.; Puig, C.G.; Pedrol, N. *Cytisus scoparius* and *Ulex europaeus* produce volatile organic compounds with powerful synergistic herbicidal effects. *Molecules* **2019**, *24*, 4539. [CrossRef] [PubMed]

81. Vokou, D.; Douvli, P.; Blionis, G.J.; Halley, J.M. Effects of monoterpenoids, acting alone or in pairs, on seed germination and subsequent seedling growth. *J. Chem. Ecol.* **2003**, *29*, 2281–2301. [CrossRef]

82. Barney, J.N.; Sparks, J.P.; Greenberg, J.; Whitlow, T.H.; Guenther, A. Biogenic volatile organic compounds from an invasive species: Impacts on plant–plant interactions. *Plant Ecol.* **2009**, *203*, 195–205. [CrossRef]

83. Campbell, M.; Segear, E.; Beers, L.; Knauber, D.; Suttle, J. Dormancy in potato tuber meristems: Chemically induced cessation in dormancy matches the natural process based on transcript profiles. *Funct. Integr. Genom.* **2008**, *8*, 317–328. [CrossRef] [PubMed]

84. Hosoki, T.; Hiura, H.; Hamada, M. Breaking bud dormancy in corms, tubers, and trees with sulfur-containing compounds. *HortScience* **1985**, *20*, 290–291.

85. Hosoki, T. Breaking bud dormancy in corms and trees with sulfide compounds in garlic and horseradish. *HortScience* **1986**, *21*, 114–116.

86. Kubota, N.; Yamane, Y.; Toriu, K.; Kawazu, K.; Higuchi, T.; Nishimura, S. Identification of active substances in garlic responsible for breaking bud dormancy in grapevines. *J. Jpn. Soc. Hortic. Sci.* **1999**, *68*(6), 1111–1117. [CrossRef]

87. Shukla, S.; Pandey, S.S.; Chandra, M.; Pandey, A.; Bharti, N.; Barnawal, D.; Chanotiya, C.S.; Tandon, S.; Darokar, M.P.; Kalra, A. Application of essential oils as a natural and alternate method for inhibiting and inducing the sprouting of potato tubers. *Food Chem.* **2019**, *284*, 171–179. [CrossRef]

88. Hartmans, K.J.; Diepenhorst, P.; Bakker, W.; Gorris, L.G.M. The use of carvone in agriculture: Sprout suppression of potatoes and antifungal activity against potato tuber and other plant diseases. *Ind. Crop. Prod.* **1995**, *4*, 3–13. [CrossRef]

89. Song, X.; Bandara, M.S.; Tanino, K.K. Potato dormancy regulation: Use of essential oils for sprout suppression in potato storage. *Fruit Veg. Cereal Sci. Biotechnol* **2009**, *2*, 110–117.

90. Finger, F.L.; Santos, M.M.d.S.; Araujo, F.F.; Lima, P.C.C.; Costa, L.C.d.; França, C.d.F.M.; Queiroz, M.d.C. Action of essential oils on sprouting of non-dormant potato tubers. *Braz. Arch. Biol. Technol.* **2018**, *61*. [CrossRef]

91. Owolabi, M.S.; Olowu, R.A.; Lajide, L.; Oladimeji, M.O.; Padilla-Camberos, E.; Flores-Fernández, J.M. Inhibition of potato tuber sprouting during storage by the controlled release of essential oil using a wick application method. *Ind. Crop. Prod.* **2013**, *45*, 83–87. [CrossRef]

92. Komai, K.; Tang, C.-S. A chemotype of Cyperus rotundus in Hawaii. *Phytochemistry* **1989**, *28*, 1883–1886. [CrossRef]

93. Neri, F.; Mari, M.; Brigati, S.; Bertolini, P. Fungicidal activity of plant volatile compounds for controlling *Monilinia laxa* in stone fruit. *Plant Dis.* **2007**, *91*, 30–35. [CrossRef] [PubMed]

94. Da Silva, A.C.; de Souza, P.E.; Amaral, D.C.; Zeviani, W.M.; Brasil Pereira Pinto, J.E. Essential oils from *Hyptis marrubioides, Aloysia gratissima* and *Cordia verbenacea* reduce the progress of Asian soybean rust. *Acta Sci. Agron.* **2014**, *36*, 159–166. [CrossRef]

95. Rienth, M.; Crovadore, J.; Ghaffari, S.; Lefort, F. Oregano essential oil vapour prevents *Plasmopara viticola* infection in grapevine (*Vitis Vinifera*) and primes plant immunity mechanisms. *PLoS ONE* **2019**, *14*, e0222854. [CrossRef] [PubMed]

96. Asthana, A.; Tripathi, N.N.; Dixit, S.N. Fungitoxic and phytotoxic studies with essential oil of *Ocimum adscendens*. *J. Phytopathol.* **1986**, *117*, 152–159. [CrossRef]

97. Chaturvedi, R.; Dikshit, A.; Dixit, S.N. *Adenocalymma allicea*, a new source of a natural fungitoxicant. *Trop. Agric.* **1987**, *64*, 318–322.

98. Dube, S.; Upadhyay, P.D.; Tripathi, S.C. Antifungal, physicochemical, and insect-repelling activity of the essential oil of Ocimum basilicum. *Can. J. Bot.* **1989**, *67*, 2085–2087. [CrossRef]

99. Wang, C.; Zhang, J.; Chen, H.; Fan, Y.; Shi, Z. Antifungal activity of eugenol against *Botrytis cinerea*. *Trop. Plant Pathol.* **2010**, *35*, 137–143. [CrossRef]

100. Rao, P.V.; Gan, S.H. Cinnamon: A multifaceted medicinal plant. *Evid Based Complement Altern. Med.* **2014**, *2014*, 642942. [CrossRef]

101. Quintana-Rodriguez, E.; Rivera-Macias, L.E.; Adame-Alvarez, R.M.; Torres, J.M.; Heil, M. Shared weapons in fungus-fungus and fungus-plant interactions? Volatile organic compounds of plant or fungal origin exert direct antifungal activity in vitro. *Fungal Ecol.* **2018**, *33*, 115–121. [CrossRef]

102. Mohammad, S.F.; Woodward, S.C. Characterization of a potent inhibitor of platelet aggregation and release reaction isolated from *Allium sativum* (Garlic). *Thromb. Res.* **1986**, *44*, 793–806. [CrossRef]

103. Yang, F.; Liu, X.; Wang, H.; Deng, R.; Yu, H.; Cheng, Z. Identification and allelopathy of green garlic (*Allium sativum* L.) volatiles on scavenging of cucumber (*Cucumis sativus* L.) reactive oxygen species. *Molecules* **2019**, *24*, 3263. [CrossRef] [PubMed]

104. Sekine, T.; Sugano, M.; Majid, A.; Fujii, Y. Antifungal effects of volatile compounds from black zira (*Bunium persicum*) and other spices and herbs. *J. Chem. Ecol.* **2007**, *33*, 2123–2132. [CrossRef] [PubMed]

105. Mandal, S.; Mandal, M. Coriander (*Coriandrum sativum* L.) essential oil: Chemistry and biological activity. *Asian Pac. J. Trop. Biomed.* **2015**, *5*, 421–428. [CrossRef]

106. Friberg, M.; Schwind, C.; Roark, L.C.; Raguso, R.A.; Thompson, J.N. Floral scent contributes to interaction specificity in coevolving plants and their insect pollinators. *J. Chem. Ecol.* **2014**, *40*, 955–965. [CrossRef] [PubMed]

107. Yazaki, K.; Arimura, G.-i.; Ohnishi, T. 'Hidden' terpenoids in plants: Their biosynthesis, localization and ecological roles. *Plant Cell Physiol.* **2017**, *58*, 1615–1621. [CrossRef] [PubMed]

108. Alborn, H.; Turlings, T.; Jones, T.H.; Stenhagen, G.; Loughrin, J.; Tumlinson, J. An elicitor of plant volatiles from beet armyworm oral secretion. *Science* **1997**, *276*, 945–949. [CrossRef]

109. Degenhardt, J.; Hiltpold, I.; Köllner, T.G.; Frey, M.; Gierl, A.; Gershenzon, J.; Hibbard, B.E.; Ellersieck, M.R.; Turlings, T.C.J. Restoring a maize root signal that attracts insect-killing nematodes to control a major pest. *Proc. Natl. Acad. Sci. USA* **2009**, *106*, 13213–13218. [CrossRef] [PubMed]

110. Li, H.-H.; Nishimura, H.; Hasegawa, K.; Mizutani, J. Allelopathy of *Sasa cernua*. *J. Chem. Ecol.* **1992**, *18*, 1785–1796. [CrossRef]

111. Cheng, F.; Cheng, Z. Research progress on the use of plant allelopathy in agriculture and the physiological and ecological mechanisms of allelopathy. *Front. Plant Sci.* **2015**, *6*. [CrossRef]
112. Weaver, T.; Klarich, D. Allelopathic effects of volatile substances from *Artemisia tridentata* Nutt. *Am. Midl. Nat.* **1977**, *92*, 508–512. [CrossRef]
113. Einhellig, F.A. *Mechanism of Action of Allelochemicals in Allelopathy*; ACS Publications: Washington, DC, USA, 1995.
114. Kohli, R.K.; Batish, D.R.; Singh, H.P. Eucalypt oils for the control of Parthenium (*Parthenium hysterophorus* L.). *Crop Prot.* **1998**, *17*, 119–122. [CrossRef]
115. Einhellig, F.A.; Rasmussen, J.A. Effects of three phenolic acids on chlorophyll content and growth of soybean and grain sorghum seedlings. *J. Chem. Ecol.* **1979**, *5*, 815–824. [CrossRef]
116. Sharkey, T.D.; Yeh, S. Isoprene emission from plants. *Annu. Rev. Plant Biol.* **2001**, *52*, 407–436. [CrossRef]
117. Kaur, S.; Singh, H.P.; Batish, D.R.; Kohli, R.K. Chemical characterization and allelopathic potential of volatile oil of *Eucalyptus tereticornis* against *Amaranthus viridis*. *J. Plant Interact.* **2011**, *6*, 297–302. [CrossRef]
118. Tsubo, M.; Nishihara, E.; Nakamatsu, K.; Cheng, Y.; Shinoda, M. Plant volatiles inhibit restoration of plant species communities in dry grassland. *Basic Appl. Ecol.* **2012**, *13*, 76–84. [CrossRef]
119. Yang, X.; Deng, S.; De Philippis, R.; Chen, L.; Hu, C.; Zhang, W. Chemical composition of volatile oil from *Artemisia ordosica* and its allelopathic effects on desert soil microalgae, *Palmellococcus miniatus*. *Plant Physiol. Biochem.* **2012**, *51*, 153–158. [CrossRef] [PubMed]
120. Zhao, J.; Yang, L.; Zhou, L.; Bai, Y.; Wang, B.; Hou, P.; Xu, Q.; Yang, W.; Zuo, Z. Inhibitory effects of eucalyptol and limonene on the photosynthetic abilities in *Chlorella vulgaris* (Chlorophyceae). *Phycologia* **2016**, *55*, 696–702. [CrossRef]
121. Tsai, C.W.; Yang, J.J.; Chen, H.W.; Sheen, L.Y.; Lii, C.K. Garlic organosulfur compounds upregulate the expression of the pi class of glutathione S-transferase in rat primary hepatocytes. *J. Nutr.* **2005**, *135*, 2560–2565. [CrossRef]
122. Hsiung, Y.-C.; Chen, Y.-A.; Chen, S.-Y.; Chi, W.-C.; Lee, R.-H.; Chiang, T.-Y.; Huang, H.-J. Volatilized myrcene inhibits growth and activates defense responses in rice roots. *Acta Physiol. Plant.* **2013**, *35*, 2475–2482. [CrossRef]
123. Mutlu, S.; Atici, Ö.; Esim, N.; Mete, E. Essential oils of catmint (*Nepeta meyeri* Benth.) induce oxidative stress in early seedlings of various weed species. *Acta Physiol. Plant.* **2011**, *33*, 943–951. [CrossRef]
124. Jin, P.; Wu, X.; Xu, F.; Wang, X.; Wang, J.; Zheng, Y. Enhancing antioxidant capacity and reducing decay of Chinese bayberries by essential oils. *J. Agric. Food Chem.* **2012**, *60*, 3769–3775. [CrossRef]
125. Baldwin, I.T.; Halitschke, R.; Paschold, A.; von Dahl, C.C.; Preston, C.A. Volatile signaling in plant-plant interactions: "Talking trees" in the genomics era. *Science* **2006**, *311*, 812–815. [CrossRef]
126. Heil, M.; Silva Bueno, J.C. Within-plant signaling by volatiles leads to induction and priming of an indirect plant defense in nature. *Proc. Natl. Acad. Sci. USA* **2007**, *104*, 5467–5472. [CrossRef]
127. Kessler, A.; Halitschke, R.; Diezel, C.; Baldwin, I.T. Priming of plant defense responses in nature by airborne signaling between *Artemisia tridentata* and *Nicotiana attenuata*. *Oecologia* **2006**, *148*, 280–292. [CrossRef]
128. Ton, J.; D'Alessandro, M.; Jourdie, V.; Jakab, G.; Karlen, D.; Held, M.; Mauch-Mani, B.; Turlings, T.C.J. Priming by airborne signals boosts direct and indirect resistance in maize. *Plant J.* **2007**, *49*, 16–26. [CrossRef]
129. Heil, M.; Ton, J. Long-distance signalling in plant defence. *Trends Plant Sci.* **2008**, *13*, 264–272. [CrossRef]
130. Biedrzycki, M.L.; Bais, H.P. Kin recognition in plants: A mysterious behaviour unsolved. *J. Exp. Bot.* **2010**, *61*, 4123–4128. [CrossRef]
131. Broz, A.K.; Broeckling, C.D.; De-la-Peña, C.; Lewis, M.R.; Greene, E.; Callaway, R.M.; Sumner, L.W.; Vivanco, J.M. Plant neighbor identity influences plant biochemistry and physiology related to defense. *BMC Plant Biol.* **2010**, *10*, 115. [CrossRef] [PubMed]
132. Chen, B.J.; During, H.J.; Anten, N.P. Detect thy neighbor: Identity recognition at the root level in plants. *Plant Sci. Int. J. Exp. Plant Biol.* **2012**, *195*, 157–167. [CrossRef] [PubMed]
133. Karban, R.; Shiojiri, K. Self-recognition affects plant communication and defense. *Ecol. Lett.* **2009**, *12*, 502–506. [CrossRef] [PubMed]
134. Masclaux, F.; Hammond, R.L.; Meunier, J.; Gouhier-Darimont, C.; Keller, L.; Reymond, P. Competitive ability not kinship affects growth of Arabidopsis thaliana accessions. *New Phytol.* **2010**, *185*, 322–331. [CrossRef]
135. Milla, R.; Forero, D.M.; Escudero, A.; Iriondo, J.M. Growing with siblings: A common ground for cooperation or for fiercer competition among plants? *Proc. R. Soc. B Biol. Sci.* **2009**, *276*, 2531–2540. [CrossRef] [PubMed]
136. Yang, X.-F.; Li, L.-L.; Xu, Y.; Kong, C.-H. Kin recognition in rice (*Oryza sativa*) lines. *New Phytol.* **2018**, *220*, 567–578. [CrossRef] [PubMed]
137. Torices, R.; Gómez, J.M.; Pannell, J.R. Kin discrimination allows plants to modify investment towards pollinator attraction. *Nat. Commun.* **2018**, *9*, 2018. [CrossRef] [PubMed]
138. Brooks, G.T. Comprehensive insect physiology, biochemistry and pharmacology: Edited by G. A. Kerkut and L. I. Gilbert. Pergamon Press, Oxford. 1985. 13 Volumes. 8200 pp approx. £1700.00/$2750.00. ISBN 0 08 026850 1. *Insect Biochem.* **1985**, *15*, i–xiv. [CrossRef]
139. Kimparis, A.; Siatis, N.; Daferera, D.; Tarantilis, P.; Pappas, C.; Polissiou, M. Comparison of distillation and ultrasound-assisted extraction methods for the isolation of sensitive aroma compounds from garlic (*Allium sativum*). *Ultrason. Sonochem.* **2006**, *13*, 54–60. [CrossRef]

140. Lee, S.N.; Kim, N.S.; Lee, D.S. Comparative study of extraction techniques for determination of garlic flavor components by gas chromatography-mass spectrometry. *Anal. Bioanal. Chem.* **2003**, *377*, 749–756. [CrossRef]

141. Mutarutwa, D.; Navarini, L.; Lonzarich, V.; Compagnone, D.; Pittia, P. GC-MS aroma characterization of vegetable matrices: Focus on 3-alkyl-2-methoxypyrazines. *J. Mass Spectrom.* **2018**, *53*, 871–881. [CrossRef]

142. Sgorbini, B.; Cagliero, C.; Liberto, E.; Rubiolo, P.; Bicchi, C.; Cordero, C. Strategies for accurate quantitation of volatiles from foods and plant-origin materials: A challenging task. *J. Agric. Food Chem.* **2019**, *67*, 1619–1630. [CrossRef]

143. Verpoorte, R.; Choi, Y.H.; Kim, H.K. NMR-based metabolomics at work in phytochemistry. *Phytochem. Rev.* **2007**, *6*, 3–14. [CrossRef]

144. Marshall, T.L.; Chaffin, C.T.; Makepeace, V.D.; Hoffman, R.M.; Hammaker, R.M.; Fateley, W.G.; Saarinen, P.; Kauppinen, J. Investigation of the effects of resolution on the performance of classical least-squares (CLS) spectral interpretation programs when applied to volatile organic compounds (VOCs) of interest in remote sensing using open-air long-path Fourier transform infrared (FT-IR) spectrometry. *J. Mol. Struct.* **1994**, *324*, 19–28. [CrossRef]

145. Stierlin, É.; Nicolè, F.; Fernandez, X.; Michel, T. Development of a headspace solid-phase microextraction gas chromatography-mass spectrometry method to study volatile organic compounds (VOCs) emitted by lavender roots. *Chem. Biodivers.* **2019**, *16*, e1900280. [CrossRef] [PubMed]

146. Danner, H.; Samudrala, D.; Cristescu, S.M.; Van Dam, N.M. Tracing hidden herbivores: Time-resolved non-invasive analysis of belowground volatiles by proton-transfer-reaction mass spectrometry (PTR-MS). *J. Chem. Ecol.* **2012**, *38*, 785–794. [CrossRef] [PubMed]

147. Capozzi, V.; Lonzarich, V.; Khomenko, I.; Cappellin, L.; Navarini, L.; Biasioli, F. Unveiling the molecular basis of mascarpone cheese aroma: VOCs analysis by SPME-GC/MS and PTR-ToF-MS. *Molecules* **2020**, *25*, 1242. [CrossRef]

148. Lebanov, L.; Ghiasvand, A.; Paull, B. Data handling and data analysis in metabolomic studies of essential oils using GC-MS. *J. Chromatogr. A* **2021**, *1640*, 461896. [CrossRef]

149. Majchrzak, T.; Wojnowski, W.; Lubinska-Szczygeł, M.; Różańska, A.; Namieśnik, J.; Dymerski, T. PTR-MS and GC-MS as complementary techniques for analysis of volatiles: A tutorial review. *Anal. Chim. Acta* **2018**, *1035*, 1–13. [CrossRef]

150. Hare, J.D. Ontogeny and season constrain the production of herbivore-inducible plant volatiles in the field. *J. Chem. Ecol.* **2010**, *36*, 1363–1374. [CrossRef]

horticulturae

Article

Transcriptomics Analysis of Heat Stress-Induced Genes in Pepper (*Capsicum annuum* L.) Seedlings

Fei Wang [†], Yanxu Yin [†], Chuying Yu, Ning Li, Sheng Shen, Yabo Liu, Shenghua Gao, Chunhai Jiao * and Minghua Yao *

Hubei Key Laboratory of Vegetable Germplasm Enhancement and Genetic Improvement, Cash Crops Research Institute, Hubei Academy of Agricultural Sciences, Wuhan 430070, China; wangfei_raymond@163.com (F.W.); yinyanxu2008@126.com (Y.Y.); yuchuying@126.com (C.Y.); n.li@msn.com (N.L.); shengshen199711@163.com (S.S.); liu13125048135@163.com (Y.L.); gaoshenghua1986@126.com (S.G.)
* Correspondence: jiaoch@hotmail.com (C.J.); yaomh_2008@126.com (M.Y.); Tel.: +86-27-87380819 (M.Y.)
† These authors contributed equally to this work.

Abstract: Pepper (*Capsicum annuum* L.) is one of the most economically important crops worldwide. Heat stress (HS) can significantly reduce pepper yield and quality. However, changes at a molecular level in response to HS and the subsequent recovery are poorly understood. In this study, 17-03 and H1023 were identified as heat-tolerant and heat-sensitive varieties, respectively. Their leaves' transcript abundance was quantified using RNA sequencing to elucidate the effect of HS and subsequent recovery on gene expression. A total of 11,633 differentially expressed genes (DEGs) were identified, and the differential expression of 14 randomly selected DEGs was validated using reverse-transcription polymerase chain reaction. Functional enrichment analysis revealed that the most enriched pathways were metabolic processes under stress and photosynthesis and light harvesting during HS and after recovery from HS. The most significantly enriched pathways of 17-03 and H1023 were the same under HS, but differed during recovery. Furthermore, we identified 38 heat shock factors (Hsps), 17 HS transcription factors (Hsfs) and 38 NAC (NAM, ATAF1/2, and CUC2), and 35 WRKY proteins that were responsive to HS or recovery. These findings facilitate a better understanding of the molecular mechanisms underlying HS and recovery in different pepper genotypes.

Keywords: pepper; transcriptomics; heat stress; transcription factor

Citation: Wang, F.; Yin, Y.; Yu, C.; Li, N.; Shen, S.; Liu, Y.; Gao, S.; Jiao, C.; Yao, M. Transcriptomics Analysis of Heat Stress-Induced Genes in Pepper (*Capsicum annuum* L.) Seedlings. *Horticulturae* **2021**, *7*, 339. https://doi.org/10.3390/horticulturae7100339

Academic Editor: Yuyang Zhang

Received: 24 August 2021
Accepted: 21 September 2021
Published: 24 September 2021

Publisher's Note: MDPI stays neutral with regard to jurisdictional claims in published maps and institutional affiliations.

Copyright: © 2021 by the authors. Licensee MDPI, Basel, Switzerland. This article is an open access article distributed under the terms and conditions of the Creative Commons Attribution (CC BY) license (https://creativecommons.org/licenses/by/4.0/).

1. Introduction

Pepper (*Capsicum annuum* L.) is an important member of the Solanaceae family and is one of the most important spice and vegetable crops in many countries [1]. It is rich in capsaicin, capsanthin, and vitamins, which can improve appetite and health [2]. Pepper grows well in warm climates but is sensitive to high temperatures, with the suitable temperature range for growth and development being 20–30 °C [3]. When the temperature exceeds 35 °C, the plant will suffer from heat stress (HS) and show symptoms of high temperature injury in the whole growth stage, which will adversely affect the plant morphology, physiological and biochemical metabolic processes, and other aspects [3,4]. With the intensification of the greenhouse effect, global temperatures have risen, impacting the growth and development of crops and presenting a severe challenge for many agricultural regions in the world, and leading to a drastic reduction in economic yields and quality [5]. Therefore, investigating the molecular mechanisms underlying the response of pepper to HS is imperative for developing varieties that are better adapted to more hostile conditions.

HS affects plant cell structure, protein denaturation, and lipid transport, resulting in the destruction of the plasma membrane structure and the death of specific cells or tissues. HS causes plant transpiration water loss, decreased photosynthetic rate, and abnormal metabolism, which affect the growth and development of plants [4]. Photosynthesis is a very heat-sensitive physiological process and is easily inhibited by HS, affecting almost all photosynthetic processes, including photosystem II, photosystem I, electron transport chain, adenosine triphosphate (ATP) synthesis, and carbon fixation [6,7]. In addition to the decrease in net photosynthetic efficiency and photosystem activity, reactive oxygen species (ROS) accumulate, resulting in the destruction of D1 protein and antenna pigment in serious cases, and thus reducing the ability of plants to absorb and utilize light energy and sequester carbon [8,9]. Additionally, ROS accumulation caused by HS in plants results in oxidative damage to cells. High temperatures cause metabolic imbalances and production of ROS in plants, which aggravate lipid peroxidation and protein denaturation of the cell membrane, thus affecting the structure and function of biofilms; severe cases can lead to cell damage and plant death [10]. High temperatures also greatly effect plant metabolism; for example, most of the genes in the anthocyanin biosynthesis pathway of eggplant are induced and downregulated under high temperatures, resulting in a decrease in anthocyanin accumulation [11]. Many abiotic stresses, including HS, directly or indirectly affect the synthesis, concentration, metabolism, transport, and storage of sugars. As a potential signal molecule, soluble sugars interact with light, nitrogen, and abiotic stresses to regulate plant growth and development [12–14].

Thermal signal perception and transduction are important parts of plant stress resistance, involving a number of signal transduction pathways, including calcium-dependent protein kinases (CDPKs) and mitogen-activated protein kinases (MAPK/MPKs), signal molecules (such as ROS), and plant hormones, which play important roles in various cellular signaling networks, by transmitting extracellular stimuli to generate intracellular responses. Thermal signal perception and transduction actively regulate gene expression and protein function under various stresses and ultimately cause adaptation to environmental stresses [15–19]. For example, CRISPR/Cas9-mediated *SlMAPK3* tomato mutants were more heat-tolerant than wild-type plants, showing less plant wilting and membrane damage, a lower ROS content, higher antioxidant enzyme activities, and higher transcriptional levels [20]. The heat-induced 47 kD MBP-phosphorylated protein SlMPK1 negatively regulates the heat tolerance of tomato by mediating antioxidant protection and redox metabolism; SlSPRH1, a protein homolog rich in serine and proline, is the target protein of SlMPK1 and can be phosphorylated by SlMPK1. Overexpression of SlSPRH1 reduces the heat tolerance and antioxidant capacity of plants and is related to SlSPRH1 phosphorylation. The SlMPK1-SlSPRH1 module negatively regulates the high-temperature signal in the high-temperature response process and cooperates with the antioxidant stress system [21]. Evidence shows that HS is accompanied by a certain degree of oxidative stress, and there is a crosstalk between the signals of heat and oxidative stresses. A study showed that H_2O_2 erupts after a short period of time under HS, owing to the activity of NADPH oxidase [22]. This outbreak was related to the induction of HS response genes [23]. H_2O_2 or menadione pretreatment can also improve heat tolerance in plants [24]. BZR1, the key regulator of brassinoid (BR) response, regulates the HS response of tomato through RBOH1-dependent ROS signaling; at least in part through the regulation of FER2 and FER3 [25].

Plant heat shock transcription factors (Hsfs) are important regulatory factors of signal transduction, which mediate the transcription of heat shock factors (HSPs) and other HS-induced genes [26]. *HsfA1a* regulates the initial response, and *HsfA1a* and *HsfB1* are constitutively expressed at a steady-state low abundance of mRNA. Under HS, the accumulation of *HsfA2* mRNA and protein is strongly induced, and *HsfA2* becomes the most abundant Hsf, regulating heat tolerance during recovery or after repeated HSs [27–29]. Under non-stress conditions, overexpression of *HsfB1* stimulates the co-activation of *HsfB1*, which promotes the accumulation of HS-related proteins and enhances heat tolerance [30,31]. Hsps are regulated by Hsfs, including *Hsp100/ClpB*, *Hsp90/HtpG*, *Hsp70/DnaK*, *Hsp60/GroEL*, and small *Hsp* (*sHsp*), which are generally considered to be important molecular chaperones for maintaining and/or restoring protein homeostasis, which plays a vital role in plant survival under HS [32,33]. In addition to Hsfs, other large families of transcription factors in plants are also involved in HS responses, such as WRKY, bZIP, MYB, and NAC. As a downstream negative regulator of the H_2O_2-mediated HS response, *CaWRKY27* prevents improper responses during HS and recovery [3]. *CabZIP63*, a member of the bZIP family in pepper, directly or indirectly regulates the expression of *CaWRKY40* at the transcriptional and post-transcriptional levels and forms a positive feedback loop with *CaWRKY40* during the response of pepper to ralstonia solanacearum inoculation or high temperature–high humidity [34]. Overexpression of *SlAN2* induced the upregulation of the expression of several structural genes in the anthocyanin biosynthesis pathway and caused anthocyanin accumulation in tomato, which enhanced the tolerance to HS [35].

In nature, when plants are subjected to HS, their ability to recover is important, as the stronger the ability to recover, the faster the plant can restore their metabolic balance and maintain their normal growth. However, the regulatory molecular mechanisms and networks of pepper have not yet been reported. Therefore, in this study, we performed transcriptome analysis of the heat-tolerant variety 17-03 and the heat-sensitive variety H1023 during HS recovery, to identify candidate genes that had altered transcription levels in the pepper leaves. Collectively, our findings provide a theoretical basis for the cultivation of high-quality, heat-resistant varieties.

2. Materials and Methods

2.1. Plant Materials and Heat Treatments

Two pepper varieties, heat-tolerant 17-03 and heat-sensitive H1023, were obtained from the Hubei Key Laboratory of Vegetable Germplasm Enhancement and Genetic Improvement, Hubei Academy of Agricultural Sciences, for transcriptome analysis. Seeds were sown in 50-hole trays and grown in a growth chamber under cool white fluorescent lights (approximately 200 $\mu mol/m^2/s$) at 25 ± 2 °C with a photoperiod of 16 h light/8 h dark and 70–80% relative humidity. At two weeks, seedlings with uniform growth were transplanted to plastic pots ($10 \times 10 \times 10$ cm), containing peat, vermiculite, and soil (v/v/v = 1:1:1), until they reached the stage of 6–8 true leaves. For heat treatment, seedlings were well watered and cultivated at 42 °C for 3 d before recovering at 25 °C for 1 d. The control seedlings were placed at 25 °C for 4 d. The lighting conditions and humidity were not changed. The first fully expanded leaf from the top of each plant was sampled from eight seedlings (three replicates) in the control and treatment groups. All samples were immediately frozen in liquid nitrogen and stored at -80 °C for RNA sequencing. The samples were named CK1 (HT1_1-HT1_3) (control group of 17-03), T1 (HT2_1-HT2_3) (heat treatment group of 17-03), M1 (HT3_1-HT3_3) (recovered group of 17-03), CK2 (HS1_1-HS1_3) (control group of H1023), T2 (HS2_1-HS2_3) (heat treatment group of H1023), and M2 (HS3_1-HS3_3) (recovered group of H1023).

2.2. Measurement of Relative Electrolyte Leakage and Proline Content

Relative electrolyte leakage, which measures cellular membrane integrity, is frequently used to evaluate plant stress tolerance [36]. The upper third of the fully expanded leaves from treated and non-treated plants of 17-03 and H1023 were excised and used to generate leaf discs (9 mm in diameter). Three replicates were used for each line, with 20 leaf discs per replicate. The leaf discs were placed into 50 mL centrifuge tubes containing 25 mL of distilled deionized water and shaken at 60 rpm for 12 h in the dark at 25 °C. The electrolyte leakage (R1) of the solution was measured using a portable magnetic conductivity meter (DDB-303A, Shanghai, China) at 25 °C. The solutions were boiled for 30 min and then cooled to room temperature. The electrolyte leakage in the boiled solution (R2) was then determined using the same method. The relative electrolyte leakage (%) was calculated as (R1/R2) $\times$ 100.

The proline content was measured as described by Ben et al. [37]. Briefly, 200 mg of ground leaf sample was extracted with 2 mL of 3% sulfosalicylic acid in boiling water for 10 min. The samples were cooled and centrifuged at 4000 rpm for 15 min. Then, 1 mL of the supernatant was transferred to a new 15 mL test tube. Then, 1 mL of glacial acetic acid and 1 mL of acid-ninhydrin reagent were added. The mixture was boiled for

30 min and cooled in an ice bath to terminate the reaction. Next, the reaction mixture was partitioned by adding toluene (3 mL). After static delamination, the upper liquid was centrifuged at 4000 rpm for 15 min. The absorbance of the organic phase was measured in a spectrophotometer (Schimadzu, Japan) at a 520 nm wavelength. The proline content of the leaf samples was calculated using a standard curve constructed with known amounts of proline.

2.3. RNA Extraction, Library Construction, and Transcriptome Sequencing

According to the manufacturer's instructions, total RNA was extracted from 18 pepper leaf samples using TRIzol reagent (Life Technologies, California, CA, USA). The mRNA with poly (A) in the total RNA was enriched using Oligo (dT) magnetic beads and divided into fragments approximately 300 bp in length using ion interruption. First-strand cDNA was synthesized using a M-MuLV reverse transcriptase system, using these RNA fragments as templates and random hexamer primers, while the second strand cDNA was synthesized using the first-strand cDNA as a template. Subsequently, the cDNA libraries were constructed after polymerase chain reaction (PCR) amplification and selected according to the fragment length of 450 bp. Then, the quality of the cDNA libraries was checked using an Agilent 2100 Bioanalyzer system (Agilent Technologies, Inc., Santa Clara, CA, USA). Based on the Illumina sequencing platform, the qualified libraries were sequenced using a double terminal (paired-end, PE) sequencer (Illumina, Foster, CA, USA).

2.4. Read Alignment and Mapping Reads to the Reference Genome

To analyze the sequencing results effectively and accurately, low-quality raw data or connectors in the sequencing data were filtered. Cutadapt was used to remove the 3′ sequencing adapter (https://cutadapt.readthedocs.io/en/stable/) (accessed on 10 March 2021), and reads with an average mass fraction lower than Q20 were removed [38]. The high-quality clean reads from each library were mapped to the pepper reference genome CM334 (https://ftp.solgenomics.net/genomes/Capsicum_annuum/C.annuum_cvCM334/) (accessed on 10 March 2021) using HISAT2 software (http://ccb.jhu.edu/software/hisat2/index.shtml) (accessed on 10 March 2021). The read count value of each gene was mapped using HTSeq as the original expression of the gene [39]. Fragments per kilobase of transcript per million mapped reads (FPKM) was used to standardize the gene expression levels based on Cufflinks software [40].

2.5. Functional Enrichment Analysis of Differentially Expressed Genes

Genes with an absolute value of | log2fold change | > 1 and a false discovery rate (FDR) < 0.05 were identified as representing significantly differentially expressed genes (DEGs), using DESeq2 in the four comparisons of CK1_vs_T1, CK1_vs_M1, CK2_vs_T2, and CK2_vs_M2 [41]. To study the putative functions and pathways of the DEGs in the above four comparisons, gene ontology (GO) functional enrichment analysis was conducted using Blast2GO (version 3.0; https://www.blast2go.com/) (accessed on 10 March 2021) [42]. Kyoto Encyclopedia of Genes and Genomes (KEGG) pathway annotation of DEGs was performed using Cytoscape software (version 3.2.0) (https://cytoscape.org/) (accessed on 10 March 2021) with the ClueGO plugin using a hypergeometric test and Benjamini-Hochberg FDR correction (FDR ≤ 0.05) [43]. Transcription factors were predicted using PlantTFDB [44].

2.6. Reverse Transcription and Quantitative Reverse-Transcription Polymerase Chain Reaction Analysis

RNA was first treated with DNase I and then reverse-transcribed to cDNA using a HiScript II 1st Strand cDNA Synthesis Kit (+gDNA wiper), according to the manufacturer's instructions (Vazyme, Nanjing, China). Then, the concentration of cDNA was diluted to 100 ng/μL, and the internal reference gene *CaUBI-3* forward and reverse primers were used for PCR to detect whether the reverse transcription was successful [45]. For the quantitative reverse-transcription polymerase chain reaction (qRT-PCR) assay, all primers were designed using Primer3 (http://primer3.ut.ee/) (accessed on 10 March 2021), and the specificity of the designed fragment was tested using SGN (https://solgenomics.net/) (accessed on 10 March 2021) (Table S8). qRT-PCR was performed using SYBR® Premix Ex Taq™ (Vazyme), with three technical replicates in 10 μL volumes containing 5 μL Fast SYBR™ Green Master Mix (2×), 2 μL cDNA template (100 ng/μL), 0.2 μL of each primer (10 μM), and 2.6 μL ddH₂O. The PCR cycling conditions were as follows: 95 °C for 30 s, followed by 40 cycles of 95 °C for 5 s, and finally 60 °C for 20 s. The relative expression level of each gene was calculated using the $2^{-\Delta\Delta CT}$ method [46].

3. Results

3.1. Phenotypic and Physiological Responses of 17-03 and H1023 under HS

To accurately evaluate the heat resistance of pepper in the seedling stage under HS, 17-03 (heat-tolerant variety) and H1023 (heat-sensitive variety) were treated at 42 °C for 3 d and then recovered for 1 d at 25 °C. Compared with the leaves of seedlings of 17-03 and H1023 that did not undergo HS (Figure 1a,d), the leaves of seedlings 17-03 after heat treatment were slightly bent (Figure 1b), while the leaves of H1023 were severely sagged, and the lower leaves were severely wilted. Additionally, the growth points of the plants were necrotic after treatment at 42 °C for 3 d (Figure 1e). After recovering at 25 °C for 1 d, the down-bent leaves of 17-03 were completely extended (Figure 1c), while those of H1023 only partially recovered and did not stretch out completely. Additionally, the edges of the leaves had different degrees of withering (Figure 1f).

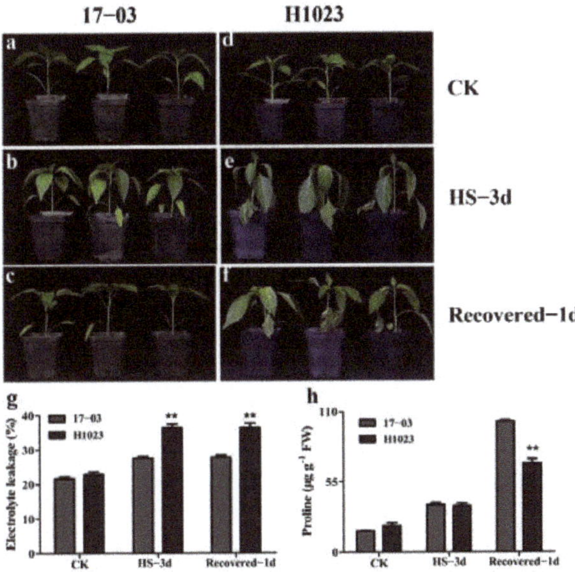

Figure 1. Phenotypic and physiological responses of 17-03 and H1023 under heat stress (HS). Seedlings of 17-03 and H1023 in control (**a,d**), treated at 42 °C for 3 d (**b,e**), and recovered for 1 d after 3 d of heat treatment (**c,f**). Relative electrolyte leakage (**g**) and proline content (**h**) in leaves treated at 42 °C for 0 d and 3 d and recovered for 1 d. Three independent biological replicates were used in each treatment, with 9 plants (6–8 true leaves) per replicate. Data are presented as mean ±SEM of three independent biological replicates. Asterisks indicate statistically significant differences between tolerant and sensitive genotypes. **, $p < 0.01$, Student's *t* test.

In this study, relative electrolyte leakage and proline content were measured in the treated plants to evaluate heat tolerance. Under normal growth conditions, there was no significant difference in the relative electrolyte leakage and proline content between the two varieties (Figure 1g,h). However, after 3 d of heat treatment at 42 °C and 1 d recovery at 25 °C, there was a remarkable increase in relative electrolyte leakage in the two varieties, with levels being significantly lower in 17-03 than in H1023 (Figure 1g). The proline content in plants increased after HS and recovery in both varieties (Figure 1h). After 3 d of heat treatment, the proline content increased, but there was no significant difference between the two varieties. However, the proline content further increased significantly in the recovery stage, and the proline content in 17-03 was obviously higher than that in H1023. These results show that 17-03 was more heat-tolerant than H1023, as the cell membranes were protected from damage, and osmotic stress was alleviated by increasing the levels of proline, an important osmotic protectant.

3.2. Overview of Transcriptomic Data for 17-03 and H1023

Transcriptome sequencing yielded 815.2 M raw reads (Table S1). After filtering, 724.31 M valid reads were obtained in 18 libraries (Table S1). The average effective data obtained from each sample were 6.03 G, accounting for 88.85% of the original data (Table S1). The Q30 base percentage of each library was above 92.81% (Table S1). The results showed that the sequencing quality was reliable, and the data were suitable for the subsequent analyses. The clean reads (82.23–85.35%) were mapped to the pepper reference genome CM334 (Table S1), indicating that the data could be used for subsequent analysis.

3.3. Identification of Differentially Expressed Genes (DEGs)

Based on the RNA-seq experiment, 24,448 expressed genes were identified in the pepper leaves (Table S2). Among these expressed genes, 11,633 DEGs were identified in 17-03 and H1023 among the four groups (Figure 2a; Table S3). Of the 11,633 DEGs, 7327 (3435 upregulated and 3892 downregulated) and 7778 (4025 upregulated and 3753 downregulated) DEGs were identified in groups CK1_vs_T1 and CK2_vs_T2, respectively (Figure 2a,b); 5185 DEGs were common between the two groups (Figure 2a); however, 2142 and 2593 DEGs were specially differentially expressed in CK1_vs_T1 and CK2_vs_T2, respectively (Figure 2a). Approximately, 3338 (2021 upregulated and 1317 downregulated) and 4822 (2256 upregulated and 2566 downregulated) DEGs were identified in groups CK1_vs_M1 and CK2_vs_M2, respectively (Figure 2b); there were 1934 common DEGs and 1404 and 2888 DEGs specially differentially expressed DEGs in CK1_vs_M1 and CK2_vs_M2, respectively (Figure 2a). Interestingly, 1229 common DEGs were identified among all four groups (Figure 2a).

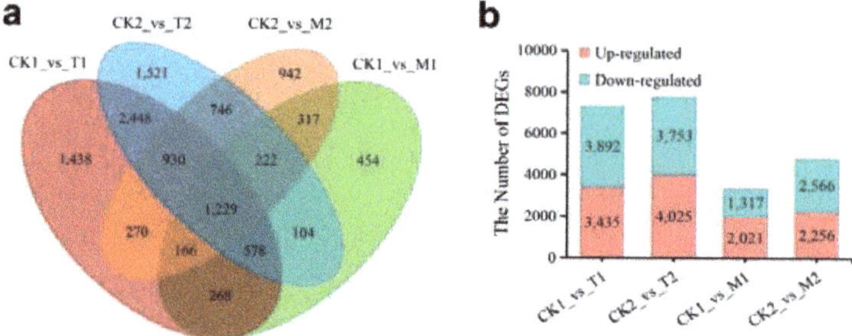

Figure 2. Expression analysis of differentially expressed genes (DEGs) in 17-03 and H1023 leaves after 3 d heat treatment at 42 °C and 1 d recovery at 25 °C. Numbers of DEGs in 17-03 and H1023 at different times (**a**). Numbers of up- and down regulated DEGs in 17-03 and H1023 at different times (**b**).

3.4. Validation of RNA-Seq Data Using qRT-PCR

To confirm the accuracy of the RNA-seq data, transcriptional levels of 14 randomly selected DEGs, representing a wide range of expression levels and patterns, were detected using qRT-PCR. All 14 DEGs participated in the process of HS response, including small heat shock protein (CA03g21390), HS transcription factor (CA03g16300), bZIP (CA08g12820), MYB (CA04g16680, CA06g27890), WRKY genes (CA03g32070, CA08g08240, and CA09g11940), NAC genes (CA05g04410, CA07g18020, CA09g12970, and CA11g04440), EG45-like domain-containing protein (CA07g00930), and universal stress protein A-like protein (CA11g00890) (Figure 3). The fold changes varied in the RNA-Seq and qRT-PCR analyses. Generally, the expression patterns determined using qRT-PCR were consistent with those obtained using RNA-Seq (Figure 3), which confirmed the accuracy of the RNA-Seq results reported in this study.

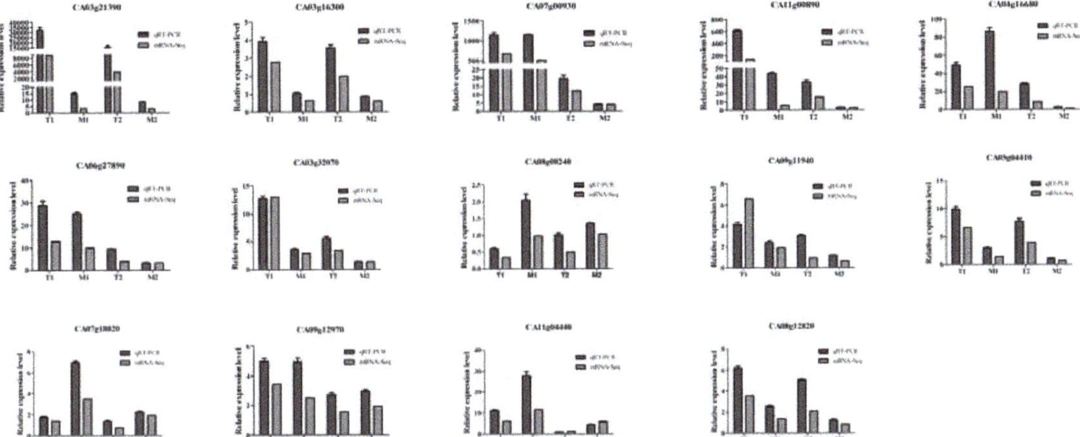

Figure 3. Quantitative reverse-transcription polymerase chain reaction (qRT-PCR)-based validation of differentially expressed genes (DEGs) in response to heat stress (HS) at different time intervals. Ordinate represents fold changes of RNA-Seq data and the relative expression level of qRT-PCR. The relative expression level of each gene under stress at each time point was compared with that under normal conditions. qRT-PCR data are presented as mean ± SEM of three independent technical replicates.

3.5. Functional Enrichment Analysis of DEGs

To explore the biological functions of DEGs in the four groups, GO enrichment analysis was performed. In total, 5133 of 11,633 DEGs in the four comparisons were annotated with GO terms and assigned to three categories: molecular function (MF), biological process (BP), and cellular component (CC) (Figure 4; Table S3). Under HS, the GO terms "photosynthesis, light harvesting" (GO:0009765), "DNA conformation change" (GO:0071103), "nucleosome assembly" (GO:0006334), "DNA packaging" (GO:0006323), "chromatin assembly" (GO:0031497), "nucleosome organization" (GO:0034728), and "chromatin assembly or disassembly" (GO:0006333) were the most commonly enriched components in the BP category in 17-03 and H1023 (Figure 4; Table S4). The most commonly enriched components were "cell wall" (GO:0005618) and "external encapsulating structure" (GO:0030312) in the CC category and "nucleosome binding" (GO:0031491) in the MF category in 17-03 and H1023 (Figure 4; Table S4). However, "DNA replication" (GO:0006260), "DNA replication initiation" (GO:0006270), "protein folding" (GO:0006457), and "protein-DNA complex assembly" (GO:0065004) in the BP category, "protein-DNA complex" (GO:0032993) in the CC category, and "nucleosomal DNA binding" (GO:0031492) in the MF category were only enriched in 17-03 (Table S4). For the recovery stage after HS, the most significantly enriched GO terms differed between 17-03 and H1023. In 17-03, DEGs were most enriched in "translation" (GO:0006412), "peptide biosynthetic process" (GO:0043043), "peptide metabolic process" (GO:0006518), "amide biosynthetic process" (GO:0043604), "cellular amide metabolic process" (GO:0043603), and "organonitrogen compound biosynthetic process" (GO:1901566) in the BP category (Table S4). "Ribosome" (GO:0005840), "non-membrane-bounded organelle" (GO:0043228), and "intracellular non-membrane-bounded organelle" (GO:0043232) were the top three CC categories (Table S4). The terms of the MF category were "structural constituent of ribosome" (GO:0003735) and "structural molecule activity" (GO:0005198) (Table S4).

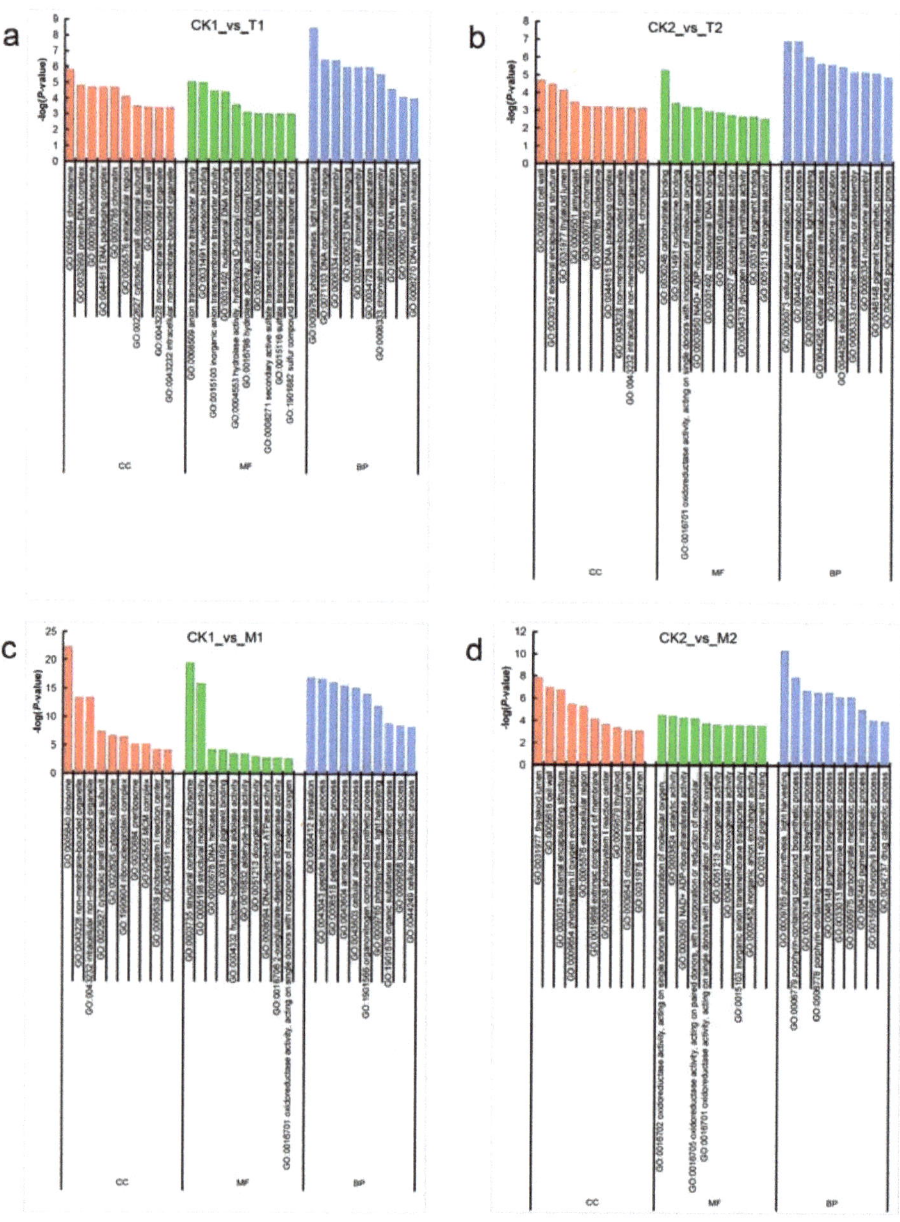

Figure 4. Gene ontology (GO) classifications of differentially expressed genes (DEGs) in the comparison groups CK1_vs_T1 (**a**), CK2_vs_T2 (**b**), CK1_vs_M1 (**c**), and CK2_vs_M2 (**d**). The DEGs were assigned to three categories: biological process (BP), cellular component (CC), and molecular function (MF). The X-axis indicates the top ten most significantly enriched BP, CC, and MF categories. The Y-axis indicates -log10 (*p*-value).

We performed KEGG pathway analysis to examine the pathways in which DEGs were involved. The significantly enriched pathways involved in HS and recovery responses are shown in Figure 5 and Table S5. The common significantly enriched pathways under HS were identified, such as "photosynthesis–antenna proteins" (cann00196), "ribosome biogenesis in eukaryotes" (cann03008), "fatty acid elongation" (cann03008), "anthocyanin biosynthesis" (cann00942), and "glycine, serine and threonine metabolism" (cann00260) (Figure 5 and Table S5). However, the significantly enriched KEGG pathways during recovery were different. The term "ribosome" (cann03010) was the most significantly enriched in 17-03, whereas "photosynthesis–antenna proteins" (cann00196), "carbon fixation in photosynthetic organisms" (cann00710), "porphyrin and chlorophyll metabolism" (cann00860), and "carotenoid biosynthesis" (cann00906) were the most significantly enriched in H1023 (Figure 5 and Table S5).

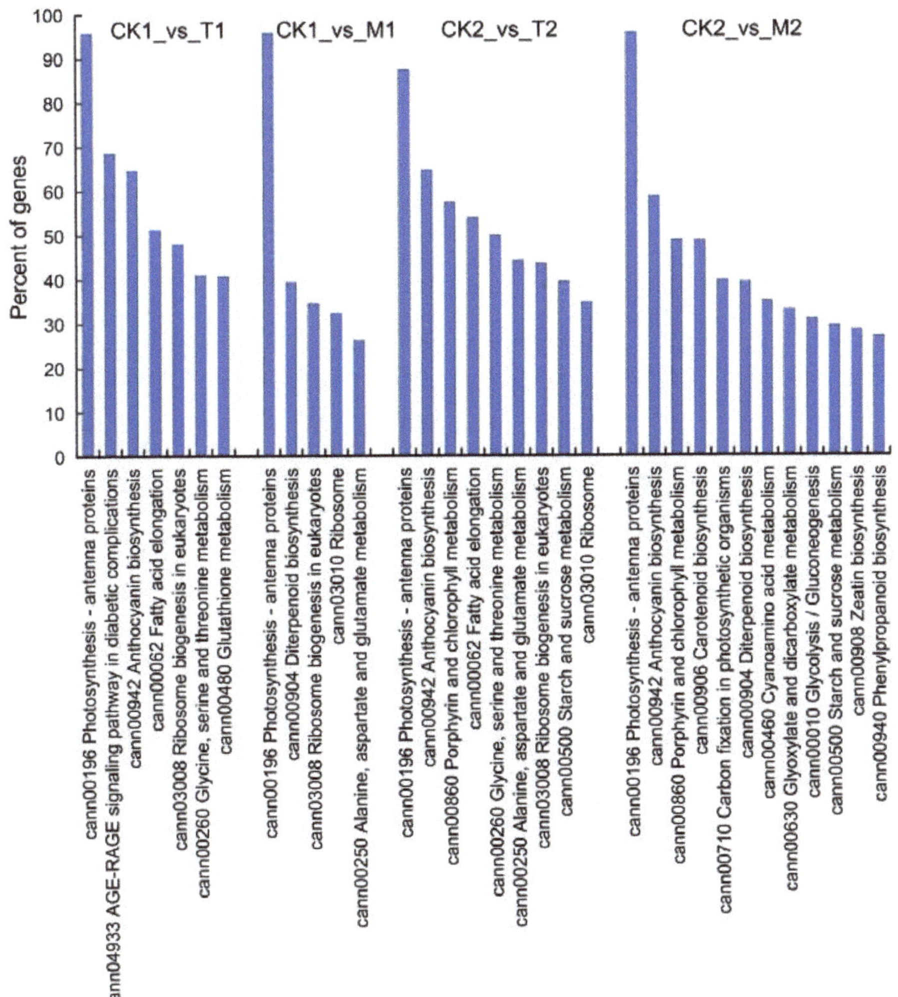

Figure 5. The significantly enriched pathways in comparison groups CK1_vs_T1, CK2_vs_T2, CK1_vs_M1, and CK2_vs_M2. The *X*-axis indicates the significantly enriched Kyoto Encyclopedia of Genes and Genomes (KEGG) pathways, and the *Y*-axis indicates the percentage of a specific category of genes in the main category.

3.6. Analysis of HS-Responsive Heat Shock Proteins and Heat Transcription Factors

Hsps and Hsfs are sensitive to HS, indicating that they play important roles in the HS response. Based on the results of transcriptome sequencing, 47 Hsp members were identified as differentially expressed in at least one of the four comparison groups (Figure 6; Table S6). Most of these were dramatically upregulated in pepper leaves after 3 d of heat treatment and were more highly expressed in H1023 than in 17-03 (Figure 6). Moreover, seven Hsps were expressed at higher levels after recovery from HS. CA01g13220, CA02g11030, CA09g06120, and CA10g10840 were highly expressed in H1023 cells (Figure 6), and CA01g31330 was highly expressed in 17-03 and H1023 cells (Figure 6). Furthermore, the expression levels of five Hsps (CA04g02800, CA09g03220, CA09g06120, CA11g13160, and CA11g13170) markedly decreased after 3 d of heat treatment (Figure 6). Similarly, 17 significantly differentially expressed Hsfs were identified, most of which were upregulated during and after recovery from HS (Figure 6). Among them, six (CA02g11030, CA03g06850, CA05g00840, CA06g08710, CA07g15920, and CA10g20440) were significantly highly expressed in H1023 after recovery from HS (Figure 6). These results indicate that these significantly differentially expressed Hsps and Hsfs might play an important role in plant protection in the long-term HS response of pepper.

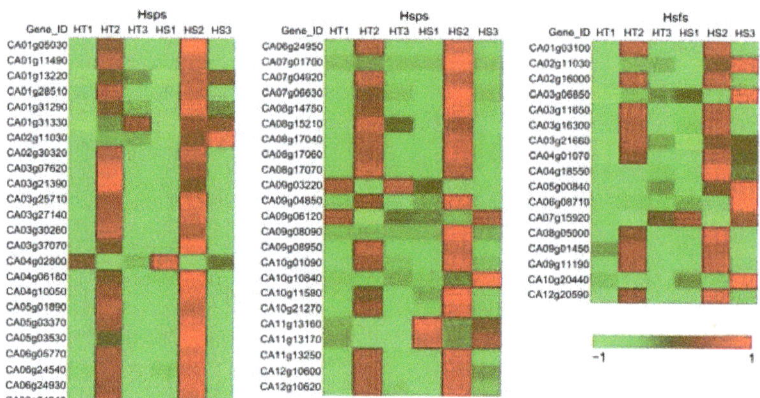

Figure 6. Heatmap of differentially expressed genes (DEGs) encoding Hsps and Hsfs in the comparison groups CK1_vs_T1, CK2_vs_T2, CK1_vs_M1, and CK2_vs_M2. The color gradient represents the normalized fragments per kilobase of transcript per million mapped reads (FPKM) value (Z-score) of DEGs. The redder the bars, the higher the gene expression level.

3.7. Analysis of HS-Responsive Transcription Factors

Transcription factors (TFs) play an important role in plant growth and development, as well as in biotic and abiotic stress response networks. Transcriptome analysis showed that many TFs in pepper were regulated by high temperatures and participated in plant recovery. A total of 49 TF families of 635 TFs were differentially expressed during heat treatment and recovery in 17-03 and H1023, including HSF, NAC, WRKY, ERF, bHLH, MYB, C2H2, B3, GRAS, bZIP, and HD-ZIP (Table S7). In this study, 38 DEGs encoding NAC proteins were identified. Among them, most were upregulated during and after recovery from HS. Moreover, the expression levels of most NAC TFs were higher in H1023 than in 17-03 during and after recovery from HS (Figure 7). WRKY proteins have also been reported to play important roles in heat response. Here, 35 DEGs encoding WRKY proteins were identified, with almost half of them upregulated in both H1023 and 17-03 during and after recovery from HS (Figure 7). Moreover, five WRKY genes (CA01g01280, CA01g34460, CA09g05110, CA09g11940, and CA12g09290) were upregulated in 17-03 but downregulated in H1023 after recovery from HS (Figure 7). Furthermore, some WRKY genes, such as CA11g05370, CA02g18540, CA09g08120, CA11g03750, and CA01g22410, were significantly highly expressed in H1023 after recovery from HS, but not at other times in H1023 or in 17-03 (Figure 7).

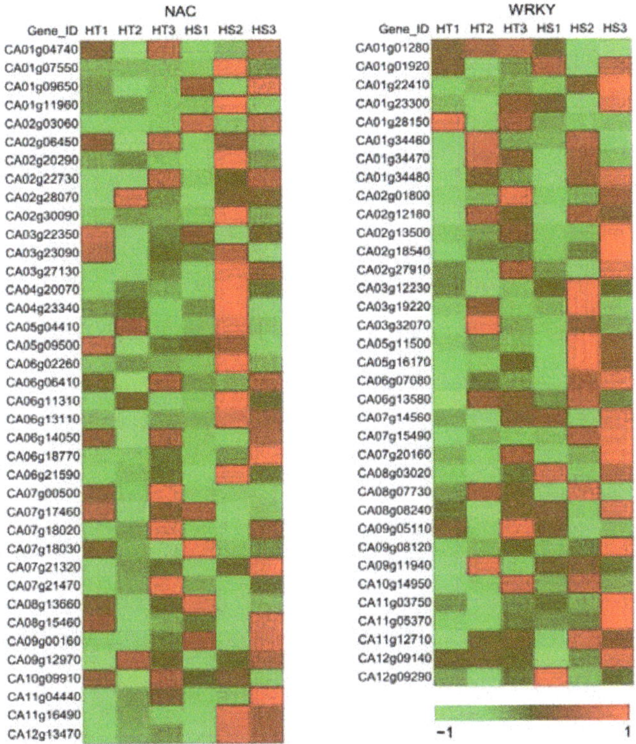

Figure 7. Heatmap of differentially expressed genes (DEGs) encoding NAC and WRKY proteins in the comparison groups CK1_vs_T1, CK2_vs_T2, CK1_vs_M1, and CK2_vs_M2. The color gradient represents the normalized fragments per kilobase of transcript per million mapped reads (FPKM) value (Z-score) of DEGs; the redder the bars, the higher the gene expression level.

4. Discussion

High temperature is one of the key climatic parameters affecting plant growth and development, resulting in crop yield losses [47]. HS can restrict photosynthesis, increase photorespiration and transpiration rate through stomatal regulation, and reduce plant biomass [7]. Pepper is a highly temperature-sensitive crop [3]. Although the physiological effects of HS on pepper have been widely studied, changes in pepper at the molecular level in response to HS and subsequent recovery are poorly understood. Therefore, to better understand the HS response in pepper, it is necessary to uncover the mechanisms underlying it. In the present study, we investigated the phenotypic and physiological changes in pepper seedlings of two varieties during and after recovery from HS. Furthermore, we comparatively analyzed heat-induced transcriptomic changes to obtain a global view of HS responses in pepper.

Under high temperature stress, the structure and function of the cell protoplasmic membrane are initially damaged, resulting in an increase in cell membrane permeability and intracellular electrolyte leakage and finally leading to an increase in electrolyte leakage of tissue leachate [48]. Therefore, the degree of electrolyte extravasation and high temperature injury can be determined by measuring the relative electrical conductivity of the tissue extract. In this study, 17-03 and H1023 were verified as high temperature-resistant and high temperature-sensitive varieties, respectively (Figure 1b,c,f,g), and were used to explore the responses and recovery patterns of pepper to HS and the possible mechanisms of the different heat resistances. There was a remarkable increase in relative electrolyte leakage in both 17-03 and H1023 after HS, but the levels were significantly lower in 17-03 than in H1023 (Figure 1h), indicating that 17-03 could alleviate damage to cellular membranes during HS. However, there was little change in the two varieties from HS to recovery (Figure 1h), indicating that the damage to the cell membrane caused by HS is irreversible. Proline, an amino acid and a compatible solute, has been widely reported to accumulate in response to various abiotic stresses, such as high temperatures [37]. After 3 d of heat treatment and 1 d of recovery, proline levels were significantly increased in both 17-03 and H1023. While there was no significant difference in proline content between 17-03 and H1023 during heat treatment (Figure 1h), the proline content increase in 17-03 during recovery was higher than that in H1023 (Figure 1h), indicating that the self-repairing ability of 17-03 after HS was stronger than that of H1023. Based on these data, we conclude that 17-03 is more heat-tolerant, as it protects cell membranes from damage and alleviates osmotic stress by increasing the proline levels.

Moreover, we obtained accurate data from transcriptome analyses based on RNA-seq and analyzed the genes of metabolic pathways that were significantly affected by HS and participated in the process of plant restoration. In 17-03 and H1023, there were more DEGs in the HS stage than in the recovery stage (Figure 2a,b), indicating that the regulatory mechanism of HS response was more active at a transcriptional level. There were significantly more upregulated DEGs than downregulated DEGs in CK1_vs_M1 (Figure 2b). However, in the other groups, the number of up- and downregulated DEGs was almost the same (Figure 2b). Moreover, the DEGs after heat treatment were mostly different from those during recovery in 17-03 and H1023 (Figure 2a). These results indicate that the defense and recovery mechanisms of pepper may have common regulatory pathways, and that there are different pathways for response, resistance, and repair.

Out of a total of 11,633 DEGs (Table S3), 5133 were assigned a GO classification (Table S3). GO enrichment analysis showed that DEGs after heat treatment were commonly enriched in "photosynthesis, light harvesting" (GO:0009765), "cellular glucan metabolic process" (GO:0006073), and "nucleosome assembly" (GO:0006334) in 17-03 and H1023 (Table S4). These findings are similar to the HS response in sweet maize (*Zea mays* L.) [49]. DEGs were also significantly enriched in "DNA replication" (GO:0006260), "DNA replication initiation" (GO:0006270), "protein folding" (GO:0006457), "protein-DNA complex assembly" (GO:0065004), "protein-DNA complex" (GO:0032993), and "nucleosomal DNA binding" (GO:0031492) in 17-03 after heat treatment (Figure 4; Table S4), which may confer increased resistance to high temperatures. The DEGs of H1023 and 17-03 during recovery were enriched with different GO terms and KEGG pathways (Figures 4 and 5, Tables S4 and S5), indicating that the repair pathways were different, which is likely due to the different degrees of high-temperature damage. In the KEGG pathway analysis, DEGs involved in HS response were predicted to function in metabolic pathways and the biosynthesis of secondary metabolites in 17-03 and H1023, which is similar to the results of previous studies [50,51].

Hsps, which are involved in multiple biological processes, such as signal transduction during HS, and have deduced functions, such as being chaperones, the folding and unfolding of cellular proteins, and the protection of functional sites from the adverse effects of high temperature, range in molecular mass from 10 to 200 kDa [52]. Hsps have functions as molecular chaperones that affect protein quality and were initially identified as proteins that were upregulated during heat treatment [51]. Many Hsps have been detected as heat response factors in tomato [53] and grape [54] plants. In this study, a total of 47 Hsps were significantly differentially expressed in the four groups. Among them, 45 Hsps were common differentially expressed in 17-03 and H1023 after heat treatment (Figure 6; Table S6). Hsps can accumulate rapidly in sensitive organs and play important roles in protecting the metabolic apparatus of cells, thus acting as a key factor in the adaptation of plants to high temperatures [55]. In this study, almost all DEGs encoding Hsps were upregulated in the four groups, the expression of which was the highest after heat treatment (Figure 6), which may play a role in protection under and after HS in 17-03 and H1023.

In addition to Hsps, various other TF genes, such as genes of Hsfs, NAC, and WRKY TFs, were also affected by HS [56]. Hsfs combine with cis-acting Hsps to play important roles in both basal and acquired thermotolerance [26,51]. Here, 17 DEGs encoding Hsfs were identified and most were upregulated in the four groups (Figure 6; Table S6). Moreover, some Hsfs were significantly highly expressed in H1023 after recovery from HS; such as CA02g11030, CA03g06850, CA05g00840, CA06g08710, CA07g15920, and CA10g20440 (Figure 6). These significantly expressed Hsfs could play important roles in the long-term HS response of pepper, by combining with the cis-acting regulatory elements, called heat shock elements, in the promoter regions of Hsps.

Plant NAC TFs have been reported to play an important role in modulating HS responses. For example, overexpression of Arabidopsis ANAC042 leads to significant thermotolerance in transgenic plants [57]. Our data indicated that the expression of 38 TFs encoded by NAC domain-containing genes was also heat-regulated (Figure 7; Table S7). Interestingly, most upregulated NAC TFs were more highly expressed in H1023 than in 17-03, during and after recovery from HS (Figure 7). WRKY TFs are one of the largest TF families in plants and have also been reported to participate in regulating plant HS response [58]. In this study, 35 WRKY TFs responded to HS in the four groups, and almost half of them positively regulated thermotolerance (Figure 7). Moreover, some WRKY TFs negatively regulated thermotolerance, such as CA01g01920 and CA01g23300 (Figure 7).

5. Conclusions

We verified the pepper varieties 17-03 and H1023 as being heat-resistant and heat-sensitive varieties and used RNA-seq to elucidate the effects of HS and the subsequent recovery on the expression of genes regulating the HS response and thermotolerance mechanisms. A total of 11,633 DEGs were identified in the four groups, with 1229 common DEGs among all four groups. Functional enrichment analysis showed that in 17-03 and H1023, DEGs were most enriched in metabolic processes under stress and photosynthesis and light harvesting during HS and after recovery from HS. The most significantly enriched pathways in 17-03 and H1023 were the same under HS, but differed during recovery. Furthermore, 38 Hsps, 17 Hsfs, 38 NAC TFs, and 35 WRKY TFs were identified as participating in the HS or recovery responses. These findings facilitate a better understanding of the molecular mechanisms underlying HS and recovery after HS in different pepper genotypes.

Supplementary Materials: The following are available online at https://www.mdpi.com/article/10.3390/horticulturae7100339/s1, Table S1: Statistical analysis of pepper clean reads in 18 libraries for RNA-seq. Table S2: Gene read count, FPKM value, annotation, and functional enrichment. Table S3: Detailed list of DEGs in 17-03 and H1023 under HS (42 °C) for 3 d and recovery (25 °C) for 1 d relative to the control. Table S4: Significantly enriched GO terms of DEGs in 17-03 and H1023 in the four groups (FDR ≤ 0.05). Table S5: Significantly enriched KEGG pathway in 17-03 and H1023 under HS (42 °C) for 3 d and recovery (25 °C) for 1 d (FDR ≤ 0.05). Table S6: DEGs encoding Hsps in 17-03 and H1023 at the heat treatment and recovery stages. Table S7: Differentially expressed transcription factors in 17-03 and H1023 at the heat treatment and recovery stages. Table S8: qRT-PCR primers used for validation of RNA-Seq data.

Author Contributions: Conceptualization, M.Y. and C.J.; methodology, M.Y. and C.J.; validation, F.W., Y.Y., C.Y., N.L., S.S., Y.L., S.G., C.J. and M.Y.; formal analysis, F.W., Y.Y., C.Y., N.L. and S.G.; investigation, S.S. and Y.L.; resources, F.W. and M.Y.; data curation, C.J. and M.Y.; writing—original draft preparation, F.W. and Y.Y.; writing—review and editing, C.J. and M.Y.; visualization, C.J. and M.Y.; supervision, C.J. and M.Y.; project administration, C.J. and M.Y.; funding acquisition, M.Y. All authors have read and agreed to the published version of the manuscript.

Funding: This work was supported by grants from the Natural Science Foundation of Hubei Province, China (2020CFA010), the earmarked fund for the China Agriculture Research System (CARS-23-G28), Key R & D Program Projects in Hubei Province (2020BBA037), China Postdoctoral Science Foundation (2017M620305), and Youth Fund of Hubei Academy of Agricultural Sciences (2021NKYJJ04).

Institutional Review Board Statement: Not applicable.

Informed Consent Statement: Not applicable.

Data Availability Statement: The data used for the analysis in this study are available in the article and Supplementary Materials.

Acknowledgments: We thank Personal Biotechnology Co., Ltd., Shanghai, China for RNA-Sequencing.

Conflicts of Interest: The authors declare no conflict of interest.

References

1. Truong, H.T.H.; Kim, K.T.; Kim, S.; Cho, M.C.; Kim, H.R.; Woo, J.G. Development of gene-based markers for the *Bs2* bacterial spot resistance gene for marker-assisted selection in pepper (*Capsicum* spp.). *Hortic. Environ. Biotechnol.* **2011**, *52*, 65–73. [CrossRef]
2. Wang, J.; Liang, C.; Yang, S.; Song, J.; Li, X.; Dai, X.; Wang, F.; Juntawong, N.; Tan, F.; Zhang, X.; et al. iTRAQ-based quantitative proteomic analysis of heat stress-induced mechanisms in pepper seedlings. *PeerJ* **2021**, *9*, e11509. [CrossRef]
3. Dang, F.; Lin, J.; Xue, B.; Chen, Y.; Guan, D.; Wang, Y.; He, S. CaWRKY27 negatively regulates H_2O_2-mediated thermotolerance in pepper (*Capsicum annuum*). *Front. Plant Sci.* **2018**, *9*, 1633. [CrossRef]
4. Belehradek, J. Physiological aspects of heat and cold. *Annu. Rev. Physiol.* **1957**, *19*, 59–82. [CrossRef] [PubMed]
5. Schiermeier, Q. Droughts, heatwaves and floods: How to tell when climate change is to blame. *Nature* **2018**, *560*, 20–22. [CrossRef]
6. Mathur, S.; Agrawal, D.; Jajoo, A. Photosynthesis: Response to high temperature stress. *J. Photochem. Photobiol. B* **2014**, *137*, 116–126. [CrossRef] [PubMed]
7. Wang, Q.; Chen, J.; He, N.; Guo, F. Metabolic reprogramming in chloroplasts under heat stress in plants. *Int. J. Mol. Sci.* **2018**, *19*, 849. [CrossRef] [PubMed]
8. Lu, T.; Meng, Z.; Zhang, G.; Qi, M.; Sun, Z.; Liu, Y.; Li, T. Sub-high Temperature and high light intensity induced irreversible inhibition on photosynthesis system of tomato plant (*Solanum lycopersicum* L.). *Front. Plant Sci.* **2017**, *8*, 365. [CrossRef]
9. Pan, C.; Ahammed, G.J.; Li, X.; Shi, K. Elevated CO_2 improves photosynthesis under high temperature by attenuating the functional limitations to energy fluxes, electron transport and redox homeostasis in tomato leaves. *Front. Plant Sci.* **2018**, *9*, 1739. [CrossRef]
10. Rogers, H.; Munné-Bosch, S. Production and scavenging of reactive oxygen species and redox signaling during leaf and flower senescence: Similar but different. *Plant Physiol.* **2016**, *171*, 1560–1568. [CrossRef]
11. Zhang, S.; Zhang, A.; Wu, X.; Zhu, Z.; Yang, Z.; Zhu, Y.; Zha, D. Transcriptome analysis revealed expression of genes related to anthocyanin biosynthesis in eggplant (*Solanum melongena* L.) under high-temperature stress. *BMC Plant Biol.* **2019**, *19*, 387. [CrossRef]
12. Kang, S.G.; Price, J.; Lin, P.C.; Hong, J.C.; Jang, J.C. The arabidopsis bZIP1 transcription factor is involved in sugar signaling, protein networking, and DNA binding. *Mol. Plant* **2010**, *3*, 361–373. [CrossRef]
13. Obertello, M.; Krouk, G.; Katari, M.S.; Runko, S.J.; Coruzzi, G.M. Modeling the global effect of the basic-leucine zipper transcription factor 1 (*bZIP1*) on nitrogen and light regulation in Arabidopsis. *BMC Syst. Biol.* **2010**, *4*, 111. [CrossRef]
14. Osuna, D.; Usadel, B.; Morcuende, R.; Gibon, Y.; Bläsing, O.E.; Höhne, M.; Günter, M.; Kamlage, B.; Trethewey, R.; Scheible, W.R.; et al. Temporal responses of transcripts, enzyme activities and metabolites after adding sucrose to carbon-deprived Arabidopsis seedlings. *Plant J.* **2007**, *49*, 463–491. [CrossRef] [PubMed]
15. Jagodzik, P.; Tajdel-Zielinska, M.; Ciesla, A.; Marczak, M.; Ludwikow, A. Mitogen-activated protein kinase cascades in plant hormone signaling. *Front. Plant Sci.* **2018**, *9*, 1387. [CrossRef]
16. Kong, F.; Wang, J.; Cheng, L.; Liu, S.; Wu, J.; Peng, Z.; Lu, G. Genome-wide analysis of the mitogen-activated protein kinase gene family in Solanum lycopersicum. *Gene* **2012**, *499*, 108–120. [CrossRef]
17. Pitzschke, A.; Schikora, A.; Hirt, H. MAPK cascade signalling networks in plant defence. *Curr. Opin. Plant Biol.* **2009**, *12*, 421–426. [CrossRef] [PubMed]
18. Sangwan, V.; Orvar, B.L.; Beyerly, J.; Hirt, H.; Dhindsa, R.S. Opposite changes in membrane fluidity mimic cold and heat stress activation of distinct plant MAP kinase pathways. *Plant J.* **2002**, *31*, 629–638. [CrossRef] [PubMed]
19. Xu, J.; Zhang, S. Mitogen-activated protein kinase cascades in signaling plant growth and development. *Trends Plant Sci.* **2015**, *20*, 56–64. [CrossRef] [PubMed]
20. Yu, W.; Wang, L.; Zhao, R.; Sheng, J.; Zhang, S.; Li, R.; Shen, L. Knockout of *SlMAPK3* enhances tolerance to heat stress involving ROS homeostasis in tomato plants. *BMC Plant Biol.* **2019**, *19*, 354. [CrossRef]
21. Ding, H.; He, J.; Wu, Y.; Wu, X.; Ge, C.; Wang, Y.; Zhong, S.; Peiter, E.; Liang, J.; Xu, W. The tomato mitogen-activated protein kinase SlMPK1 is as a negative regulator of the high-temperature stress response. *Plant Physiol.* **2018**, *177*, 633–651. [CrossRef]

22. Vacca, R.A.; de Pinto, M.C.; Valenti, D.; Passarella, S.; Marra, E.; De Gara, L. Production of reactive oxygen species, alteration of cytosolic ascorbate peroxidase, and impairment of mitochondrial metabolism are early events in heat shock-induced programmed cell death in tobacco Bright-Yellow 2 cells. *Plant Physiol.* **2004**, *134*, 1100–1112. [CrossRef]

23. Volkov, R.A.; Panchuk, I.I.; Mullineaux, P.M.; Schöffl, F. Heat stress-induced H_2O_2 is required for effective expression of heat shock genes in Arabidopsis. *Plant Mol. Biol.* **2006**, *61*, 733–746. [CrossRef] [PubMed]

24. Larkindale, J.; Huang, B. Thermotolerance and antioxidant systems in Agrostis stolonifera: Involvement of salicylic acid, abscisic acid, calcium, hydrogen peroxide, and ethylene. *J. Plant Physiol.* **2004**, *161*, 405–413. [CrossRef] [PubMed]

25. Yin, Y.; Qin, K.; Song, X.; Zhang, Q.; Zhou, Y.; Xia, X.; Yu, J. BZR1 transcription factor regulates heat stress tolerance through feronia receptor-like kinase-mediated reactive oxygen species signaling in tomato. *Plant Cell Physiol.* **2018**, *59*, 2239–2254. [CrossRef]

26. Kotak, S.; Larkindale, J.; Lee, U.; von Koskull-Döring, P.; Vierling, E.; Scharf, K.D. Complexity of the heat stress response in plants. *Curr. Opin. Plant Biol.* **2007**, *10*, 310–316. [CrossRef] [PubMed]

27. Fragkostefanakis, S.; Simm, S.; Paul, P.; Bublak, D.; Scharf, K.D.; Schleiff, E. Chaperone network composition in Solanum lycopersicum explored by transcriptome profiling and microarray meta-analysis. *Plant Cell Environ.* **2015**, *38*, 693–709. [CrossRef]

28. Hahn, A.; Bublak, D.; Schleiff, E.; Scharf, K.D. Crosstalk between Hsp90 and Hsp70 chaperones and heat stress transcription factors in tomato. *Plant Cell.* **2011**, *23*, 741–755. [CrossRef]

29. Mishra, S.K.; Tripp, J.; Winkelhaus, S.; Tschiersch, B.; Theres, K.; Nover, L.; Scharf, K.D. In the complex family of heat stress transcription factors, *HsfA1* has a unique role as master regulator of thermotolerance in tomato. *Genes Dev.* **2002**, *16*, 1555–1567. [CrossRef]

30. Fragkostefanakis, S.; Simm, S.; El-Shershaby, A.; Hu, Y.; Bublak, D.; Mesihovic, A.; Darm, K.; Mishra, S.K.; Tschiersch, B.; Theres, K.; et al. The repressor and co-activator HsfB1 regulates the major heat stress transcription factors in tomato. *Plant Cell Environ.* **2019**, *42*, 874–890. [CrossRef]

31. Bharti, K.; Von Koskull-Döring, P.; Bharti, S.; Kumar, P.; Tintschl-Körbitzer, A.; Treuter, E.; Nover, L. Tomato heat stress transcription factor HsfB1 represents a novel type of general transcription coactivator with a histone-like motif interacting with the plant CREB binding protein ortholog HAC1. *Plant Cell.* **2004**, *16*, 1521–1535. [CrossRef] [PubMed]

32. Waters, E.R.; Vierling, E. Plant small heat shock proteins—Evolutionary and functional diversity. *New Phytol.* **2020**, *227*, 24–37. [CrossRef] [PubMed]

33. Yang, J.; Sun, Y.; Sun, Q.; Yi, Y.; Qin, J.; Li, H.; Liu, J. The involvement of chloroplast HSP100/ClpB in the acquired thermotolerance in tomato. *Plant Mol. Biol.* **2006**, *62*, 385–395. [CrossRef]

34. Shen, L.; Liu, Z.; Yang, S.; Yang, T.; Liang, J.; Wen, J.; Liu, Y.; Li, J.; Shi, L.; Tang, Q.; et al. Pepper CabZIP63 acts as a positive regulator during Ralstonia solanacearum or high temperature-high humidity challenge in a positive feedback loop with CaWRKY40. *J. Exp. Bot.* **2016**, *67*, 2439–2451. [CrossRef]

35. Meng, X.; Wang, J.; Wang, G.; Liang, X.; Li, X.; Meng, Q. An R2R3-MYB gene, *LeAN2*, positively regulated the thermo-tolerance in transgenic tomato. *J. Plant Physiol.* **2015**, *175*, 1–8. [CrossRef]

36. Campos, P.S.; Quartin, V.; Ramalho, J.C.; Nunes, M.A. Electrolyte leakage and lipid degradation account for cold sensitivity in leaves of Coffea sp. plants. *J. Plant Physiol.* **2003**, *160*, 283–292. [CrossRef]

37. Ben Rejeb, K.; Abdelly, C.; Savouré, A. Proline, a multifunctional amino-acid involved in plant adaptation to environmental constraints. *Biol. Aujourdhui* **2012**, *206*, 291–299. [CrossRef]

38. Chen, Y.; Chen, Y.; Shi, C.; Huang, Z.; Zhang, Y.; Li, S.; Li, Y.; Ye, J.; Yu, C.; Li, Z.; et al. SOAPnuke: A MapReduce acceleration-supported software for integrated quality control and preprocessing of high-throughput sequencing data. *Gigascience* **2018**, *7*, 1–6. [CrossRef]

39. Anders, S.; Pyl, P.T.; Huber, W. HTSeq—A Python framework to work with high-throughput sequencing data. *Bioinformatics* **2015**, *31*, 166–169. [CrossRef]

40. Trapnell, C.; Williams, B.A.; Pertea, G.; Mortazavi, A.; Kwan, G.; van Baren, M.J.; Salzberg, S.L.; Wold, B.J.; Pachter, L. Transcript assembly and quantification by RNA-Seq reveals unannotated transcripts and isoform switching during cell differentiation. *Nat. Biotechnol.* **2010**, *28*, 511–515. [CrossRef] [PubMed]

41. Love, M.I.; Huber, W.; Anders, S. Moderated estimation of fold change and dispersion for RNA-seq data with DESeq2. *Genome Biol.* **2014**, *15*, 550. [CrossRef] [PubMed]

42. Götz, S.; García-Gómez, J.M.; Terol, J.; Williams, T.D.; Nagaraj, S.H.; Nueda, M.J.; Robles, M.; Talón, M.; Dopazo, J.; Conesa, A. High-throughput functional annotation and data mining with the Blast2GO suite. *Nucleic Acids Res.* **2008**, *36*, 3420–3435. [CrossRef] [PubMed]

43. Bindea, G.; Mlecnik, B.; Hackl, H.; Charoentong, P.; Tosolini, M.; Kirilovsky, A.; Fridman, W.H.; Pagès, F.; Trajanoski, Z.; Galon, J. ClueGO: A Cytoscape plug-in to decipher functionally grouped gene ontology and pathway annotation networks. *Bioinformatics* **2009**, *25*, 1091–1093. [CrossRef] [PubMed]

44. Jin, J.; Tian, F.; Yang, D.; Meng, Y.; Kong, L.; Luo, J.; Gao, G. PlantTFDB 4.0: Toward a central hub for transcription factors and regulatory interactions in plants. *Nucleic Acids Res.* **2017**, *45*, D1040–D1045. [CrossRef] [PubMed]

45. Wan, H.; Yuan, W.; Ruan, M.; Ye, Q.; Wang, R.; Li, Z.; Zhou, G.; Yao, Z.; Zhao, J.; Liu, S.; et al. Identification of reference genes for reverse transcription quantitative real-time PCR normalization in pepper (*Capsicum annuum* L.). *Biochem. Biophys. Res. Commun.* **2011**, *416*, 24–30. [CrossRef] [PubMed]

46. Schmittgen, T.D.; Livak, K.J. Analyzing real-time PCR data by the comparative C(T) method. *Nat. Protoc.* **2008**, *3*, 1101–1108. [CrossRef]

47. Chaudhry, S.; Sidhu, G.P.S. Climate change regulated abiotic stress mechanisms in plants: A comprehensive review. *Plant Cell Rep.* **2021**. [CrossRef]

48. Demidchik, V.; Straltsova, D.; Medvedev, S.S.; Pozhvanov, G.A.; Sokolik, A.; Yurin, V. Stress-induced electrolyte leakage: The role of K^+-permeable channels and involvement in programmed cell death and metabolic adjustment. *J. Exp. Bot.* **2014**, *65*, 1259–1270. [CrossRef]

49. Shi, J.; Yan, B.; Lou, X.; Ma, H.; Ruan, S. Comparative transcriptome analysis reveals the transcriptional alterations in heat-resistant and heat-sensitive sweet maize (*Zea mays* L.) varieties under heat stress. *BMC Plant Biol.* **2017**, *17*, 26. [CrossRef]

50. Mangelsen, E.; Kilian, J.; Harter, K.; Jansson, C.; Wanke, D.; Sundberg, E. Transcriptome analysis of high-temperature stress in developing barley caryopses: Early stress responses and effects on storage compound biosynthesis. *Mol. Plant* **2011**, *4*, 97–115. [CrossRef]

51. Li, T.; Xu, X.; Li, Y.; Wang, H.; Li, Z.; Li, Z. Comparative transcriptome analysis reveals differential transcription in heat-susceptible and heat-tolerant pepper (*Capsicum annum* L.) cultivars under heat stress. *J. Plant Biol.* **2015**, *58*, 411–424. [CrossRef]

52. Huang, B.; Xu, C. Identification and characterization of proteins associated with plant tolerance to heat stress. *J. Integr. Plant Biol.* **2008**, *50*, 1230–1237. [CrossRef] [PubMed]

53. Frank, G.; Pressman, E.; Ophir, R.; Althan, L.; Shaked, R.; Freedman, M.; Shen, S.; Firon, N. Transcriptional profiling of maturing tomato (*Solanum lycopersicum* L.) microspores reveals the involvement of heat shock proteins, ROS scavengers, hormones, and sugars in the heat stress response. *J. Exp. Bot.* **2009**, *60*, 3891–3908. [CrossRef] [PubMed]

54. Liu, G.; Wang, J.; Cramer, G.; Dai, Z.; Duan, W.; Xu, H.; Wu, B.; Fan, P.; Wang, L.; Li, S. Transcriptomic analysis of grape (*Vitis vinifera* L.) leaves during and after recovery from heat stress. *BMC Plant Biol.* **2012**, *12*, 174. [CrossRef] [PubMed]

55. Wahid, A.; Gelani, S.; Ashraf, M.; Foolad, M.R. Heat tolerance in plants: An overview. *Environ. Exp. Bot.* **2007**, *61*, 199–223. [CrossRef]

56. Qin, D.; Wu, H.; Peng, H.; Yao, Y.; Ni, Z.; Li, Z.; Zhou, C.; Sun, Q. Heat stress-responsive transcriptome analysis in heat susceptible and tolerant wheat (*Triticum aestivum* L.) by using wheat genome array. *BMC Genom.* **2008**, *9*, 432. [CrossRef] [PubMed]

57. Shahnejat-Bushehri, S.; Mueller-Roeber, B.; Balazadeh, S. Arabidopsis NAC transcription factor JUNGBRUNNEN1 affects thermomemory-associated genes and enhances heat stress tolerance in primed and unprimed conditions. *Plant Signal. Behav.* **2012**, *7*, 1518–1521. [CrossRef]

58. Zhang, C.; Wang, D.; Yang, C.; Kong, N.; Shi, Z.; Zhao, P.; Nan, Y.; Nie, T.; Wang, R.; Ma, H.; et al. Genome-wide identification of the potato WRKY transcription factor family. *PLoS ONE* **2017**, *12*, e0181573. [CrossRef]

MDPI
St. Alban-Anlage 66
4052 Basel
Switzerland
Tel. +41 61 683 77 34
Fax +41 61 302 89 18
www.mdpi.com

Horticulturae Editorial Office
E-mail: horticulturae@mdpi.com
www.mdpi.com/journal/horticulturae

www.ingramcontent.com/pod-product-compliance
Lightning Source LLC
LaVergne TN
LVHW070102170726
843515LV00007B/1595